MULTIFORMITY OF SCIENCE

POZNAŃ STUDIES
IN THE PHILOSOPHY OF THE SCIENCES AND THE HUMANITIES

VOLUME 79

Poznań Studies is partly sponsored by Adam Mickiewicz University

The address: prof. L. Nowak, Cybulskiego 13, 60-247 Poznań, Poland.
Fax: (061) 8477-079 or (061) 8471-555
E-mail: przybysz@main.amu.edu.pl

MULTIFORMITY OF SCIENCE

Jan Such

Amsterdam - New York, NY 2004

The paper on which this book is printed meets the requirements of "ISO 9706:1994, Information and documentation - Paper for documents - Requirements for permanence".

ISSN 0303-8157
ISBN: 90-420-0938-1
©Editions Rodopi B.V., Amsterdam - New York, NY 2004
Printed in The Netherlands

Contents

PREFACE

This volume contains 37 of my essays, mainly devoted to philosophical reflections on science in general, and on physical science (physics, astronomy and cosmology) in particular.

Of the six sections in which the essays have been included, only the sixth section contains works that do not immediately deal with philosophy of science but rather with the theory of reality.

The vast majority of works in this volume were published earlier in various journals and collected works, beginning in 1973, either in Polish or in English (this is indicated in footnotes for every article or study that was published before). Here I wish to express my gratitude to the publishers and editors of my previously published papers for their kind permission to republish them in this volume.

Most of the texts in this volume were translated by Andrzej Pietrzak, unless indicated otherwise.

I would like to acknowledge the kind and expert assistance of Dr Katarzyna Paprzycka in editing and improving the language of my book. Any remaining mistakes and imprecisions are my exclusive responsibility and that of the translators.

Part I
Contributions to the Idealizational Theory of Science

IDEALIZATION AND CONCRETIZATION
IN THE NATURAL SCIENCES*

I. The Method of Idealization in Ancient Science

In this study, assumptions and the conceptual apparatus of the theory of idealization are used as developed by Leszek Nowak and published in his numerous works on the subject (see e.g. Nowak 1970, 1971, 1972, 1974).

We know that the Egyptian and Babylonian civilisations abstracted geometrical figures from physical objects and that the Greeks made a decisive step towards something that might be conceived as the idealization of geometrical concepts, thus creating geometry as a branch of mathematics. Hence, mathematics was the first science in which the method of idealization was applied. In the natural sciences, it was the Greeks, too, who brought about a breakthrough by applying this method.

Archimedes was probably the first scientist who was aware of applying the idealizational method in the natural sciences. While developing his theory of a lever, he assumed that an arm of a lever or scales, in spite of being supported at just one point (in the middle) is perfectly flat. He assumed that thin ropes supporting weights were perfectly perpendicular to the arm and thus precisely parallel to each other. Still, he knew that bodies fall in the direction of the centre of the round Earth, i.e. that the lines of gravitation forces meet concentrically, as do radiuses in a sphere's centre. This was a successful application of the method of idealization in statics.

The fact that statics is considered to be the only strict branch of the natural sciences which was worked out by the ancient thinkers indicates a close connection between the method of idealization and science. They appeared together due to the efforts of the same person. The maker of the first strict branch of the science of nature was, at the same time, the maker of the idealizational method in the natural sciences. This is not strange, since Marx considered the method of idealization and gradual concretization to be the proper method of all developed empirical sciences, both natural and social. In this sense, it was not Aristotle with his physics based on direct common experience who was the pioneer of contemporary science. Rather, it was Archimedes with his statics. He assumed

---

* First published in English in *Poznań Studies in the Philosophy of the Sciences and the Humanities*, vol. **4**, 1978, 49-73.

consciously and was aware of contradictions between the "common observation" and the idealizing assumptions.

Eratosthenes, a great astronomer and Archimedes' contemporary, living in the 3rd century B.C., was the second outstanding representative of the method of idealization in Ancient science. He was the first to correctly calculate the Earth's circumference. As the basis, he assumed the angular difference in the angles under which the sun rays fall on the Earth's surface in Alexandria and Syene and he accepted an idealizing assumption that the sun rays falling on different spots of the Earth's surface (their latitude notwithstanding) were perfectly parallel to each other.

Thus, physics (statics) and astronomy were the first natural sciences in which the Ancient scientists successfully applied the method of idealization.

II. The Method of Idealization and the Beginning of Modern Science

Contemporary natural sciences also owe their successful start and quick growth, abounding in theoretical and practical achievements, to the method of idealization. The 17th century owes it its name as "the age of scientific revolution," an age of fundamental transformation in the very mode of thinking on the part of the scientists. Galileo applied it, fully aware of its superiority to the previous modes of investigation, quoting its success in Archimedes' statics. The unusual productivity and universality of its applications in contemporary natural sciences resulted from two basic premises of scientific development, experiment and mathematics, which were ready at the time. It is only within the context of experimental research and an ability to analyze and work out its results with the aid of mathematical means that its real power in explaining and predicting natural phenomena could be demonstrated. Also, it was the only context which could demonstrate the superiority of the approach over the previous method based on common experience and on the far-fetched metaphysical assumptions characteristic of the physics of the Peripatetics (Aristotle's followers). On the other hand, it is worth mentioning that the very application of mathematics in a given branch of science is possible only because some simplifying (idealizing) assumptions are accepted. These in turn allow one to transform the problems from a given branch of science into idealized ones that may be formulated in the language of numbers and geometry as mathematical problems.

While Aristotle based his physics on common experience and metaphysical (speculative) assumption, Archimedes and Galileo based their physics on a scientific experiment and idealizing assumptions which enabled them to apply mathematics. In this way, modern physics gained relative independence of metaphysics but relied heavily on mathematics which started playing a principal role in physical research.

There arose an argument concerning the validity of this new method of inquiry between Galileo, who may be considered to be the founder of the method of idealization in the modern natural sciences, and the Peripatetics. In this discussion, Galileo faced a seemingly paradoxical charge of "betraying experience" with the application of the new method: instead of sticking to direct observation he assumed statements which contradicted the results of experiments and were supposed to be lost in "mathematical subtleties."

Contrary to what is usually believed, however, it turns out that the main methodological argument between Galileo and the Peripatetics (i.e. between the followers of the old Aristotelian physics and the followers of the new Galilean one) did not concern the relation of inquiry to experience, i.e. it was not an argument about the role of experiment in physical inquiry. Instead, it was a dispute about the meaning and validity of the method of idealization in physics. From the viewpoint of the Peripatetic school, Galileo's major error consisted in accepting idealizing assumptions. Moreover, Galileo faced the above-mentioned charge because he used the method, which the Peripatetics viewed as strange. They accused him of leaving experience out, contradicting the experimental method and overestimating the meaning of the mathematical method.

The Galilean law of inertia obviously contradicted common experience in which we observe that bodies generally tend to stop when the force which has been applied ceases to influence them. Hence, in order to keep a body in a uniform movement, even on a horizontal plane, we have to apply some force to it. In a similar way, a heavy body remaining on the surface of the Earth (for instance a stone) may not move at all, in spite of being influenced by some force, which seemed to contradict Galileo's law while supporting Aristotle's law of inertia. Aristotle was therefore right in the sense of support by direct observation when he wrote in his *Physics* with complete assurance: "Truly all things will be put in a motion simultaneously with the primary source of movement and shall cease moving if this primary source of movement ceases to move them" (Aristotle 1968, p. 299).

Galileo, conducting more thorough experiments than those of Aristotle and his followers (the Peripatetics), observed that the results of observation (the conduct of bodies) were due to the fact that the forces of friction and air resistance were generally present. If those forces were removed, the bodies would continue their movement. He noticed that friction diminished by sliding or rolling those bodies on very smooth surfaces, and air resistance diminished with the provision of more appropriate shapes, thus removing part of air from the space in which they moved in. Thus, one can achieve greater compliance of experimental results with the law of inertia which states that the observable tendency of the moving bodies to stop is not, as Aristotle thought, their natural tendency. By contrast, the tendency is caused by second-rate reasons of external nature from which one should abstract when formulating the law of inertia.

III. The Application of the Method of Idealization by Galileo to the Analysis of Inertial Movement

Let us first consider how Galileo arrived at the principle of inertia, subsequently named after him, and at the belief that it is correct. In this method, he used both real and thought experiments. The latter was somehow a complement to the former in the case when the former could not be carried out, as its continuation in an ideal situation. Galileo's thought experiment was an essential part of the procedure of idealization.

In the *Dialogue*, where the fundamental principle is the principle of inertia, Salviati (representing Galileo's attitude) recalled previously conducted experiments. He considered intellectually two "cases of what happens to a moving body on two planes" while assuming that a "body" was a perfectly round sphere and "planes" were perfectly smooth surfaces (boards). The first case occurs when "on a steep plane a heavy body moves downward by itself, constantly increasing its velocity," so that "one has to apply force in order to keep it resting" (Galileo 1962, p. 157). "But on a plane which goes upwards, where the second case occurs, one has to apply force in order to move it up and even in order to stop it while the movement enforced slows down constantly and finally eases completely" (p. 157). Moreover, "in both cases there arise differences depending more or less on steep planes going upwards or downwards, where the greater slope downwards causes greater velocity and reversely, on upward going plane the same body pushed with the same force goes further if the slope is smaller" (p. 157).

After these conclusions were drawn from the real experiments, Galileo could state a new question, fundamental to his new method of idealization (which decided the difference between his method and the methods of Aristotle applied in physical research): "And now tell me – what would happen with this body which moved on a plane that went neither up nor down?" (p. 157). Asking this question, Galileo obviously made the reader conduct the same thought experiment which was based on four idealizing assumptions: (1) a rolling body is perfectly spherical, (2) a plane is ideally smooth, (3) a plane is placed in a perfectly horizontal position, (4) all resistance caused by the environment is excluded, i.e. the sphere moves in absolute vacuum. According to Galileo, these assumptions were equivalent to the assumption that no external force influences a moving body (p. 158).

He led his listeners to believe that if "all coincidental and external obstacles" were removed in this way the movement would last "as long as the plane extended, going neither upwards nor downwards" (p. 158). "If space was therefore infinite, movement would also last without limits, hence, it would be infinite" since in an existing situation "there is no reason for slowing down nor for acceleration, since there is neither slope nor hill" (p. 158). In an assumed situation, a body could accelerate its movement just by going downwards, i.e. moving closer to the centre of the Earth, and slow down – only by going up – i.e. moving away from the Earth's centre. In the same way Galileo wrote in *Colloquies*: "If a body moves on a horizontal plane without any obstacle then from all that has been previously presented *in extenso* it follows that this movement is uniform and constant when the plane stretches infinitely."[1]

Although, as in the above passages, one could speak of a possibility of infinite inertial movement with a uniform velocity on a plane, Galileo (mistakenly) thought that movement of a free falling body occurs not along a straight line (or plane, in a geometrical meaning of the term, i.e. on a flat, not curved surface) but along a circular line equally distant from the centre of the Earth in every point. As he wrote: "a plane which moves neither upward nor downward would have to be equally distant from the centre of the Earth in all its parts" (Galileo 1962, p. 158). Hence a question "do these planes exist anywhere in the world?" was being answered by one of the men participating in the discussion as follows: "we are not short of these – even the surface of our globe – if it was quite smoothed, not as it is now – hilly and uneven – or rather the surface of the water when it is calm and motionless" (p. 158).

This instance clearly indicates that Galileo arrived at an erroneous formulation of the principle of inertia since he accepted fewer idealizing assumptions than were needed (or, which is the same, the ones that he assumed were too weak).

Let us formulate the Galilean assumptions more precisely:

p_1: a tested body moves in absolute vacuum (no air friction);

p_2: a moving body is a perfectly rigid sphere (a sphere with no deformations);

p_3: a body moves along a perfectly smooth surface (no friction with a plane a body moves on);

p_4: a body moves on a horizontal plane (not on a plane: while moving, it does not change its distance from the centre of the Earth).

---

[1] Galileo (1930), p. 176. Both in that volume and elsewhere, the author uses misleading names such as "horizontal plane" and "flat plane" which are, strictly speaking, inherently contradictory.

It turns out that these assumptions were not equivalent to the assumptions which occurred in a correct formulation of this law supplied by Newton. This new assumption (of course, also of an idealizing kind) is the following: no force influences a body or, in other words, a body is in a free state. Instead of assuming the movement of a body along a horizontal curve close to the Earth, Galileo should have imagined (if he wanted to arrive at a correct formulation of the principle of inertia), as Newton did later on, the movement of a body in a space devoid of any other bodies (including our planet), hence of any influence of these forces. It would be a uniform and isotropic space, i.e. one in which no point and no direction would be distinguished. In this space, the movement of a body could be considered accelerated (i.e. requiring an application of force) also when velocity had a constant numerical value but changed direction. Galileo lacked the understanding of the fact that in order to keep a body moving along a curve it was necessary (according to Newton's second principle of dynamics: $F = ma$) to apply the force which is directed to the centre of a curve of the size equal to mv^2/r (whereby m is the mass of the body, v – its velocity, r – the radius of the curve), since it is then that the body's acceleration is v^2/r (this force is equal, according to the third principle of dynamics, "equality of action and reaction," to the centrifugal force, i.e. the force which the body exerts on what keeps it on a curve). Having no adequate formulation of the law of inertia, Galileo could not arrive at a proper view of gravity, even in a qualitative aspect. He did not assume that the Earth attracts those bodies; instead, he assumed that those bodies "tend" to the centre of the Earth. Contrary to the Aristotelian view, this was not "tending" towards the centre of the Universe. It would not have made Galileo much happier if he had learned that this approach to the problem of inertia was an anticipation of the general theory of relativity. In his concept of circular inertia, he assumed, without being quite aware of it, that movement in a gravitational field is also an inertial movement. This "corresponds" to a general theory of relativity and a generalized theory of non-relativity formulated on the basis of the former, according to which the forces of gravitation do not exist and gravitation itself demonstrates so many similarities to inertia that bodies moving within the gravitational field may be considered to remain in a free fall, i.e. to be in inertial movement.

Nevertheless, in spite of accepting the improper idealizing assumptions, which caused the law of inertia to be formulated erroneously, the application of the law to the analysis of the movements of earthly bodies (which can be found in Galileo's works, primarily in *Dialogue* and in *Colloquies*, his most outstanding work, in which the foundations of modern physics were created) turns out to be correct in almost all cases. This is connected with the following "lucky" circumstance. When Galileo applied his law to the bodies which were not influenced by any greater forces (those which could not be neglected when the problem was considered), he assumed that bodies move uniformly and linearly, and correctly assumed that linear and circular movement (around a great circle like the Earth's circumference) coincides sufficiently for small distances which we deal with in local movements of the bodies. That gave approximately correct results for small distances but when the distances of the size comparable to the dimensions of the Earth were concerned, this approach was quite incorrect.

In the case when Galileo considered (also accepting some idealizing assumptions) the movement of bodies which move along curves influenced by the forces which cannot be neglected (for instance the movement of missiles and thrown bodies), he did not use, not always being aware of it, his own formulation of the law of inertia, which was incorrect. Instead, he used the law of inertia which stated that if no force influences a body then it will move with a uniform and linear movement. In this case, dividing movement along a curve into "inertial" components and a remaining fragment (for instance a gravitational one which causes a change of velocity or a bending of a track), Galileo had a correct formulation of the principle of inertia in mind, i.e. he assumed that the "inertial" component of the movement along a curve is always a movement along a line and the remaining component accounted for any deviation from linear movement. This was what he did when discussing the power of centrifugal force ("the force of rejection movement") acting on the surface of the Earth and caused by its revolution around its own axis, and when discussing the force which "might have caused stones to deviate from tangents" (Galileo 1962, p. 234), he grasped the inertial component of the movement of bodies on the Earth as the movement along a tangent to its surface as a movement along a straight line. In those cases, Galileo usually noted that this was an approximate approach which did not account for the fact that inertial movement is, strictly speaking, a circular one.

In the first case, Galileo applied the procedure of (approximate) concretization of the law of inertia in his own (erroneous) formulation. In the latter case, however, he employed the procedure of concretization (often a strict one) of the law of inertia in the (correct) formulation provided by Descartes and Newton, and then the procedure of approximate concretization "eliminating" the latter's deviance from the law of inertia in his own formulation. Galileo did not forget, even then, that according to his law of inertia a linear movement is only approximately inertial.[2]

---

[2] If applications which Galileo made in *Discorsi* concerning the thrown body and on other occasions, led to correct results, then this was because he treated the respective part of the arc as a straight line in movements occurring within a small area. Nonetheless, when he justifies doing it in some other passages, saying this is only supposedly correct, he points out that he did not have a correct formulation of the law of inertia (Galileo 1962, p. 503, footnotes).

Galileo tried to confirm his law of inertia in three ways. First, he attempted to do so by means of isolating the body in the best possible way, i.e. by creating experimental conditions similar to the ones mentioned in the precedent of the law. But the movement of celestial bodies was, according to Galileo, most similar to the ideal state of free fall. Hence, Galileo saw his astronomical observations of the planets as excellent direct confirmation of his principle of inertia, which was concerned primarily with space movement along curves, and was seemingly disqualified by earthly observations (which is why the Galilean principle of inertia is often called the principle of cosmic, or circular, inertia).

In a sense, experimental procedure was, according to Galileo, a complement of the procedure of concretization. The former created the conditions that were as close to free fall as possible for earthly bodies or, generally speaking, conditions that were similar to those formulated, among others, due to idealizing assumptions in the precedents of this or that law. In order to achieve this, Galileo selected planes that were as smooth and even as possible. His followers placed moving bodies in a relative vacuum, etc. That allowed the employment of the law of inertia while investigating earthly bodies as well, accounting just for small corrections resulting from the non-ideal nature of the surrounding vacuum.

Secondly, Galileo thought that the principle of inertia is directly confirmed by all these cases when the forces influencing a body are balanced, i.e. offset one another. A typical case is illustrated by a body falling from a great height in ordinary conditions (i.e. in the air) which originally increases the velocity of fall but only up to the moment when the air resistance equals the force of gravity influencing the body: then both forces offset each other, their resultant equals zero and the body falls (moves) with a uniform and linear movement according to the principle of inertia.

Thirdly, Galileo looked for a confirmation of the law of inertia by means of its concretization, i.e. accounting for forces which influence a body and introducing proper corrections into the consequent of the law. A typical example is provided by dissolving real movement into its vector components, i.e. inertial and remaining (for instance, gravitational) components. These result respectively from the fact that a body possesses inertial mass (and hence apparent forces of inertia act upon it), and from the fact that it also possesses gravitational mass (and hence gravitational forces influence it). In this case, Galileo applied both strict and approximate concretization.

It is worth mentioning that Galileo clearly employed the principle of the superposition of forces and movements in his analysis (i.e. of velocities and accelerations) and even clearly formulated it. In *Colloquies*, we read: "if a plane is limited and a heavy body moves along it, then upon reaching the brim of the plane it will still move and to the primary, uniform, unceasing movement there will be added a complex movement caused by gravity and the one called missile movement which is composed of uniform horizontal and uniform accelerated ones" (Galileo 1930, p. 176). We are dealing here with a division of the investigated movement into two independent ones: uniform movement along a curve caused by inertia and perpendicular to it, uniformly accelerated, movement caused by gravity. Real movement occurs along the diagonal of parallelogram made up of the two (vectorial) components of this movement. This real movement turns out to be a movement along a parabola (assuming it occurs in idealized conditions). Every element of a parabola is considered by Galileo to be an infinitely small diagonal of an infinitely small parallelogram composed by uniform horizontal movement and acceleration which is perpendicular to it. Although the thought of linking (adding) some force preserving initial acceleration to a force of gravity was already expressed before Galileo (among others by Nicolaus of Cuza and Leonardo da Vinci), here is the first expression of an idea of a continuous adding of the forces of primary acceleration and the force of gravity in the whole movement of a thrown body. All this is based on an assumption that the horizontal and vertical movement of the physical bodies may be considered as two independent movements[3] and that a body may be subjected to two forces which are independent of each other (Aristotle's physics did not allow for this possibility: it did not allow for a falling body moving along a parabola, since in that case it would go farther from the centre of the Earth, at least in its horizontal movement).

IV. Galileo's Employment of the Method of Idealization in the Analysis of Accelerated Movement

Let us consider another classic example of Galileo's employment of the method of idealization which led – this time – to a fully correct result (within the required limits): the law of free fall of the bodies, which is also named after him. In Galileo's formulation, the law took the following form: freely falling body (in vacuum) covers the distance $S = gt^2/2$.

Formulating this law, Galileo accepted four following assumptions, of which one (third) labelled as p_1 he knew to be an idealizing assumption:

---

[3] Indicating on p. 179 of *Colloquies...* that a movement composed of a uniform linear one and the one which is perpendicular to it, and also uniformly accelerated results in parabolic motion of the falling body, Galileo remarks that this reasoning is based on "the assumption that a diagonal movement is always uniform, that a naturally accelerated movement is proportional to the squares of time and that these movements and their velocities mix but do not interfere with each other, do not change and do not slow down" (p. 180). Further on (pp. 185-190), Galileo presents the way of marking at every moment of the velocity and direction of a complex movement on the basis of its components, i.e. the inertial and gravitational component.

1. Bodies fall when in the vicinity of the Earth: $h(x) \ll R_e$ (where $h(x)$ symbolizes the height from which x falls, "$\ll$" the relation "considerably smaller than" and R_e – the radius of the Earth).
2. In a moment preceding the moment a body began falling, the same body remained resting (with respect to the Earth): $V_0(x) = 0$, where $V_0(x)$ is the starting velocity of the falling body x.
3. (p_1) No force influences the body except for the Earth's gravity (for Galileo, this was equal to an assumption that the body falls in a perfect vacuum: $F_{tr}(x) = 0$, where $F_{tr}(x)$ stands for forces of friction, i.e. resistance of the environment which influences the body x).
4. The force of the Earth's gravity is constant and proportional to the mass of the falling body, i.e. it causes permanent acceleration equal for all masses: $g = const$ (where g stands for the Earth's acceleration, equal to 9.8 meters per second^2).

Should these assumptions occur in the precedent of the law in its complete formula, the law itself would assume the following form: if no force apart from constant force of the Earth's gravitation influences a body remaining in rest in the vicinity of the Earth, then this body shall be constantly accelerated so that its falling distances will be equal to the constant of the Earth's gravitation (i.e. earthly acceleration) g multiplied by a square of falling time t^2 and divided by two: $S = gt^2/2$.

Let us briefly discuss Galileo's formulations concerning the precedent and the consequent of the law. The first assumption was accepted by Galileo since he started with the right idea that bodies falling in the vicinity of the some celestial body other than the Earth (e.g. the Sun or the Moon) would be falling onto the nearest celestial body, not necessarily the Earth. The second assumption is indispensable since if a body had been moving before it started to fall, and not resting, then the distance of the falling s would be determined according to the formula $s = V_0t + gt^2$, whereby V_0 is the starting velocity of the body.[4] The third assumption, p_1, of an idealizing nature, aims at excluding the environmental resistance (the air) and possibly other forces influencing a body apart from the Earth's gravitation. This assumption was indispensable, among others, in order to understand that the common sense division into "heavy" and "light" bodies

---

[4] Galileo notes that deviation from his law of free fall is at its greatest for the bodies with the greatest starting velocity (for instance: missiles fired vertically downwards), since the air resistance is then so great that their movement is not uniformly accelerated even in the first approximation. . . : on the contrary, this is a slowed-down motion, since the starting velocity of the missile is greater than velocity with which it reaches the surface of the Earth.

Since lifting the above assumption of non-idealizing nature (realistic one) also calls for introducing a respective correction (namely the member V_0t), a more general remark is suggested, i.e. that also lifting some non-idealizing assumptions is being accomplished via the procedure of concretization (strict or approximate), or, strictly speaking, analogous with some respect to the procedure of concretization.

(whereby heavy bodies tend downwards in the direction of the Earth's or the Universe's centre, whereas light bodies tend upwards) results from failing to account for the role of the air and in fact divides bodies not into "heavy" and "light" ones in an absolute sense but into those that are heavier and those that are lighter than air (i.e. bodies whose proper weight is greater or smaller than the proper weight of air).

In a space devoid of air or any other kind of matter (i.e. in a vacuum), all objects fall down in a "natural way"; since they tend toward the Earth, they are heavy. From the commonsensical point of view and from the viewpoint of Aristotle's physics based on it, the assumption of vacuum is the most important one. It abolishes not only the distinction between heavy and light bodies but also the closely related concept of the bodies falling in a quicker or slower way – that is, strictly speaking, the concept according to which the acceleration of falling bodies is directly proportional to their heaviness. The fact that this concept has been confirmed by common-sense experience shows that this law of Galileo was also contrary to common sense and experience. The fourth assumption, which is also of an idealizing nature in view of Newton's physics, was not considered to be such by Galileo. He did not realize what the nature of gravitation was like. In particular, he was unaware that the gravitational force is reversely proportional to the square of distance and that, accordingly, the Earth's gravitation $g = GM/r^2$ (whereby G is a gravitational constant, M – the mass of the Earth, and r – the distance between the attracted mass and the centre of the Earth) was constant only if the Earth's mass remained constant, too, its distribution and distance between the attracted body and the centre of the Earth undergoing no change. In order to account for the above condition, Galileo would have to formulate his law of free fall in a more complex form:

$$S = \frac{\frac{GM}{r^2}t^2}{2}$$

It is worth mentioning that Galileo was not only unaware of the idealizing nature of the fourth assumption but he was also not quite aware of the idealizing nature of the third one, since he did not know that by accepting this assumption he abstracted not only from the forces of friction but also from all other forces influencing a falling body, especially from the gravitational forces stemming from celestial bodies other than the Earth.

As far as the consequent of the law of free fall is concerned, it establishes the fact that the velocity of a falling body is proportional to the time of its fall (so that after a second its velocity is 9.8 meters per second, after two seconds it is doubled and amounting to 19.6 meters per second, after three seconds it is three times as great and amounts to 29.4 meters per second, and so forth), whereas the distance of falling is proportional to the square of the time of fall (hence after the first second a body covers the distance (a point achieved is placed below the starting one) of 4.9 meters, after two seconds it covers the distance of $4.9 \cdot 2^2 = 19.6$ meters, after three seconds $4,9 \cdot 3^2 = 44.1$ meters, and so forth) which means that the movement of a falling body is uniformly accelerated.

Galileo, and physicists in general, rarely mention all idealizing assumptions and are generally explicit with respect to those that are significant and essential for the relation described by the consequent of the law, i.e. the ones that they want subsequently to lift while performing a (strict) concretization of the law in order to explain and predict the phenomena which really occur.

For both the principle of inertia and for the law of free fall, this significant idealizing assumption consisted in assuming the existence of a perfect vacuum. It may therefore be said that the intellectual premise of modern physics was presenting a vacuum. Among the predecessors of Newton and Galileo, no one succeeded in formulating the principle of inertia of the uniform movement just because they were unable to present this movement in a perfect vacuum, in a situation when a body encounters no obstacles on its route. Hence, it is no wonder that Galileo and his followers (Torricelli and others) opposed Aristotle's arguments against vacuum's existence since this physical fiction is indispensable for a simplified but profound analysis of the course of physical phenomena.[5]

The theoretical mechanics of Galileo and Newton considered environmental resistance to be an insignificant aspect of movement (dynamics) of a solid body (naturally, this does not concern ballistics and applied mechanics in general). This was thus in fact a transformation of empirically (practically) significant factors, which often considerably influenced the course of directly observable events, into theoretically insignificant ones, i.e. those which may be bypassed in theoretical analysis which considers phenomena in their "pure form." This "transformation" is nothing else but the acceptance of respective idealizing assumptions. Proving theoretical insignificance of environmental resistance for dynamics of solid bodies was at the same time a proof of the indispensability of accepting idealizing assumption concerning vacuum.

In some situations, Galileo mentioned the rules that he had been guided with when accepting some idealizing assumptions, i.e. the rules which determined what kind of factors one could abstract from in physics formulating idealizational laws. Here is one of the rules applied. Galileo wrote that as far as obstacles "resulting from environmental resistance are concerned, they, ... because of their great diversity, cannot be included in

---

[5] It is interesting in the history of discoveries that the correct formulation of the principle of inertia has first been provided by Descartes, the scientist who did not believe in the existence not just of absolute but also of relative vacuum (in the sense of diffused matter).

strict rules, for when we consider air resistance, it interferes with all movements in different manners, depending on infinitely various changes of shapes, weights and velocity of the bodies. ... For these infinitely different cases, weights, and shapes, one may not elaborate any strict theory. One should therefore abstract in order to treat this subject scientifically" (Galileo 1930, p. 181-182).

Accepting idealizing assumptions, we should, then, according to Galileo, abstract first of all from these factors interfering with the course of an investigated phenomenon the influence of which may not be described in a simple and, at the same time, general way. Further development of physics showed that in many cases it was also convenient to accept idealizing assumptions which concerned minor influences since their effect would not be univocally and generally described.

We shall see now how Galileo tested and applied the law of free fall by concretizing it. Galileo, lifting the idealizing assumption p_1, concerning vacuum, recalled a well known law of Archimedes which says that a body submerged in fluid loses as much weight as the weight of fluid forced out by this body. Applying this law to the bodies "submerged in the air" Galileo assumed that a body submerged in the air loses as much weight as the weight of the air this body forced out. He assumed then that environmental resistance diminishes the weight of a falling body with a magnitude equal to the weight of the part of environment which has been forced out, i.e. he assumed a principle of concretization which states that the distance covered by a falling body in the air equals the distance of freely falling body minus a magnitude resulting from the air resistance.[6]

This principle allows, on the basis of the law of free fall, to deduce the aerodynamic law of bodies falling in the air. Analogically, Galileo constructed other laws of aero- and hydro-dynamics referring to bodies falling in different liquid and volatile environments. He also performed a reverse step: from some actual law of aero- or hydro-dynamics he went to an idealizational law referring to freely falling bodies, i.e. bodies falling in vacuum, the law which "acts" in empty space, and which says: if a body does not force out any weighing environment (of the volume equal to the volume of the body) then its weight is not diminished, and, consequently, all bodies fall in vacuum with the same acceleration making their distance

---

[6] Galileo writes: since "the environment subtracts as much weight from the body moving within it, as is the weight of the ousted part of environment, if one diminishes in this proportion the velocity of the falling body, which would be constant in a non-resistant environment (as has been supposed) we shall obtain what we are looking for. Assuming that – for instance lead is 10 000 times heavier than air, and ebony only 1000 times, we shall subtract one grade from 10 000 (lead) and ten from 10 000 (one from 1000) in case of lead ... from the velocity of these bodies, which would be equal in non-resisting environment. Reasoning according to this principle I think that we shall find experience corresponding to this calculation more than to the Aristotelian one. In a similar way one may find the relation between velocities of the same body in various fluid environments" (Galileo 1930, p. 63).

of falling equals $S = gt^2/2$. Galileo asserted that in order to reconcile the law of free fall with experiments (in order to test it) one should find the way to measure air whose resistance "disfigures" the law. With respect to this he undertook a number of experiments which aimed at establishing the proper weight of air, and of "coincidental" forces resulting from the air resistance, which should have been accounted for applying the law of free fall to real bodies or from which one should have abstracted when formulating the basic law.

The law obtained in the course of the procedure of concretization was not yet a factual law in a strict sense of the term: Galileo, however, thought it to be just that, since he was unaware of the idealizing nature of other assumptions, which he accepted or which he should have accepted.

However, applying it directly to bodies falling in the air, he obtained good approximations, which are within the limits of error of his measuring technique.

Nevertheless, the concretization which he performed was an approximate rather than a strict one. Introducing a "correction," referring to air resistance, to the law of free fall allowed him to explain the fact that light bodies fell more slowly and roughly to predict the behaviour of falling bodies but did not allow for precise quantitative testing (in Galileo's times). Galileo's measurements of the proper weight displaced by a falling body in the air were so incorrect that he was unable to determine whether quantitative relations assumed by this law tested positively or not.

The fact that the forces of environmental resistance (in this case, the forces of air friction) are the function of velocity (the smaller the velocity, hence a distance of fall, the smaller friction forces, hence they may be neglected when a small distance of fall is the case) gave Galileo an idea that this law might be tested by application, without any major corrections, to bodies falling from a small distance.[7] In this case, he attempted a

---

[7] Galileo knew, for instance, that for objects falling from considerable altitudes the air resistance, once they achieve some velocity, becomes so considerable that it counterbalances the force of gravity and then bodies fall with the same velocity (the so-called border velocity), hence without acceleration (according to contemporary data, this velocity equals 60 m/sec for a human with a closed parachute, and 5-6 m/sec for a human with an open one). He did not, however, know the numerical data characterizing this relation, i.e. he did not know that for small velocities the environmental resistance increases proportionally to velocity, for medium velocities – proportionally to the square of velocities, and for big ones even to the cube of the latter. Galileo knew, of course that if a starting velocity of a falling body surpasses its border velocity, deviations from his law of free fall become even more obvious: in spite of being accelerated, a body falls with a diminishing velocity.

The passage to the above way of testing this law of free fall was also conditioned by the fact that the greater an altitude (hence velocity) of falling, the greater an influence of an environment on a light body (calculated per unit of mass) compared to the analogous

procedure of approximate concretization. The difficulty that he encountered then was unsolvable at the time. A short time of fall could not have been precisely measured (Galileo had no good clock). We may see that troubles and difficulties in testing the law of free fall which Galileo encountered resulted from great velocity and great acceleration of falling bodies.

These difficulties made Galileo switch from experiments with bodies falling in the air to experiments with bodies rolling along inclined plane, which slowed down the processes caused by the Earth's gravitation (that might have been made conveniently small by diminishing the angle of inclination). That meant that he gave up testing the law of free fall in a more direct way and passed on to a direct testing of the behaviour of bodies on an inclined plane, which also made him hope to achieve precise testing of the basic law. Galileo did not just want to make the movement of the bodies slower, and hence accessible to a direct measurement. He also wanted to diminish the influence of the minor factor – air resistance (the more slowly a body moves, the smaller environmental resistance is). Galileo writes: "in order to deal with a possibly slow movement when the resistance of environment modifying the phenomenon caused by a simple gravitational force is diminishing, I have thought of forcing a body to move along an inclined plane, placed under a small angle towards the horizon."

Experiments with inclined planes really allowed Galileo to calculate the velocity of falling bodies by means of a direct measurement of the time and distance of falling. In this way, he learned that the time necessary to cover the divisions equal to 1, 4, 9, 16 of the units of distance increased as a root of these numbers, i.e. assumed respectively the value 1, 2, 3, and 4 that was a direct confirmation of the law of free fall which claimed that the distance covered was proportional to the square of time.[8] But here, too, a

---

influence on a heavier body. It may be seen that with a small altitude both heavy and light bodies fall with the same velocity (almost), whereas with a great altitude big differences in velocity occur. Moreover, environment slows heavy bodies more than it does lighter ones, since in the process of diminishing the body's volume (and weight) falls more quickly than its surface – and the bigger the surface of the body, the bigger the environmental resistance (assuming that the shape of the body does not change) (see Galileo 1930, pp. 69-74).

[8] Measuring the time with an amount of water flowing from a broad vessel through a small hole, Galileo decided that a sphere rolling down an inclined plane covers four times of its single distance when a double portion of water is being collected. Of course a sphere rolls down an inclined plane more slowly than falling straight down in a free fall. Galileo, abstracting from air resistance and friction on the surface of the plane, demonstrated that the force F_w, accelerating a rolling body is equal to the Earth's gravitation F_g, multiplied by a sinus of an angle of inclined plane – $sin\alpha$ = the height of an inclined plane / length of the inclined plane. It means that "the time of falling down an inclined plane is related to the time of vertical fall in the same way as the length of inclined plane to its height" (Galileo 1930, p. 138) and that times of falling along differently inclined plane of the same height are proportional to the lengths of these planes (p. 139). In this way, placing inclined plane

difficulty arose, which Galileo was unable to overcome: the trouble with
the friction of a body against the surface of the plane which was hard to
estimate.[9]

The above difficulty made Galileo discover the fourth and, in his
times, the most perfect way of testing the law of free fall, i.e. experiments
with the pendulum. This was also a convenient way to deal with
environmental resistance, since, as it turned out, the motion of the
pendulum was not so strongly influenced by air resistance as the
movement of bodies rolling down an inclined plane. With respect to other
features, the motion of the pendulum was very similar to the movement
along inclined plane: the bob of the moving pendulum could be treated as
movement along an inclined plane which at every point was inclined at a
different angle to the horizon. This kind of curved plane is termed a
runway.

Pendula, made of different materials but of the same length, covered
the distance with the same velocity, i.e. the time of their swings was the
same, no matter what the amplitude or mass was. This confirmed the fact
that acceleration of falling bodies does not depend on the kind of material
the pendula are made of, nor on their weight. This method also allowed for
a quantitative test of the law in question, it led Galileo towards
discovering an idealizational law concerning the motion of the pendulum
according to which "with the pendula of different lengths times are in a
relation of square roots of their lengths, or, in other words, distances are
related in the same way as squares of the times of swings, if one pendulum
is to move twice as slow as the other, it must be four times longer."[10]

---

under this or that angle Galileo could "reduce" the sphere's acceleration with any
magnitude. For instance, on a plane of 30° inclination, the acceleration of a body is equal to
half of the Earth's acceleration, i.e. it is 4.9 m/sec².

[9] Formulating the laws of the fall along inclined plane Galileo naturally assumed some
idealizing assumptions. In particular, those laws turn out to be true "... assuming always
that all coincidental, external obstacles are removed, and that planes are hard and smooth,
and a body has a perfectly round shape, so that neither a body nor a plane are rugged"
(1930, p. 126). In order to bring the conditions of inclined plane closer to the ideal ones,
Galileo made a groove on a plane and glued it with a "smooth and clean" parchment, a
sphere rolled was "smoothly made and of a very hard brass" (p. 133). This was also a
rolling sphere, not a sliding disc which also diminished the influence of friction, although
Galileo realized that this calls for another idealizing assumption and creates additional
problem: in the case of a sliding disc, the whole force of gravitation accelerates it along the
plane, while in the case of a rolling sphere, a part of it goes for accelerating its revolutions.

[10] Galileo (1930, p. 78). One should stress that Galileo did not realize that two further
idealizing assumptions had to be accepted in order to make the law of the period of the
pendulum true. First, the amplitude of the swings (the biggest swing) is sufficiently (strictly
speaking, infinitely) small or the bob of a pendulum is attracted by the Earth with a constant
force no matter how far it deviates from the position of balance. Second, a curve along
which the bob swings is not an arc of a circle but of a cycloid. As Huygens demonstrated by
introducing the concept of the ideal mathematical pendulum and constructing a precise

The physicists of the next generation, including his disciples such as Torricelli and others, tested Galileo's law in yet another (fifth) way: by means of creating a vacuum, hence the conditions closer to the ones described in the precedent of the law. Galileo knew of this way but had no means of observing falling bodies in highly diffused air.

The method turned out to be more perfect than the one available to Galileo. It did not call for any concretizations of the law in question, apart from an approximate one, which stemmed from the fact that Torricelli's vacuum was not an ideal one.

Newton finally applied the sixth and the most perfect method of learning whether the law in question is true. While the method was less direct and more theoretical than the others, it was more precise at the same time. It applied a deduction from more general and strict laws (primarily from the law of universal gravitation). The conclusions concerned the behaviour of the bodies which fell in gravitational fields of celestial bodies including the Earth. The method made it clear that the law is correct only with some approximation.

Newton was also the first scientist who was fully aware of the idealizing nature of the third and fourth assumption, and of the necessity to accept some further idealizing assumptions in order to make the formulation of the law of free fall more adequate.

V. The Idealizational Nature of the Law of Free Fall in Classical Physics

Below is a complete list of assumptions which are necessary if Galileo's law of free fall is to be adequately formulated (in the view of contemporary physics, this list should be completed with some further idealizing assumptions).

Realistic assumptions (non-idealizing ones):

1. $M(x) \neq 0$ – a falling body is an object with a mass (strictly speaking two masses – inertial and gravitational one, which coincide numerically).[11]

---

pendulum clock, the period of swings of a pendulum is independent of the amplitude of swings only when it moves along a cycloid, i.e. a curve described by a constant point on a circumference of a circle which rolls without sliding on a plane (because of this property, the cycloid is also often called the isochronic curve or, more briefly, the isochrone or tautochrone and the pendulum the period of which does not depend on amplitude H-isochronic pendulum). The time of a heavy body falling along a cycloid does not depend on the point of the cycloid from which the body starts. More precisely, an ideal material point which swings without friction on a cycloid under the influence of gravity has a period of swings independent of the amplitude of movement (this is, as Jean Bernoulli has pointed out, the shortest period, which is why we call the cycloid "the path of the quickest descent" – the brachistochrone).

[11] If a body is influenced by the force of the Earth's gravitation F_g (called the weight of the body) then according to the second principle of Newton's dynamics a body acquires

2. $V_0(x) = 0$ – a falling body starts from the resting position.

Idealizing assumptions:

3. (p_1) No force apart from the Earth's gravitation influences the body:
   a) $F_{tr}(x) = 0$ – a body falls in vacuum (the force of environmental resistance and other mechanical forces are equal to zero),
   b) $F_{gr}(x) = 0$ – no gravitational forces apart from the Earth's gravitation influence the body,
   c) $F_{el}(x) = 0$ – no forces of other nature influence the body (in particular no electrodynamics force).

4. (p_2) The force of the Earth's gravitation is constant (strictly speaking, it is proportional to the mass of the body, does not depend on the distance and is a central force directed from the centre of the body to the centre of the Earth, i.e. along the line linking both bodies):

   a) $W_e = 0$, or $\vec{V_e} = const \wedge P_e = const \wedge W_e = 4/3\pi R^3 \wedge \omega e = 0$,

   where:

   (a_1) $W_e = 0$ – the Earth is a material point (i.e. its volume is equal to zero) or behaves in a way indicating its mass to be concentrated in the centre,

   (a_2) $\vec{V_e} = const$ – the Earth is in an inertial state (its vector of velocity is constant),

   (a_3) $P_e = const$ – the consistence of the Earth is constant (distribution of mass is equal),

   (a_4) $W_e = 4/3\pi R^3$ – the Earth is an ideal sphere (not just an approximate ellipsoid),

   (a_5) $\omega e = 0$ – the angular velocity of the Earth equals zero (the Earth does not revolve, centrifugal forces and Coriolis' forces acting on its surface are equal to zero),

   b) $W(x) = 0$ or $P(x) = const \wedge W(x) = 4/3\pi R^3$, whereby:

   (b_1) $W(x) = 0$ – a falling body is a material point,

   (b_2) $P(x) = const$ – a body has constant consistence,

   (b_3) $W(x) = 4/3\pi R^3$ – a body is an ideal sphere,

---

acceleration $a = F_g/m_b$ (where m_b is an inertial mass of the body). Assuming such a unit for m_g (gravitational mass) of the body that $F_g = m_b g$ and putting it into the first formula, we obtain $a = gm_g/m_b$. Hence, if gravitational and inertial mass did not coincide numerically, we would have:

$$S = \frac{m_g g t^2}{2m_b} .$$

Since they coincide, we obtain $m_g/m_b = 1$ which leads to the formula of Galileo: $S = gt^2/2$.

   c) $g(x) = const - 981 \text{cm/sec}^2$ – earthly acceleration is constant
   (the mass of the Earth is constant – both with respect to the
   magnitude and distribution and acceleration – it does not
   depend on the distance from the centre of the Earth, or,
   strictly speaking, a body falls in a homogeneous gravitational
   field, i.e. non-relative value of the difference of potentials
   between two random points on the way of the falling body is
   equal to zero, an approximate equivalent of this assumption is
   to assume that bodies fall close to the surface of the Earth),
   d) $g_e = 0$ – the constant of acceleration is equal to zero (a body
   does not accelerate the Earth in the least) since $g_e = g(x)$
   $m(x)/M_e$ (where $m(x)$ is the mass of the body, M_e – the mass
   of the Earth), hence the assumption in question is equal to an
   assumption that $m(x) = 0$ or $M_e = \infty$ since $g_e = 0 \cong M_e = \infty$ or
   $m(x) = 0$.

It is assumed that the assumption (a_1) is approximately equal to the
conjunction of the assumptions: (a_2), (a_3), (a_4), (a_5), i.e. (a_1) $\cong$ (a_2) $\wedge$ (a_3) $\wedge$
(a_4) $\wedge$ (a_5). Newton already demonstrated that if a sphere is homogeneous
(has equal consistence in each point) then it attracts the external body as if
its whole mass were concentrated in the centre. (For the sake of simplicity,
this is not considered, although even Newton knew that a sphere with
concentric distribution of the mass acts gravitationally in an analogous
way, i.e. the distribution which makes all points equally distant from the
centre uniformly dense. The condition of constant consistence can
therefore be replaced with a weaker condition of a concentric distribution
of matter.)

Assuming that the Earth is an ideal sphere with a constant distribution
of mass and assuming that it is neither in a revolving motion around its
own axis, nor in an accelerated movement around the Sun, one assumes
what may be considered approximately equivalent, based on Newton's
theory of gravitation, to the assumption (a_1), which states that the Earth is
a material point, i.e. that its whole mass is concentrated in the middle.

Similarly, it is assumed that the assumption (b_1) is – based on the same
theory – approximately equivalent to the conjunction of the assumptions:
(b_2), (b_3), hence: (b_1) $\cong$ (b_2) $\wedge$ (b_3).

The law of free fall in its full formulation within classical mechanics
should therefore assume the following form:

$$(x) \ \{M(x) \neq 0 \land V_0(x) = 0 \land F_{tr}(x) = 0 \land F_{gr}(x) = 0 \land F_{el}(x) = 0 \land$$

$$\land [W_e = 0 \lor (\vec{V_e} = const \land P_e = const \land W_e = 4/3\pi R^3 \land \omega e = 0)] \land$$

$$\land [W(x) = 0 \lor (P(x) = const \land W(x) = 4/3\pi r^3) \land g(x) = const \land g_e = 0]$$

$$\rightarrow S(x) = g(x) t^2(x)/2 \}.$$

Why do physicists who formulate this law never provide all the idealizing assumptions that matter? Physicists are explicitly formulating only those idealizing assumptions which they consequently want to waive by way of strict concretization. It is not hard to demonstrate that it is impossible to waive some of the above assumptions by means of strict concretizations. For instance, it is impossible to calculate (even with a good approximation), a joint effect of all gravitational forces from cosmic masses on the falling bodies. This fact makes neither law of free fall nor Newton's law of universal gravitation accessible to the final concretizations, i.e. factual laws derived by way of strict concretization of the starting laws (only the so-called approximate laws are accessible).[12]

Newton realized this difficulty perfectly well. It leads to the impossibility of a precise mathematical description of the movement of bodies. From the viewpoint of the law of universal gravitation, a precise determination of the movement of a given celestial body (e.g. the Earth) is impossible, since it is being influenced gravitationally by all other bodies in the universe. By means of distinguishing gravitational forces acting between two of these bodies from this universal gravitational link (the Earth and the Moon) Newton succeeded in determining the gravitational forces in question.[13] This law, however, has never been tested precisely

---

[12] It is worth noting that it is impossible to concretize finally all the idealizational laws in which formulation irrational numbers occur, like Archimedes' π or Euler's e. One may not lift idealizing assumptions formulated with the help of irrational numbers by way of strict concretization but exclusively by way of approximate concretization (approximation). This concerns also those laws in the formulation of which there occur rational numbers which may not be expressed with a finite decimal fraction, for instance the number 1/3. This is connected with the following circumstance: all results of actually conducted experiments may be presented with decimal fractions (which are only approximations of irrational numbers), hence there is always an "error," i.e. the difference between the results of measurement and precise theoretical value of the measured magnitude expressed with numbers which are not (finite) decimal fractions. This error is greater if there are fewer decimal positions of the fraction bringing us closer to the number which we have accounted for in the calculations.

[13] It should be stressed that the gravitational forces that Newton had to abstract from are greater in this case than those which he considered (i.e. the force with which the Earth attracts the Moon). It turns out that the force of the sun's attraction for a body of a mass of one kilogram within the distance of 150 million km (i.e. the distance between the Sun and the Moon) is 599 dynes, whereas the force of the Earth's attraction for a body of the same

with respect to the fact that a movement of any body is influenced by a great (maybe infinite) amount of gravitational forces stemming from the remaining masses in the universe. On the basis of the movement of a given body one may not therefore infer that this law is strictly true.

Of course, after formulating the law of gravitation, it might be applied, theoretically speaking, to a greater number of bodies. But, as we know, a strict solution obtained with the help of known mathematical methods just finds the problem of "the movement of two bodies," while the problem of the movement of "three bodies" may only be found approximately. This is why Newton and his followers usually applied the law of gravitation to two bodies, i.e. when investigating the movement of the Earth they accounted for the mutual attraction with the Sun alone, whereas when investigating the movement of the Moon – only for mutual attraction of the Moon and the Earth, and when investigating the movement of the bodies which fall in the vicinity of the Earth – only for the Earth's gravitation, etc. They may occasionally have accounted for three bodies, i.e. they may have accounted for a joint interaction of two bodies against the third one. The passage from a precisely formulated but highly idealized case of gravitational interaction of two bodies, to the more concrete but only approximately mathematically formulated case of three bodies, elaborated by a theory of disturbances in the 19th century, brings us only slightly closer to the real situation. If we wanted to consider it precisely and universally, it would call for taking the interactions of a given gravitational mass with all the others into account.

The situation is not hopeless, though. As gravitational influences differ much with respect to intensity, the "departure from reality" is more of a quantitative than qualitative kind, for not accounting for a great number of attractions on the part of sufficiently small or sufficiently distant bodies does not result in great discrepancies between theoretically calculated and actual movement. One may therefore say that lifting idealizing assumptions which assume no attraction on the part of these bodies, we are making a legitimate approximate concretization.

In a similar way, by means of approximate concretization, physicists waive most of the above idealizing assumptions, since they apply the law of falling bodies in some concrete field. But some of these assumptions –

---

mass in a distance of 386,000 km (i.e. the distance between the moon and the Earth) is 273 dynes. It follows that the gravitational force of the sun influencing the moon is more than double the analogous force with which the Earth attracts that body. We are dealing here with an instance (in this case – Newton's one) of assuming such idealizing assumption in which one abstracts from a greater influence than the one which is being considered (investigated). As may easily be noticed, Newton might have done so (i.e. neglect the sun's attractive force when considering the gravitational force with which the Earth attracts the moon) for the sun influences the Earth with approximately the same force (per unit of mass) it attracts the moon with, thus making these bodies (approximately) equally accelerated in every moment of their movement.

different in different applications – are lifted by way of strict concretization.

Sometimes one waives assumption 4(a_5): $\omega e = 0$ and 4(a_4): $W_e = 4/3\pi R^3$ by way of strict concretization. That is why a law of falling is sometimes formulated in such a way that idealizing assumptions 4(a_2), (a_3), (a_4) and (a_5) are accounted for:

$$(x)\left\{[(\vec{V_e} = const \wedge P_e = const \wedge W_e = 4/3\pi R^3 \wedge \omega e = 0)] \right.$$
$$\left. \rightarrow S(x) = \frac{g(x)t^2(x)}{2}\right\}$$

or, in a brief form the law

$$(1)\quad S_4(x) = \frac{g_4(x)t^2(x)}{2}.$$

Lifting the idealizing assumption which says that the Earth does not revolve, the physicists introduce the following principle of concretization: The principle of concretization $K(4\text{-}3)$:

$$S_3(x) = S_4 - \frac{g(x)\left(1/289\cos^2\alpha\right)t^2(x)}{2}$$

(where α is an angle determining latitude at which a body falls).

The above principle of concretization includes a "correction" which should be placed within the consequent of law (1), if the idealizing assumption 4(a_5): $\omega e = 0$ is lifted. This correction accounts for the effect of centrifugal forces which appear in the case of the revolving movement of the Earth, and it is equal approximately to 0.34% g.[14]

From law (1) and from the principle of concretization $K(4\text{-}3)$ the following law follows:

$$(2)\quad S_3(x) = \frac{g_4(x)t^2(x)}{2} - \frac{g(x)\left(1/289\cos^2\alpha\right)t^2(x)}{2}.$$

Lifting, in turn, the idealizing assumption which assumes that the Earth is an ideal sphere, one introduces the following principle of concretization which accounts for the Earth's bulge on the equator caused by the revolutions around its own axis.

The principle of concretization is the following:

---

[14] This means that due to the centrifugal force on the surface of the Earth revolving round its axis, the bodies on the equator weigh about 0,34% less than on the pole.

$$K(3\text{-}2) \quad S_2(x) = S_3(x) - \frac{g(x)\left(98/55199\cos^2\alpha\right)t^2(x)}{2}.$$

From the law (2) and the principle of concretization $K(3\text{-}2)$ the following law follows:

$$
\begin{aligned}
(3) \quad S_2(x) = {} & \frac{g_4(x)\,t^2(x)}{2} - \frac{g(x)\left(1/289\cos^2\alpha\right)t^2(x)}{2} - \\
& - \frac{g(x)\left(98/55199\cos^2\alpha\right)t^2(x)}{2}
\end{aligned}
$$

i.e. $\quad S_2(x) = \dfrac{g_4(x)\left(1 - 1/191\cos^2\alpha\right)t^2(x)}{2}.$

Introducing the principle of approximate concretization (approximation), which lifts two remaining idealizing assumptions

$$K(2\text{-}0) \quad S_0(x) \underset{\varepsilon}{\cong} S_2(x)$$

(where ε is some constant number limiting permissible error) we obtain the following factual law:

$$(4) \quad S_2(x) = \frac{g_4(x)\left(1 - 1/191\cos^2\alpha\right)t^2(x)}{2}$$

which concerns the bodies falling (in a vacuum) in the vicinity of the Earth which revolves around its axis and has bulges on the equator.

Strictly speaking, the passage from law (1) to law (2) should also include, apart from the correction due to centrifugal force, a correction due to Coriolis force (which also appears in revolving objects and makes the falling bodies deviate from vertical fall): but as the latter is generally much smaller than the former, it is usually neglected. Nevertheless, one may say that the passage from law (1) to law (2) is being made by way of strict and approximate concretization.

On other occasions, for instance when the law is applied to the bodies falling from a great height, assumption 4(c): $g(x) = const$ is being lifted by way of strict concretization (that this is an idealizig assumption follows from the fact that $g = GM/r^2$ i.e. constancy of the Earth's gravitational force g assumes immutability of both gravitational constant G and the mass of the Earth M, and the distance of the falling body from the center of the Earth r).

Let the starting law has the following form :

$$(1) \ (x)\left[g(x) = const \rightarrow S(x) = \frac{g(x)\, t^2(x)}{2} \right]$$

briefly speaking: $S_1(x) = g_1(x)t(x)^2/2$. Introducing a correction for diminishing of the earthly attraction with the altitude we are applying the principle of concretization

$$K(1-0): \quad S_0(x) = S_1(x) - \frac{g(x)\left[1 - R^2/(R+h)^2\right]t^2(x)}{2}$$

where R is the radius of the Earth, h – altitude of the fall.

From law (1) and the principle of concretization $K(1-0)$ the following law results:

$$(2) \quad S_0(x) = \frac{g_1(x)\, t^2(x)}{2} - \frac{g(x)\left[1 - R^2/(R+h)^2\right]t^2(x)}{2}$$

i.e.

$$S_0(x) = \frac{g_1(x)\left[R^2/(R+h)^2\right]t^2(x)}{2}.$$

Sometimes in place of the correction $R^2/(R+h)^2$ the magnitude which approximates it is deduced $1-2h/R$ such that:

$$R^2/(R+h)^2 \underset{\varepsilon}{\cong} 1 - 2h/R$$

where:

$$\varepsilon = \frac{R^2}{(R+h)^2} - (1 - 2h/R) = \frac{3Rh^2 + 2h^3}{R(R+h)^2} = \frac{h^2(3R + 2h)}{R(R+h)^2}$$

one should mention the fact that ε is an approximate magnitude here, depending on both R and h.

Then we have the principle of approximation (approximate concretization): $K'(1-0)$: $S_1(x)-S_0(x) \leq \varepsilon$ from which in conjunction with the law (1) the following law is obtained:

$$(2') \quad S_0(x) \underset{\varepsilon}{\cong} \frac{g_1(x)(1 - 2h/R)t^2(x)}{2}$$

or, more precisely:

$$\frac{g_1(x)(1 - 2h/R)t^2(x)}{2} - S_0(x) \le \varepsilon .$$

For instance, when the altitude of the falling body a is $h(a) = 6.4$ km, then, as $R = 6.37$ km:

$$S_0(a) \underset{\varepsilon}{\cong} \frac{g_1(1 - 0.002)\, t^2(a)}{2}$$

or more precisely:

$$\frac{g_1(1 - 0.002)\, t^2(a)}{2} - S_0(a) \le \varepsilon$$

where $\varepsilon = 0.000007$ km.

Naturally, the above concretization is not strict but is merely an approximation. It does not establish precisely (only with some approximation) what correction should be introduced to the idealizational law when the respective idealizing assumption is being lifted.

There are two differences between the strict and approximate concretizations usually performed in physics. First, one idealizing assumption is usually lifted in strict concretization, compared with usually some more in approximate concretization (all those which have not been lifted by way of strict concretization). Second, strict concretization is regularly explicit, while approximate one is implicit, stating just that a given idealizational law is fulfilled with a good degree of approximation in a given empirical domain (although this is not a precise manner of saying it).

The procedure of concretization is always finished (or should be finished) with an approximate, not a strict approximation. This follows from the fact that the idealizing assumptions which have been necessary to make the law true are always more numerous than the ones the constructor of the law is aware of (sometimes this is linked to the fact which we have mentioned above, namely that some idealizing assumptions are of the kind which may not be lifted by way of strict concretization). Therefore, in the course of the growth of a given branch of knowledge the amount of realized idealizing assumptions grows, since the ones assumed so far turn out to be insufficient for a complete formulation of some law. The growth of physics reveals new relations, additional influences on the part of minor phenomena, which were not realized before, and which must be abstracted from when some idealizational law is being formulated.

It turns out that, on many occasions, some assumptions formulated in the precedent of some law and thought not to be idealizing ones are actually idealizing. This was the case of the constancy of earthly acceleration *g* assumed by Galileo to be non-idealizing. This was also the case of the existence of an ideal vacuum which in Newton's and Torricelli's times, and a little later, was considered to actually exist in some areas, and which turned out to be a physical fiction. The assumption which concerned it turned out to be, as Galileo rightly believed, an idealizing assumption.

I have discussed the method of idealization as understood and used by Galileo, for whom it was a method of constructing particular idealizational laws.

In the hands of Newton, it became a tool for constructing whole model theories (made of idealizational laws) and entered a new, higher stage of growth therewith. The sense of abstract concepts of classical mechanics, such as absolute time, absolute space, absolute rest, absolute movement, material point, isolated system, free fall and others, may be grasped only when the method of idealization in physics is understood. According to Einstein, Newton could not have erected the monumental of classical physics without them.

It is astonishing, though, that while Galileo was fully aware of the method that he applied, Newton was almost completely unaware of his, and taught the physicists the erroneous view that in both theoretical and experimental physics the method usually does not go beyond normal inductive methods of generalization and synthesis of what may directly be observed, which is quite inadequate to their scientific practice. This brings to mind the words of Einstein: if one wants to learn real methods that physicists apply, one should not mind what they say but see what they really do.[15]

---

[15] See Einstein (1933). The above passage from Einstein is similar in intention to a well known saying of Lenin that one should think of the philosophers based not on the signboards they erect for themselves but on the manner in which they solve the basic philosophical problems.

PLATO'S PHILOSOPHY AND THE ESSENCE
OF THE SCIENTIFIC METHOD

I. Galileo – the Founder of Modern Scientific Method

Galileo is the founder of a scientific method of idealization which has been so successfully applied in modern science. He is not only the first modern scholar to apply this method on a wide scale, but probably also the first to be fully aware of its nature.[1]

With regard to the latter point, Galileo clearly surpasses even the greatness of Newton, who – applying the method of mathematics-based idealization (or idealizing abstraction) which was virtually identical to Galileo's – was mistaken in claiming that his was the conventional method of induction consisting in generalization of experimental facts.[2]

Given all that, my presentation, which aims at contrasting Plato's philosophy with the essence of scientific method, will of necessity turn into a comparison of the kind of philosophy represented in the writings of the author of the *Republic* with Galileo's scientific method, or to be exact, with the method of idealization applied for the first time on such a massive scale by a scholar.[3]

II. Galileo's Method of Idealization

Galileo not only systematically applied idealization in his research but was also able to grasp its essence and characterize it in his works (see: Galileo 1962, pp.157-159, 219-252, 466-467; Galileo 1930, pp. 54-55, 60-63, 181-183). He was also aware of the fact that his method was superior to the method of inductive generalization and synthesis of what can be directly observed. The latter method became standard as it was used by Aristotle, particularly in his physics.

---

[1] Compare what the following have to say on the subject: Butterfield (1958, pp. 5-9), Kotarbiński (1961, pp. 327-328), Such (1978, pp. 49-74) and Nowak (2000, pp. 17-27).

[2] Newton's mistake of misrepresenting his own research method was noted by such different scholars as for instance Einstein (1933) and Engels (1953, p. 211).

[3] It is worth noting at this point that attempts at using the method of idealization go back as far as the antiquity. Occasionally, it was used by Archimedes and Eratosthenes. Cf. Nowak (1973), Such (1978).

What does Galileo's method of idealization consist in?[4]

Instead of simply generalizing experimental facts, it is based on forming idealizational laws and then, in its second stage, on making them concrete; that is the theoretical concretization stage.

What then are idealization laws? They are strictly general statements directly referring not to some empirical (observable) objects but rather to their simplified (idealized) models which are often called ideal types. Formulating laws of that kind, we often make certain assumptions, known as idealizing assumptions. It is assumed in the hypotheses (counterfactually, i.e. contrary to what we think the real state of affairs is) that certain characteristics of the objects under analysis (mass, velocity, size, and so on) reach their boundary (extreme) values or limits; they usually equal zero or, to simplify somewhat, "infinity."

At the same time, it is commonly known that characteristics of empirically existing ("real") objects never (or in any case almost never) reach such boundary values. That, in turn, means that "real" objects approximate, to various degrees, the ideal types which model them but can never be identical to them. Thus, idealizational laws are statements which satisfy no domain of reality, as objects they refer to form a null class in this domain; or to put it differently, the laws do not concern existing objects.

Thus, what are such laws formed for?[5] They are formed because they let one describe a given phenomenon or a process "in its pure form," undistorted by various accidental factors of secondary importance. Thus, for instance, in order to find out how "bodies behave when left to themselves," Galileo, Descartes, and Newton all make the same idealizing assumption, namely that no external forces have any influence on the bodies in question. On the basis of that they can arrive at the idealizational law known as the law of inertia. In Newton's terms, it states that if no external forces act upon a body, it preserves the state it was previously in, i.e. the state of rest or inertial motion.

Similarly, when Galileo formulated his law of falling bodies, he assumed counterfactually, i.e. made an idealizing assumption, that every body falls in an ideal vacuum, i.e. it falls freely and no other force (except for the gravity of the Earth, which must be constant) acts upon it.

If we are to answer the question of why science needs idealizational laws (or rather why idealization method is used in science), we should also

---

[4] Galileo himself called his method "geometrical." His decision can be explained by the fact that until Newton and Leibniz discovered the differential and integral calculi, geometry (together with trigonometry) had been the most powerful mathematical tool applied in physics and other sciences.

[5] We should remember that the majority of laws formed within theoretically well-developed empirical studies are idealizational laws; thus, within sciences of that type they are the rule rather the exception.

add at this juncture that no strict mathematical description of phenomena would be possible without idealizational procedures.

That, in turn, points to a way out of the well-known paradox of the usefulness of mathematics in empirical studies (abstract study far from reality serves various empirical studies deeply involved in this reality so well). It is reality that is to blame. It is so complex that its full description in strictly mathematical terms would be incomprehensible to us.

III. Plato's Theory of Ideas as a Quasi-Description of the Scientific Method

It seems that it was Pythagoras and Plato who first understood that well developed science does not result from applying strictly empirical observational methods plus induction but rather from mathematically based idealization and approximative modelling of phenomena. A conclusion like that could be arrived at from the fact that the two philosophers were the first to be simultaneously interested in the progress in mathematics of their times and also in the first attempts at the mathematization of the budding natural sciences of the epoch.

It should also be noted here that their interest in scientific activity and scientific method did not in any way prevent them from giving philosophical justifications of their ideas whose nature was much closer to religious or even mystical theorizing than to scientific one. I mean here Pythagoras's theory of mathematical harmony and Plato's theory of ideas.

And yet, it is impossible to disagree with two German physicists and philosophers, Heisenberg and Weizsäcker, when they claim that Plato's theory of ideas has been the first fully developed doctrine to properly grasp the essence and "nerve" of the scientific method. (cf. Heisenberg 1959, pp. 52-66; Weizsäcker 1978, Part IV, pp. 433-520). In his theory of science (or, to be exact, of infallible knowledge), the first to have come down to our times, Plato arrived at the right conclusion that science (scientific knowledge) is concerned not with observationally given phenomena but instead with the ideal models (archetypes) of the phenomena of the world. In its theoretical statements, science does not characterize – contrary to what Aristotle and even Newton claimed later on and which was to be sanctioned in the so-called inductive model of scientific reasoning – features common to the analyzed objects (its typical qualities). What it does deal with are the essential features (qualities) that are hidden under the surface of the investigated phenomena. We are able to get at them with the help of idealizing abstraction which always gets at the essence of phenomena but never with the aid of generalizing abstraction skimming over their surface. And by essential features (attributes), we usually mean those attributed to the ideal type which lacks all secondary features (attributes).

Thus, to give an example, the ideal-type greylag goose (greylag) is neither any single goose of the many wild greylag geese (greylags), nor anything of the "greylag-goose-ness" characteristic of any empirical greylag goose (greylag); nor is it any statistical mean of all empirical greylag geese (greylags) (cf. Weizsäcker 1978, pp. 448-478; also: Lorenz 1975). Thus, "the greylag goose" (the greylag) as described by a zoologist interested in its habits can never be empirically observed exactly in the same way in which it appears in his/her description (as "the ideal type" with respect to its anatomy and behaviour). Yet science deals just with that ideal type; in fact it is only possible just with respect to that type (Weizsäcker 1978, p. 450).

Ideal types are thus regularities which make greylag geese (greylags) possible; they are paragons (prototypes) that are by some distance removed from empirical grey geese. Other notions of the developed science, such as e.g. "intelligence" or "atom," are obviously of the same nature as that of "greylag goose" (greylag). Therefore, the ideal type (the idea, paradigm, prototype, regularity) as described directly in theoretical statements of the developed science has nothing to do with the structures (forms, Gestalten, types) inherent in the analyzed objects and phenomena. Instead, ideal types are certain simplified models of those phenomena and objects which it is at all possible to describe formally in some general way, for example in mathematical terms. If we accept all that, we may then agree with Weizsäcker that "Platonic ideas of things are the world's structures which make those things possible" (p. 451).

IV. Plato's vs. Aristotle's Idea of Science

There can be no doubt that Aristotle went much too far in the criticism of his teacher's and master's theory of ideas. He rejected not only the ontology of the ideas, thus denying the notion of their existence as the only true and fully subsistent entities (he seems to have been right in that respect) but at the same time he refused to accept the epistemological value and methodological importance of Plato's ideas as indispensable theoretical means of scientific analysis of empirically existing objects.

It is true that what we learn about are, in Aristotle's opinion, also universals (forms, sets of typical attributes). Yet his universals are inherent in particular things and can be "removed" from those things by means of generalizing abstraction (i.e. enumerative induction, which Fr. Bacon later termed, with a shade of disdain, "the induction of the ancients").

Plato's ideas (the archetypes) are, on the other hand, things that we learn about in a scientific way: they are not dissimilar from the rules of nature discussed in contemporary natural sciences. One cannot arrive at them directly through the senses. Along the way to knowledge one starts from, let us say, a given ball and moves to a perceptually given sphere in order to arrive at the sphere which is a mathematical idea, the latter being the true object of scientific analysis.

As a matter of fact, Plato distinguishes between two types of scientific thinking (which are simultaneously two levels of "intellect," i.e. of human understanding and knowledge). The first, that is hypothetical (or discursive) reasoning, is characteristic of mathematics (i.e. arithmetic and geometry), while the second, "higher" kind of "noetic" reasoning, needs no assumptions: in its process ideas may by directly apprehended by reason (that is, we arrive at the essence of such ideas as the Truth, Beauty, Goodness, and others). The second kind of reasoning is characteristic of dialectics (or philosophical reasoning). Thus, according to Plato, dialectics becomes the most profound and important science of all, the science of the essence (of ideas), whereas mathematics forms a kind of introduction to dialectics being the study of numbers and geometrical forms.

Plato's prescientific knowledge (called "opinion" and dealing with sense perceptions and beliefs) resembles dreaming and can only lead to error. In turn, his mathematical hypothesizing (which concerns only mathematical objects) is simply a basic step on our way to mastering the highest of sciences, i.e. dialectics, which allows us arrive at the ideas themselves and, furthermore, it may even lead us to understand "the assumption-free principle of the ideas" (the monad and the unrestricted diad).

Platonic method of "arriving at the idea" (understood in the sense of looking for the essence of a thing) is more or less the same as the way of forming scientific notions and scientific laws. Thus, it is the method of science. In the ancient world of Greece, attempts to follow the method were made by Archimedes in his mathematics and physics (his statics) and by Eratosthenes (in astronomy).

Just as Ptolemy's astronomy, however, Aristotle's physics and cosmology (related to astronomy) took a different route, which was much closer to common-sense knowledge, i.e. the path of inductive generalization from empirical facts. While the first (Platonic) path led to the discovery of what is essential in things (essentialism), the other (Aristotelian) tried to capture the general in things (phenomenalism). The former made use of the idealization method based upon the procedure of idealizing abstraction; the latter favoured the method of induction based on the procedure of generalizing abstraction.

The Middle Ages (which for all practical purposes improved neither on Aristotle's physics nor on Ptolemy's astronomy, the impetus theory and the theory of local movement aside) chose the second path, which turned out to be a dead end. Modern science of the 16th and 17th centuries, however, took the first route, which had been blazed by Plato. And it was the idealization method (used also by Copernicus and Kepler) that resulted in a successful rise of modern science and its fast progress marked by great achievements (see Nowak 1971, pp. 173-176; Krajewski 1974, pp. 3-22; Nowakowa 1975, pp 48ff; Such 1978, pp. 49-74). The 17th century owes the method its name as "the century of scientific revolution." It was the period of crucial changes in the way scientists reasoned. Galileo, for instance, made use of the method fully aware of its superiority over scientific methods applied so far (at the same time, he mentioned its previous successful applications in Archimedes's statics). The unusual fertility and versatility of its applications in modern natural science (particularly in Newton's writings) seem to have resulted from the fact that the 17th century favoured both experiments and mathematics which are two prerequisites of the success of the method. The idealization method may flourish only in the context of experiments, given that there are clever analysts capable of making use of the experiments' results with the aid of mathematics. Only then can the method show its real explanatory and predictive powers and prove its superiority over the previous method, which despite appearing simple (as being based on common sense and everyday experience) was in fact deeply implicated in metaphysical assumptions of Aristotle and his Peripatetic followers that were divorced from any common sense or experience.

Thus, while Aristotle founds his physics on everyday experience and metaphysical (i.e. speculative) assumptions, Archimedes and Galileo, on the other hand, found theirs on scientific experiments and idealizing assumptions. The latter, in turn, are rooted in the most controversial yet the most fertile of all Plato's theories, i.e. his theory of ideas.

THE IDEALIZATIONAL THEORY OF SCIENCE AND PHYSICS OF THE MICROWORLD*

I. Presentation of the Problem

Since there are various methodological differences between microphysics and classical physics (physics of the macroworld), a question arises of whether the power of the idealizational procedures, which is fully manifested in classical sciences, especially in macrophysics, can be extended onto the whole (so far existing) physics, or whether it is unexpectedly restricted in the physics of the macroworld.

Such a doubt may arise particularly when one considers the fact that in classical physics the concept of isolation was probably the most frequently used concept in the process of characterizing theoretical objects. As it is, it is a "component" of such objects which it considers as "a body in a free state," "isolated system," "inertial system," "inertial motion," "free falling," and even such as "the perfect gas" or "the ideal liquid," where it occurs in a form that is as if concealed.

On the other hand, in quantum mechanics (within it or on its foundation) a number of principles are formulated which indicate the inadmissibility of interpreting microobjects (especially elementary particles or virtual particles) as isolated objects (and that are in any case found fully isolated from their surroundings).

These principles impose restrictions in particular on the following isolation procedures (either of a theoretical or an experimental character).

(1) Isolation of a microobject from other microobjects (and their complexes); it is connected with the fact that in the microworld the fact of "internal indefiniteness" of microobjects (which consists in the fact that a given state of a microobject does not determine univocally any of its subsequent single states but only a probability of future states) is accompanied by "an increased role" of external influences that cause the future states in most cases to depend in a great measure on the surroundings.

---

* Translation of "Idealizacyjna teoria nauki a fizyka mikroświata," published in R. Egiert, A. Klawiter, P. Przybysz (eds.) (1996). *Oblicza idealizacji* [Facets of Idealization]. (*Poznańskie Studia z Filozofii Humanistyki,* **2**, 15), Poznań: Wydawnictwo Naukowe UAM, pp. 173-183.

(2) Isolation of microobjects from fields which are connected with them; in the field theories, microobjects are usually interpreted as *quanta* of definite fields and sometimes even as *singularities* of a field of one kind or another.

(3) Isolation of microobjects from (macro)measuring instruments which is connected with the fact that due to Planck's constant and the impossibility of an optional minimalization of influences there are no purely cognitive procedures (purely informational ones) that would not destabilize the state of the microsystem. It is expressed in the Bohr's principle of complementarity and in the principle of relativization to the means of observation (as formulated by Fock). One should pay attention to the fact that the impossibility of isolating in a quantum description of a microobject from macroinstrument is connected with two other impossibilities: (3a) the impossibility of isolating a microobject from the rest of the world and (3b) the impossibility of isolating the microobject from the observer. As has rightly been observed by W. Heisenberg, "the measuring device deserves this name only if it is in close contact with the rest of the world if there is an interaction between the device and the observer" (Heisenberg 1959, p. 56). Therefore, in order to perform its functions, the measuring instrument must be in contact with three systems: (1) the object researched, (2) the studying subject and (3) the rest of the world. The peculiarity of measurement in microphysics consists in the fact that the first of these systems is a microobject, which in view of the fact that the other two are always macrosystems leads to numerous complications in the quantum-mechanic description.

The peculiarities given here (and other ones) of the phenomena of the microworld cause speaking of the reality (about the existence) of microobjects in the same sense in which the objects of classical physics (stones, machines, tables, planets) are real (exist) to bring about various difficulties, which lead to the concept of the "potential reality" (possibility) as a form of existence of microobjects. According to a widespread conviction formulated by Heisenberg, "in the experiments about atomic events we have to do with things and facts, with phenomena that are just as real as any phenomena in daily life. But the atoms or the elementary particles themselves are not as real; they form a world of potentialities or possibilities rather than one of things or facts" (p. 160; emphasis J.S.). It might surely be expressed in the following way: only after a connection with the macroinstrument does a microbject obtain full reality, it becomes fully defined (as a particle or a wave). This leads to a conclusion that in the light of quantum mechanics microphenomena really exist (i.e., occur in time and space) only when they are components of the observational situation (for the concept of observational situation see below).

However, a different ontology of phenomena of the microworld (and also the fact that, perhaps in order to be adequate, their description requires application of a logic that is different from the classical one, i.e. a quantum logic)[1] at the same time causes the interpretation of a microsystem exclusively as a particle (or exclusively as a wave respectively) to be – in the light of Bohr's complementarity principle – an inadmissible idealization, in spite of the fact that macrophenomena of classical physics are successfully interpreted by it (idealizationally, i.e. counterfactually) as (exclusively) corpuscular objects or (exclusively) as waves. The inadmissibility of this idealization in the sphere of microphysics itself indicates also at least partially potential nature of existence of microobjects as it is when revealing at the present time (in a given experimental situation) its wave properties, a microobject is able to reveal (in another experimental situation) also the corpuscular properties somehow inherent in it in the potential. After all, Heisenberg's principle of indeterminacy does not allow the possibility of interpreting microobjects as simply particles or waves either, since e.g. attributes ("the initial conditions") of particles which determine their state are (closely determined) locations and momenta (velocities), which is precisely what is precluded in the case of microobjects.

(4) In my view, one more restriction in the applicability of the concept of isolation in microphysics is indicated by the fact that the quantum-mechanics description of the state of a microobject is never a pure and fully objective description of the state of a microobject but is at the same time a metadescription of our knowledge concerning this state (the description that cannot be separated from it). This is one of the most critical differences between classical physics and quantum physics. The difference resulted in physics (in any case, physics of the microworld) being forced to renounce the ideal of a fully objective description, which was the guiding principle of classical physics, and with which, as is well known, researchers of the calibre of Planck or Einstein could not agree. This is a proof that in the quantum-mechanics description a microobject one cannot be isolated (even if only in thinking) not only from its physical situation but also from "factors" which are at least partly mental, such as observation procedures

---

[1] According to C.F. von Weizsäcker's opinion, which is also shared by Heisenberg, quantum logic – just as intuitionist logic – must give up the law of the excluded middle. Along with true and false utterances, it introduces "complementary utterances" concerning the so-called co-existing states which do not have their own classical equivalent. "The coexisting states" are such quantum states which – through "the interference of probability" – constitute "a mixture" of mutually contradictory (mutually exclusive) classical states. Heisenberg points out that contradiction may be avoided here also on the grounds of classical logic but then the word "state" should be considered as "describing some potentiality rather than a reality ... the concept of 'coexistent potentialities' is quite plausible, since one potentiality may involve or overlap other potentialities" (Heisenberg 1959, p. 159. See also C.F. Weizsäcker 1978, pp. 193-199: "Logika kwantowa i wielokrotne kwantowanie" [Quantum Logic and Multiple Quantification]).

or the state of knowledge of the subject who describes it. The fact of the inadmissibility of this isolation is emphasized by the concept of observational situation introduced by Heisenberg.[2]

The concept of an observational situation assumes the inseparability of the object from the subject of observation, i.e. their unity. It is only their interpretation as one whole which makes it possible to include in the description, among other things, the knowledge which the subject has on the object.[3]

As can be seen, an observer appears twice in the considerations on the interpretation of quantum mechanics: once in conjunction with the role of the measuring macroinstrument which s/he manipulates, and the second time in connection with the role of knowledge that participates in the quantum mechanics description. Heisenberg explains these two functions of an observer as follows:

> The transition from the "possible" to the "actual" takes place during the act of observation. If we want to describe what happens in an atomic event, we have to realize that the word "happens" can apply only to the observation . . . , and we may say that the transition from the "possible" to the "actual" takes place as soon as the interaction of the object with the measuring device, and thereby with the rest of the world, has come into play; it is not connected with the act of registration of the result by the mind of the observer. The discontinuous change in the probability function, however, takes place with the act of registration, because it is the discontinuous change of our knowledge in the instant of registration that has its image in the discontinuous change of the probability function (Heisenberg 1959, p. 54).

Therefore, the first function of the observer consists in "physical action" (that is why it can be replaced by an analogical function of the measuring device) while the other on "mental influence" (influence of knowledge).

Because of the four restrictions indicated which must be taken into consideration in all isolation procedures in microphysics, no quantum-mechanics description is a full description and that the full description would have to concern the whole Universe. C.F. Weizsäcker pays particular attention to it when he writes:

> Mutual interaction described by means of the language of quantum mechanics is an internal dynamism of a combined object consisting of mutually interacting objects; the primary considered object "got lost" in this combined object. The measurement on the initial object occurs only then when on the objects which interact with it, which we call a measuring instrument, occurs an irreversible process. The irreversibility is not, however, a quantum theory feature of the description of states; it signifies a transition

---

[2] See Heisenberg (1979, pp. 255-256). After all, Heisenberg claims that the concept of an observational situation is indeed "lacking in Newtonian physics" but it occurs also in "the Gibbsian interpretation of thermodynamics"; therefore, the need for it was present already on classical grounds.

[3] See the translator's remark on p. 256 of Heisenberg's work (1965). The translator (K. Wolicki) stresses that "the notion of an observational situation encompasses more than the notion of observational conditions" and, in contrast to the latter, is not a concept with an exclusively objective content.

to the classical description, description of knowledge of finite beings on finite things. Thus, by necessity it dispenses with some portion of the possible quantum theory information on the combined system (dependencies of phases between objects and a measuring instrument) and therefore, a unity of this system. One could say that spatial definitions become possible only when some part of the quantum theory unity is lost.

Let us now apply it to the Universe. Actually, the description of any object in the world as the isolated One is after all always unjustified. An object would not be an object in the world if it were not connected with it by mutual interaction. However, in such a case, strictly speaking, there are no objects. If something could surely be a quantum theory object then only at best it could be the whole world (Weizsäcker 1978, pp. 503-504).

Let us bear Heisenberg's opinion, which is similar to the one above:

It must be observed that the system which is treated by the methods of quantum mechanics is in fact a part of a much bigger system (eventually the whole world); it is interacting with this bigger system; and one must add that the microscopic properties of the bigger system are (at least to a large extent) unknown. This statement is undoubtedly a correct description of the actual situation. Since the system could not be the object of measurements and of theoretical investigations, it would in fact belong to the world of phenomena if it had no interactions with such a bigger system of which the observer is a part (Heisenberg 1959, pp. 153-154).

It can be shown that also a whole number of other idealizing assumptions (apart from the one on isolation) that are widely used in the physics of the macroworld (and in astronomy and cosmology) encounter significant restrictions in the physics of the microworld. In particular, the assumption that the objects studied are points belongs here. As is well known, the "point" interpretation of elementary particles in quantum electrodynamics leads to the so-called divergencies in the process of determining their mass (and, respectively, energy) or a load: these magnitudes assume an infinite value. It is only the application of some procedures of "renormalization" that allows ascribing to these particles the real values of these magnitudes. And the concept of "a material point" previously appeared to be very fruitful not only in mechanics (mechanics of material points) but also in astronomy (celestial bodies as material points), kinetic theory of gases ("the perfect gas"), and in many other fields as "the material point" constituted a perfect idealized model of macrobodies (as well as atoms and molecules).

As far as quantum theory is concerned, the point interpretation of microobjects is – according to the principle of complementarity – permitted to a much larger extent in some experimental situations than in others, namely when a microobject reveals its corpuscular properties. On the other hand, in situations when it occurs primarily as a wave, its properties stray far away from the properties of point creations. One reservation must be made at this juncture. It is obvious that the principle of indeterminacy imposes limitations on the point interpretation of a microobject also in such situations in which it reveals exclusively corpuscular properties because of a canonical connection of position and velocity (momentum).

II. The Problem of Isolation in Microphysics

Thus, what is the possibility of applying isolation procedures in microphysics on the level of theory and practice?

It is beyond any doubt, at least in my mind, that those procedures are used, albeit to a lesser extent than in the older (classical) disciplines of physics such as classical mechanics, classical thermodynamics, Faraday-Maxwell electrodynamics or theory of relativity. After all, also some other fields of modern physics use isolation procedures on a lesser scale than their classical counterparts. Modern thermodynamics, is defined as "thermodynamics of open systems," while the classical thermodynamics of Boltzmann-Gibbs was, as is well known, a "thermodynamics of closed systems." The openness of the systems studied made modern thermodynamics find just its fundamental application in biology and also in synergetics and other theories of self-organization, since living creatures and other systems capable of self-organization (dissipative systems in Prigogine's terminology) are always open systems.

It is just the classical physics which, at the cost of abstracting from the influence of the measuring instrument on the investigated object, could realize its ideal of the completely objective (and at the same time fully deterministic) description and explanation of phenomena. Quantum theory cannot do that because without taking into account the results of the mutual influence between a microobject and macroinstrument it is not possible to predict the future states of the microobject, and in this connection, the future results of the measurement. This state of affairs is not changed by an endeavour[4] to define the concept of the studied object in order to include the measuring instrument into it (see Heisenberg 1959, p. 56).

Nevertheless, we have to agree with a statement that "in all physical investigations it is assumed that the system studied undergoes only certain definite influences" (Amsterdamski 1965, p. 237). Indeed, in the opinions of Weizsäcker and Heisenberg, it is that, due to the unavoidable interaction of the microobject with the rest of the world, a fully exhaustive quantum description of any object would have to be at the same time a description of the whole Universe. However, descriptions actually realized in quantum theory (e.g. of the state of elementary particles or atoms of simpler elements) are considered to be adequate enough in spite of the fact that they are not in a position to account for the whole of these interactions. Therefore, one cannot abstract merely from some (namely, four previously mentioned) kinds of interaction in the quantum theory description of the state of microobjects. This side of quantum theory (and

---

[4] S. Amsterdamski terms this attempt "a terminological attempt" (cf. Amsterdamski 1965, p. 238).

of other fields of microphysics) appears simply a discipline of a lesser degree of abstraction (idealization) than the macroscopic physics, where one can (idealizationally) abstract also from each of the four interactions mentioned above, taking of which into consideration is unavoidable.

Taking into account the fourth (last) kind of influence (dependence of the quantum description on the state of our knowledge), Heisenberg states: "It may be said that classical physics is just that idealization in which we can speak about parts of the world without any reference to ourselves. Its success has led to the general ideal of an objective description of the world" (Heisenberg 1959, p. 55).[5]

III. The Problem of Objectivity of the Quantum Theory Description

Let us now deal with one more issue, i.e. that of the possibility of abstracting (idealization) from the subjective factors in the process of quantum description. As we know, this description is performed by means of the probability function, which "represents a mixture of two things, partly a fact and partly our knowledge of a fact" (Heisenberg 1959, p. 47).

With regard to the objective and subjective content of this function, Heisenberg writes as follows:

> The probability function combines objective and subjective elements. It contains statements about possibilities or better tendencies ("potentia" in Aristotelian philosophy), and these statements are completely objective, they do not depend on any observer; and it contains statements about our knowledge of the system, which of course are subjective in so far as they may be different for different observers. In ideal cases the subjective element in the probability function may be practically negligible as compared with the objective one. The physicists then speak of a "pure case" (p. 53).

Therefore, it appears that only in certain exceptional cases, "ideal" and "pure" cases, can one abstract from subjective elements in the description of the quantum state by means of the probability function. On the other hand, under normal conditions (when a description is concerned with "mixed" cases) abstraction (idealization) of this type is forbidden. In this context, Heisenberg poses the following question: "To what extent, then, have we finally come to an objective description of the world, especially of the atomic world?" (p. 54). Heisenberg's answer is as follows:

> In classical physics, science started from the belief – or should one say from the illusion? – that we could describe the world or at least parts of the world without any reference to ourselves. This is actually possible to a large extent. We know that the city of London exists whether we see it or not. It may be said that classical physics is just that idealization in which we can speak about parts of the world without any reference to ourselves. Its success has led to the general ideal of an objective

---

[5] According to Heisenberg (1959, p. 102) even such words as "space" and "time" in their common sense "refer to a structure of space and time that is actually an idealization and oversimplification of the real structure." Heisenberg is inclined to interpret all scientific concepts as idealizations which are devoid of an immediate relation with reality (see pp. 96-97).

description of the world. Objectivity has become the first criterion for the value of any scientific result. Does the Copenhagen interpretation of quantum theory still comply with this ideal? One may perhaps say that quantum theory corresponds to this ideal as far as possible. Certainly quantum theory does not contain genuine subjective features, it does not introduce the mind of the physicist as a part of atomic event (pp. 54-55).

While formulating a thesis on the fall of the ideal of a purely objective description of the world, Heisenberg claims that great theoretical systems which followed one after another in physics appear "to be ordered by the increasing part played by the subjective element in the set" (p. 95).

Only in classical mechanics was this element not taken into account at all. On the other hand, thermodynamics, while introducing statistical laws into consideration, and along with them probabilistic concepts, which are always combined with the incompleteness of our knowledge, actually deviates to some extent from this ideal. According to Heisenberg, the special theory of relativity, creating with the classical electrodynamics "the third system," goes even further.

Nevertheless, "the first three systems" (classical mechanics, thermodynamics and electrodynamics, along with the special theory of relativity), according to him, while advocating the ideal of the fully objective description, constitute (in some respect) a certain idealization which has been given up only by quantum theory:

> In the fourth set that of quantum theory, man as the subject of science is brought in
> · through the questions which are put to nature in the *a priori* terms of human science.
> Quantum theory does not allow a completely objective description of nature (pp. 95-96).

IV. Conclusion

The above considerations lead us to a conclusion that quantum mechanics abandons certain idealizations (idealizing assumptions) which played an important role in classical (macroscopic) physics. In particular, some assumptions variously concerning isolation of the studied objects (microobjects) do not play in it such a significant role as before, which leads even to giving up an ideal of a purely objective description.

However, this does not mean that this field – in contradistinction to all other areas of scientific research – does without any idealizational procedures; it does not even mean that idealization (idealizing abstraction) plays in it a less significant role than, e.g., in the physics of the 17th-19th centuries. One should agree with Heisenberg that "idealization is necessary for understanding [the reality]" (1959, p. 97). Heisenberg rightly observes that:

> When we represent a group of connections by a closed and coherent set of concepts, axioms, definitions and laws which in turn is represented by a mathematical scheme we have in fact isolated and idealized this group of connections with the purpose of clarification (p. 96).

THE IDEALIZATIONAL CONCEPTION
OF SCIENCE AND THE STRUCTURE
OF THE LAW OF UNIVERSAL GRAVITATION*

1. When the question is raised about the aims of idealizing assumptions (or broadly speaking: modelling assumptions) being introduced to the antecedent of a law, the laws in question should be given some attention, such as their best known representative, i.e. the Newtonian law of universal gravitation.

On reconstructing the law as an idealizational one, it is generally held that one of the necessary idealizing assumptions appearing in its correct formulation is that of the gravitational isolation between a two-element system (i.e. a system of two objects having gravitational mass) and the remaining masses (bodies) of the universe.[1]

2. The assumption, however, turns out to be superfluous for the correct formulation (and hence, for the proper reconstruction) of the law, at least within classical physics. This is due to the fact that in Newton's theory of gravitation it is assumed that the force of gravitational interaction between some (e.g. two) given objects is independent of other influence – both gravitational and otherwise – exerted upon those objects by the other objects. The assumed simple superposition (summation) of gravitational forces is fully justified due to the linear character of the gravitational field equations.

It should be remarked in this connection that things look different in the case of the general relativity theory (GRT), which is at the same time a

---

* First published in J. Brzeziński, F. Coniglione, T. Kuipers, L. Nowak (eds.) (1990), *Idealization II: Forms and Applications* (*Poznań Studies in the Philosophy of the Sciences and the Humanities*, **17**), Amsterdam-Atlanta: Rodopi, pp. 125-130.

[1] Cf. e.g. Nowak (1974, p. 93) where we can read: "The law of gravitation was formulated by Newton for material points and under the assumption that the two considered points are not affected by a third one." The role that is played in the gravitation law by the isolation assumption finds an even stronger expression in J. Kmita, who writes: ". . . the conditional, the form of which the law of universal gravitation has, does not narrow down its antecedent to the condition that the individual objects x and y may be any physical ball-shaped objects (or material points) but – above all – it defines them as members of a relatively isolated two-element system; thus – due to the abstraction – it adopts an isolation assumption that no interactions with other objects are taken into account" (Kmita 1976, pp. 42-43). A similar view is presented in my paper (1978).

modern relativistic theory of gravitation whose field equations are, however, of a non-linear character, which is evident from the fact that the gravitational field here "interacts with itself" and, accordingly, simple principles of superposition do not apply to it.[2]

3. Needless to say, all gravitational (and other) interactions exert an influence on the motion (and/or form) of investigated objects, thus influencing indirectly – by causing changes in their relative positions – the quantity of gravitational interaction, which is of our present interest. However, there is no mention in the law of universal gravitation of the motion of gravitating objects, only of their gravitational interaction which is conceived as a function of the quantity of respective masses and the square of relative distance.

In this respect, the law differs considerably from many other physical laws, such as Galileo's law of free fall. The latter defines the fall of objects (as a function of the time of fall and the force of the Earth's gravitation), whereas in reality it also depends on other interactions (both gravitational and non-gravitational). Thus, the idealizing assumptions that introduce (to the antecedent of the law) the isolation of the falling object become indispensable.

4. It seems that from the above-mentioned point of view, all physical laws can be divided into two groups: one which comprises the laws for which the condition of isolation is superfluous, and the other which consists of the laws whose correct formulation must be accompanied by that condition. The laws belonging to the former group include, next to Newton's law of gravitation, Coulomb's law, Newton's third principle of dynamics as well as other laws accounting for forces and interactions of various kinds (e.g. field interactions, in linear theories of field) as functions of position, mass, object velocity, etc. The latter group, in turn, would include, next to Galileo's law of free fall, Newton's first and second principles of dynamics, and other laws describing body actions (i.e. motion) under the influence of the applied forces and other interactions.

Due to the universal nature of the above classification (at least within physics), the case of gravitation law indicates – as one in whose correct formulation the isolation conditions do not play any significant role

---

[2] The non-linear character of the gravitation field equations of GRT has turned out to be an auspicious opportunity which made it possible to introduce logical simplification ("inner improvement," in Einstein's terms) into the theory by deriving equations of GRT motion from the (non-linear) field equations of the theory. This was accomplished in the 1930s by two unrelated groups of scientists: by Einstein, Infeld and Hoffmann (Einstein, Infeld, Hoffmann 1938, pp. 65-100 and 1940, pp. 455-464) on the one hand, and by Fock, Pietrova and their co-workers (Fock 1939, p. 81) on the other. The opportunity for such derivation, which arose due to the non-linear character of GRT field equations, was first noticed (as recorded by Einstein) in 1927 by Hans Reichenbach.

because (two or more) objects are attracted to one another with a given gravitational force depending on their mass and the distance between them, and independent of the interactions they may enter with other objects – that the question posed is of a general nature and seems to be methodologically valid.

5. Needless to say, the above discussion does not prove in any way that the laws of the first group are not idealizational in nature, or, in other words, that the introduction of idealizing (or broadly speaking: modelling) assumptions to their antecedents is unnecessary.

Owing to the very fact that strictly inertial systems simply do not exist, each law of classical physics in its usual form, i.e. not in its general covariant form in which it is strictly satisfied, is idealizational in nature, at least in inertial systems.[3]

In addition, there appear various other circumstances due to which theoretical laws of the first group are also usually idealizational. Thus, in case of the law of universal gravitation, one (but certainly not the only one) such circumstance is the fact that the law concerns, strictly speaking, material points and only such objects (which in reality do not exist), possessing the gravitational mass, whose mass distribution is exactly barycentric (concentric), which makes it mathematically tenable to accept the assumption (albeit of limited validity) that their mass is clustered in the centre; because the objects satisfying the condition of barycentricity attract other objects (in fact, only the ones lying outside) as if their mass was clustered in the centre.

6. Moreover, it turns out that with respect to the first group of laws the assumptions of the object isolation are not devoid of all significance either, even though they do not perform the function that is usually assigned to them. Thus, although they have no validity from the point of view of adequacy assessment of a ready-made law – i.e. their introduction to the antecedent of a law does not increase its adequacy – they may be crucial for the very process of the law formulation and its verification. In other words, they may be said to be devoid of any (direct) semantic values

---

[3] The situation changes again when GRT appears, which suggests to the physicists the way of "crossing beyond the narrow conceptual bounds of the inertial systems," owing to the fact that within GRT it became possible to work out a method of covariant formulation of physical laws; the method which, from the point of view of physics, is grounded in the principle of acceleration-gravitation equivalence (inertial mass – ponderable mass equivalence) that reveals the relative character of acceleration. The relative nature of acceleration is demonstrated, to a certain degree, in Einstein's thought experiment with a lift, whose actual state can be always interpreted in two different ways: either as a standstill (or a state of inertial motion) in a gravitation field of non-zero intensity, or as a state of the accelerated motion of the direction contrary to the direction of the affect of the gravitation field which is assumed by the former interpretation.

but not of pragmatic ones which, moreover, are sometimes crucial. Realizing that other factors (in addition to those which are of the immediate study) bring about similar effects as the studied ones, makes it possible to study the investigated factors in their undistorted form, i.e. free of any "side effects," by having those factors eliminated, either conceptually or experimentally.

Thus, Newton, for instance, having abstracted from mutual gravitational interactions of the planets and accepting several other simplifying (idealizing) assumptions, could derive Kepler's laws of his law of universal gravitation. As Kepler's laws had already been well confirmed, he was in a position to verify experimentally his newly discovered law of gravitation.

The above example shows that the opposition between "pragmatic" and "semantic" values should not be treated as clear-cut. Its maintenance, however, is of true significance once we realize the difference between some idealizing assumptions being introduced for the purpose of adequacy assessment of a law, on the one hand, and for the purpose of its easier application or verification, on the other.

7. In my opinion, the fact that idealizing assumptions are superfluous for the correct formulation of the gravitational law has been confirmed by a well-known historical evidence. Thus, while accounting for the gravitational interaction at first with relation to the system constituted by the Earth and the Moon, Newton disregarded the gravitational interaction occurring between that system and a certain external body (i.e. the Sun), whose interaction has turned out to be twice as strong as the interaction between the Moon and the Earth.[4]

When, in 1798, Cavendish was defining, with the aid of torsion balance, the gravitational constant k, testing experimentally the force of attraction between two lead balls, 1 kg each, the main difficulty lay in "grasping" the force of attraction free of various distortions caused by gravitation and other factors which made it impossible to study the system of two lead balls in ideal vacuum.

It goes without saying that in each of the above cases the greater gravitational force could have been disregarded, so that the smaller one had been easier to define, due to the fact that the greater force functioned as some "constant background" upon which the investigated system had been "thrown." In practical applications of the law of gravitation, this is a normal procedure that only more intensive gravitational interactions are

---

[4] It has turned out that the Sun exerts a gravitational influence upon a body of 150 millions km away from it (this is the mean distance between the Moon and the Sun) with the force of 599 dynes, while the Earth affects a body of the same mass located 386,000 km away from it (this is the mean distance between the Moon and the Earth) with the force of 273 dynes.

taken into account and the weaker ones are simply bypassed. However, it often happens in theoretical analyses and in scientific experiments, as was the case with Newton and Cavendish that gravitational interactions are studied between two or more objects seen as isolated systems, irrespectively of the disregarded forces being greater or smaller than those actually investigated.

If, therefore, the isolation assumptions were necessary conditions for the correct formulation of the law of gravitation, the situation in which – applying the law – some (much) stronger gravitational interactions are disregarded while other (much) weaker ones are taken into account would be ruled out

8. As is evident from the above discussion, the two mentioned classes of physical laws, one exemplified by Newton's law of universal gravitation and the other by Galileo's law of free fall, illustrate, so to speak, two distinct aspects of the question of adaptation and cancellation of the idealizing assumptions; for the laws of the former class, the idealizing assumptions are necessary not so much for the adequacy assessment of their (final) form as for facilitating the process of their formation, of theoretical and practical applications, as well as for facilitating the procedure of verifying them.

One might say that in the case under discussion the idealizing assumptions perform the function of some auxiliary substructure rather than that of construction blocks forming the edifice of science.

Naturally, the problem outlined above requires further thorough analysis.

Part II
The Nature of Scientific Cognition

ON KINDS OF KNOWLEDGE

I. Five Kinds of Human Knowledge

A contemporary person encounters various kinds of knowledge. Two of them are taken into consideration most often: scientific knowledge and common knowledge, also called common-sense knowledge. It seems advisable to distinguish, apart from these two, three other kinds of human knowledge: artistic and literary knowledge connected with literature and art, speculative knowledge, contained mainly in the systems of speculative philosophy and religion, and the so-called irrational knowledge, connected, for instance, with mystic cognition or some kinds of intuition. It can be accepted that the above mentioned kinds of knowledge account comprehensively for the whole of modern human knowledge.

In the course of further considerations, an attempt will be made at characterizing the above mentioned kinds of knowledge and listing basic differences between them. Let us start with scientific knowledge.

II. Scientific Knowledge

1. Planes of Interpreting Science

Scientific knowledge is an aspect of the multidimensional product of social development which we call science. As is well known, "science" is a term which has many meanings. Seven fundamental meanings of this term can be distinguished; thus, also seven most important planes of interpreting science. When speaking about science we can mean:

(1) a kind of knowledge ("scientific knowledge"),
(2) a kind of activity ("scientific research activity"),
(3) a method of proceeding ("scientific research method"),
(4) a group of people practising science ("scientific community"),
(5) a social institution ("complex of scientific research institutions"),
(6) a form of social consciousness ("scientific consciousness"),
(7) a component of the productive forces of a community ("scientific productive force").

The aspects of science (planes of its interpretation) mentioned above are the subject of study of various disciplines which deal with science and which are part of a discipline called the science of science. The most important ones are: the philosophy of science (also known as the theory of science or methodology of the sciences), the sociology of science, the psychology of science, the economics of science and the history of science.

2. Criteria of the Scientific Nature of Knowledge

One of the tasks of philosophy of science, which focuses mainly on the first three planes of interpreting science, is to distinguish scientific knowledge from other kinds of human knowledge. In order to do this, scholars formulate distinguishing factors of scientific knowledge which are usually called criteria of the scientific nature of knowledge.

Since scientific knowledge is genetically derived from common knowledge, one takes into consideration first of all that makes it different from common knowledge when establishing the criteria of scientific knowledge. However, those currents in the philosophy of science which aim at liberating science from all kinds of metaphysics and speculation, formulating criteria of the scientific nature of knowledge, are primarily concerned with drawing a distinct demarcation line between scientific knowledge and philosophical knowledge of a non–scientific nature (also known in this case as "metaphysics"). This is the way in which neopositivists of the period of the Vienna Circle (Schlick, Carnap) proceeded.

Some of the most important discriminants of scientific knowledge are the following:

i. *A high degree of logical systematization, or the ordering of statements by means of the relation of consequence.* Generally, science aims at constructing (approximate or strict) deductive systems, which are just sets of statements, ordered by means of relation of consequence. A deductive system constitutes a class of statements (theses of a system) which result logically (deductively, reliably) from a certain subset of statements called axioms (postulates, principles, and the like).

The degree of the systematization of statements within other kinds of knowledge is generally much lower; in addition, it is not a logical ordering, but, for instance, an alphabetical one (as in a telephone directory).

ii. *A high degree of informational content: logical and empirical.* Scientific statements are formulated so that they provide as much information as possible on a subject which they concern. The logical content of a statement is determined by the number of logical consequences, which result from it, while the empirical content can be measured by means of the number of consequences of an observational nature (i.e. consequences concerning states of affairs which can be directly observed, including prognostic sentences). In turn, the number of consequences of a statement depends on the level of its generality and precision: the more general and the more strict a statement, the more it tells about the world, the greater its informational content, both logical and empirical.

iii. *A high degree of the intersubjectivity of knowledge.* Intersubjective knowledge is knowledge which, firstly, may be understood by every normal, properly prepared (e.g. educated) person (this condition is termed the condition of intersubjective communicability). Secondly, it is submitted to public control, i.e. may be tested by every person (or a group of people) who is/are properly prepared or having at his/her/their disposal appropriate means (this condition is called the condition of intersubjective testability). The other kinds of knowledge fulfil the condition of intersubjectivity (especially intersubjective testability) to a much lower degree, and irrational knowledge does not fulfil it at all (or only to a very small extent).

And it is so in science that scientific discoveries are generally made by one person or a small group of people but they become part of the treasury of scientific knowledge only after they have been tested by many scientists (or groups of researchers), working independently of one another to such an extent that the control results might be considered as evidence which have their own independent value.

iv. *A high degree of self-criticism and self-control.* Scientific knowledge is characterized by the self-critical attitude to its own achievements, which is unparalleled in other kinds of knowledge. It is rightly said that science was created when people started to critically relate to all the previous achievements in the field of cognition: both achievements of their predecessors, teachers and colleagues as well as their own. On the other hand, the other kinds of knowledge are distinguished by a considerable measure of apology (this refers mainly to religious knowledge and knowledge contained in the systems of speculative philosophy): they are mostly satisfied with the state achieved, which is one of the impediments to their development. Among other things, the reason why science develops so fast is the self-critical attitude of scientists, based their assumption that no result of scientific studies is everlasting (in fact, this is a subjective factor, and one that is of secondary importance, yet worth emphasizing). That means that each one of them remains in the treasury of scientific knowledge only as long as a better one appears, and this makes them continue their search of a better (more adequate and deeper knowledge) than the one obtained so far.

Obviously, not all scientists and not always are capable of such an adequate criticism and "distancing" from their own views. But a criticism of which the initiator of new ideas him/herself is not capable is exercised within a wider scientific community, which performs (should perform) a merciless selection of newly formulated hypotheses and suppositions.

v. *A high degree of theoretization.* This is the fifth of the most important distinguishing factors of scientific knowledge which is never restricted to a set of empirical data, which are a direct result of observation, experiments and measurements, but concerns deeper characteristics and levels of reality, reaching the essence of things. Other kinds of knowledge generally provide a more superficial, "mirror-like" knowledge, which touches just the surface of phenomena, and which often concerns appearances. While using methods of modelling, abstraction and idealization, science omits those aspects of the phenomena which are less important, adventitious and incidental, in order to reach the internal mechanisms and regularities, concealed from our senses.

vi. *A high level of generality.* Scientists never cease to formulate singular statements, i.e. statements that concern individual phenomena. Singular and existential statements usually serve in science as premises for (inductive and other) generalizations as a result of which statements of a great level of generality, sometimes universal statements, are obtained. Such statements concern (or at least are supposed to concern) all the phenomena which occur in the world. For instance, a scientist who has solid foundations for supposing that "all bodies attract one another" will not formulate statements of a kind "all planets attract one another" or "all cigars attract one another" but formulates, as Newton did, "the law of universal gravitation." Even the so-called idiographic-nomological sciences (previously termed idiographic), such as history, are not restricted to formulating singular statements but search for true general statements: the laws or historical generalizations.

vii. *A high degree of precision.* Science is not satisfied with vague general statements, whose degree of generality is high but which are devoid of precision. They cannot be comprehensively and severely tested and, apart from that, they do not provide detailed prognoses, and, in this connection, do not constitute a proper theoretical basis for effective activities. Since mathematics provides the most powerful means of increasing the degree of statement precision, thanks to the fact that it transforms qualitative statements (laws) into quantitative statements (laws), it is small wonder that it is applied to a greater and greater extent in various sciences.

viii. *High prognostic power of knowledge.* Better than other kinds of knowledge, scientific knowledge provides means which serve as a reliable prediction in many fields. Prediction is an indispensable condition of effective activity, hence great practical importance of scientific knowledge.

ix. *High explanatory power of knowledge.* Science does not only allow a description of phenomena, or an answer to the "how?" question (how do the phenomena occur?). It also permit an explanation of phenomena, or an answer to the "why" question (why do the phenomena occur the way they do and not differently?). This is because scientific knowledge is to a great degree theoretical: only knowledge, reaching deeper mechanisms of reality and laws which rule it, is capable of providing the real (rather than merely apparent) explanations.

x. *High heuristic power of knowledge.* Scientific knowledge accumulated so far always constitutes an indispensable and powerful instrument for obtaining new knowledge. The greater fertility of scientific knowledge than of other kinds of knowledge is connected with values such as a high degree of systematization, generality, precision, its theoreticality, and high prognostic and explanatory power.

The most important criteria of the scientific character of knowledge mentioned above do not exhaust the whole list. Neither can they be concretized nor made more precise (e.g. by indication "how high the level" of systematization, generality, preciseness the newly formulated statements should achieve so that they can be included into science), as long as we speak about science in general. As it is, various fields of sciences on various levels of their development make unequally severe demands on the statement formulated within them as far as the ten above mentioned features are concerned. In this connection, it is said that criteria of the scientific character of knowledge are, first, historically variable, and second, relativized to a given discipline, and so they do not have an extra-temporal (extrahistorical, universal) meaning. Therefore, if one wants to make them more precise or concretize them, one should move into the realm of considerations concerning particular fields of sciences or even singular scientific disciplines.

3. Basic Kinds of Science

Let us select from amongst different classifications of science, which serve various purposes, the one which is probably most often adduced just because of its great naturalness.

It is represented in the following schema:

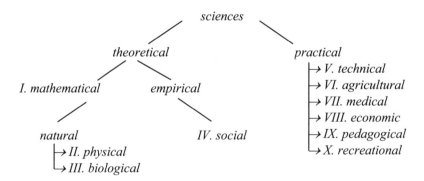

This schema specifies ten basic fields of science. Two more kinds of science should be mentioned here. These sciences are developing unusually dynamically and are typical of contemporary times. These are:

Sciences at the interface (XI) (also called the borderline sciences), which are created at the interface of two, three or more "classical" sciences (here belong such disciplines as: physical chemistry, physicochemistry, biophysics, biochemistry, bionics, and the like), and

Integrating sciences (XII) which combine in their subject matter aspects of phenomena so far explored many times not only by related disciplines but also sciences which are distant from one another (the most important ones are: cybernetics, synergetics, information theory, communication theory and general theory of systems). Integrating sciences are also borderline sciences but in a somewhat more extensive meaning of the word: they are created at the border of such fields of sciences as theoretical and practical sciences, mathematical and empirical sciences, natural and social sciences, biological and physical sciences and the like.

Let us characterize briefly theoretical sciences and practical sciences.

Theoretical sciences conduct the so-called fundamental research whose aims are purely cognitive: a description and explanation of phenomena. Thus, they allow to theoretically capture the world, i.e. to understand it. On the other hand, they pursue their practical aims mostly by means of practical sciences which, due to this, are also called applied sciences. Apart from the cognitive and practical function, theoretical sciences perform a significant world-view function by providing philosophy with premises that allow the creation of the general view of the world.

The aims of practical sciences are not only directly cognitive. They also have practical tasks in the field of effective transformation of natural and social reality. They are a "bridge" between fundamental studies and the so-called implementation and development research whose goal is the creation of new technologies and other means that serve to satisfy various non–cognitive social needs.

Practical (applied) sciences are distinguished by their comprehensive and complex character in the sense that in general they include the whole of knowledge about a given type of phenomena. Any applied science is based on several fundamental disciplines, which penetrate from various points of view the phenomena that are the object of interest of a given applied science. For instance, pedology, as one of the main disciplines of agricultural sciences, is based on such sciences as physics, chemistry, geology, biology, geography. Medical sciences take knowledge about man from such sciences as physics, chemistry, biology, psychology, sociology, cybernetics. If the high degree of generality of knowledge is a distinguishing factor of fundamental (theoretical) sciences then the high degree of wholeness and complexity of knowledge characterizes applied (practical) sciences.

III. Common Knowledge

Both in our phylogenetic and ontogenetic development, we encounter this kind of knowledge at the earliest. It is based on common sense and everyday experience of people. It is an unusually extensive type of knowledge, and at the same time an extremely inhomogeneous one since common sense is "a reservoir" of true judgements, based on people's practice in the course of their life, accumulated since the oldest times, as well as on superstitions which have been preserved through custom and tradition.

Common knowledge is superseded by scientific knowledge in many respects but not in all of them. Its characteristic features – as compared with scientific knowledge – are as follows:

1. *A low level of theoretization.* It does not directly concern models (theoretical constructs) of phenomena and objects studied; instead, it pertains to the surrounding phenomena themselves, given to man in sensual cognition (perception).

2. *A low level of generality.* Statements, formulated within it, mostly concern either particular individual objects or sets of objects consisting of a small number.

3. *A low level of precision.* Precision is not its characteristic feature, it usually formulates the so-called qualitative statements, which do not quantify the phenomena studied and which do not apply an advanced mathematical apparatus.

4. *Low informational content.* It is a consequence of the two previous features. Common statements, which are general and precise to a small extent, do not state much about the objects which they concern.

On the other hand, common knowledge is not inferior to scientific knowledge, in spite of what is sometimes thought, with respect to the degree of truthfulness and the degree of certainty.

Common statements, just because of their low level of theoretization, generality, precision and informational content, thus because of the fact that "they are close to experience" and only to a small extent exceed that which is "given directly," are mostly reliable in practical activities and are not usually inferior to scientific statements as far as truthfulness is concerned. Let us use an example of two statements on gravitation: the common and the scientific one.

The common statement which states the fact of gravitation, formulated already before Newton, and sometimes called the qualitative law of gravitation says: "All bodies attract one another."

Its scientific counterpart is the quantitative law of gravitation, formulated by Newton, who called it "the law of universal gravitation." It states: "All bodies (which possess mass) attract one another with the force determined by means of the formula $F_{gr} = Gm_1m_2/r^2$."

In the light of general relativity theory, which is a theory of relativist mechanics and at the same time a modern theory of gravitation, Newton's law of gravitation has been disproved since not all bodies attract one another according to Newton's formula, while the common law of gravitation has not been shaken, since it is still true that "all bodies gravitate."

Common statements, just because of their vagueness (a low level of exactness), and consequently limited informativeness (i.e. the fact that they say very little about the world), often appear to be more true and more certain than scientific statements, telling much about the course of phenomena and leading to precise prognoses.

However, common knowledge, while providing an enormous number of correct descriptions or correct statements on how things really are, at the same time does not have either great explanatory power or great prognostic power. The proverbial "grandmother" (or "grandfather") as someone who has great resources of common knowledge but has not had many opportunities to get in touch with scientific knowledge – can supply many true and useful descriptions of the course of various events. However, when we start asking her (or him) why things are as they are and not different, we usually obtain wrong answers. And this is just an evidence that common knowledge is capable of providing true descriptions rather than true explanations, while scientific knowledge contains both. This is connected with the generally superficial character of common knowledge with the reservation that, unlike from scientific knowledge, it does not penetrate deeper levels of reality. In particular, it usually does not reach regularities which control the phenomena, the causes which evoke them and mechanisms of their course.

Also, predicting is not very good on the grounds of common knowledge; due to its low precision and the tendency to use vague general statements, prognoses based on them are usually not very exact and very deceptive.

However, it should be emphasized that common knowledge is needed by man exactly "everywhere" and that without it none of us could normally live and function in the society. That is why the postulate of replacing common knowledge by a more perfect scientific knowledge is mistaken. It could not be carried out if only for the lack in many spheres of human life of a scientific equivalent of common knowledge. It is also redundant in such a general formulation, since where common knowledge suffices for the realization of a given aim, there is no need to involve knowledge of an admittedly higher cognitive status but which is more difficult to master and apply, and after all scientific knowledge is just such knowledge. A sensible postulate of replacing common knowledge with the scientific knowledge is that it should take place only where, on the one hand, common knowledge appears to be insufficient, and where, on the other hand, scientific knowledge ensures the efficiency of those activities that are indispensable for achieving them.

IV. Artistic/Literary Knowledge

This is the third most important kind of knowledge without which humankind could not do, and in any case could not develop normally.

Knowledge contained in literature and art is first of all knowledge about humans, their complex nature and their psychological resources, about their behaviour and outlook on the world and life. Literature and art also reach, one may assume, those levels and dimensions of human life which, due to their complexity or low accessibility, cannot so far be submitted to the exact and reliable scientific analysis, based on strictly established facts. One of such dimensions is that which is called "the spiritual image" of the human being.

Artistic–literary knowledge brings general truths about the human being and the world closer by means of the so-called presented reality showing concrete life situations. It is generally characterized also by an image and sensual way of expression due to which it is easily (and, in an illiterate society, even commonly) available.

Its value is not exclusively instrumental: it consists in providing the human being not only with information about issues of life but also with artistic experiences. While getting to know a work of art, the human being at the same time makes his/her life more pleasant; contact with a work of art is usually also a form of recreation. In this context, the kind of knowledge under discussion has excellent recreational values. Of great importance is also its participation in forming human being's personality. The artistic/literary knowledge is also a vital factor in the emergence of human worldviews (*Weltanschauung*) and the human attitudes towards life. Its educational value goes hand in hand with the above.

There is yet another function of this type knowledge which deserves attention. Contact with a work of art, in any case with the work of literary art, consists in a better definition ("concretization") of the reality depicted in the work. This is of great importance for the development of children and young people, especially of such features of their intellect as originality, fantasy and imagination.

The artistic/literary knowledge is, therefore, an irreplaceable culture-forming factor in human life, which develops all levels of personality, mainly the intellectual and emotional spheres.

The artistic/literary knowledge reveals numerous direct relations with all the other kinds of knowledge. First of all, as knowledge about person it is interrelated with the humanities, with common knowledge about person and issues of life and with the philosophical and religious knowledge about person and society.

V. Speculative Knowledge

This term refers to knowledge which is separated from experience and life practice and, at the same time, from science. The "separation from life" usually consists in the fact that the solution of the problems posed is of no great significance to human behaviour. For this reason, they are sometimes called – somewhat derisively – "academic," "scholastic" or "speculative" problems. The "separation from science" consists in turn in that it resolves its problems in a non-scientific manner, i.e. neither drawing on the results of sciences nor applying scientific methods. Thus, it is knowledge that is built on a frail material similar to sand: on fantasy, fiction and speculation. It is a product of "pure thinking" (as it is sometimes said) where what counts are results not so much in agreement with "the logic of facts" but results which correspond to our wishes, expectations and images "what things should be like."

On the whole, religious knowledge is speculative in nature, as was – and sometimes still is – the case of philosophical knowledge. These are not, however, kinds of knowledge of purely speculative character.

Religious knowledge, which, on the one hand, is knowledge concerning personal behaviour evaluated from the point of view of good and evil (or the behaviour submitted to ethical evaluations), and not only from the point of view of eschatological aims (those concerning eternity), displays numerous relationships with the human life practice. On the other hand, however, it is knowledge of an external world and of the so-called supernatural phenomena, accumulated by our very distant ancestors. This knowledge reflect their views on the world and life, opinions which are often in collision with results of scientific studies and with the views of contemporary man, superimposed on science, art and everyday practice.

As far as philosophical knowledge is concerned, during the twenty three centuries (from the middle of the 4th century B.C., i.e. since Aristotle until mid-19th century) it was generally speculative in nature, despite the process of approximating it slowly to life and science. During this period, philosophers (called "the period of speculative philosophy") constructed great systems whose goal was "to explain everything," but actually provided, in most cases, only apparent explanations. It was the period of hegemony of philosophy over sciences in which philosophy was said to be the queen of sciences, and sciences were treated as its subjects. Ignoring facts sometimes assumed in this period such drastic proportions that if some scientific theory was not compatible with facts then it was said: "so much worse for the facts!," obviously provided that it was in agreement with the prevalent system of speculative (religious or philosophical) worldviews.

During the last century and a half, however, there has been a breakthrough in philosophy as a result of which some fields of philosophy (in some interpretations) obtained the status of scientific knowledge. This does not mean, however that all speculation has been eliminated from them. It is impossible since, as shown by epistemological analyses, no kind of knowledge is totally free of speculation. It is so even within the so-called exact sciences (mathematics, logic, exact natural sciences, econometrics, mathematical linguistics), which also contain elements of speculation. For instance, Newton's great system of physics, which constitutes the first well-developed discipline of exact natural sciences, was based, among other things, on a speculative assumption that physical interactions (e.g. gravitational ones) may disperse with any, even "infinite" velocity ("momentarily"). And it is only in the following centuries that field physics and relativistic physics showed that the upper limit of velocity of dispersing of any interactions (causal ones including) is the speed of light in a vacuum (marked with symbol c).

Obviously, this does not mean that all knowledge "is speculative in character." The distinguishing factor of speculative knowledge is not simply that it is not tested knowledge (through reference to experimental facts or logico–mathematical determinations) but that this is a fundamentally untestable knowledge. Here I mean knowledge that is formulated in such a way that no facts – whether experimental or logico-mathematical in nature – are capable of either confirming or refuting it.

For this reason, the main value of speculative knowledge does not consist in providing faithful descriptions or explanations of the surrounding phenomena. As fantasy not bound by facts, it is not so much a model reflecting features of the real world but one that creates models of fictitious worlds, being a creation of our imagination and – no less importantly – our wishes and longings. Thus, through its means we "are in touch with" the possible worlds at the most, which undoubtedly expands our intellectual horizons and strengthens the power of our imagination.

It follows that it is not adequacy and truthfulness, i.e. adaptation to the real world, but originality and richness of ideas that are its main distinguishing factor. However, let us say that this is not the most sought-after value in knowledge that is why knowledge recognized as speculative does not enjoy great popularity in the contemporary society in which the authority of scientific knowledge has been strengthened.

As is shown by the analyses of the development of science, a seemingly paradoxical phenomenon is revealed here, which allows viewing speculative knowledge from a different, more favourable, angle. Namely, it appears that it is indispensable material, a starting point which cannot be replaced by anything, and a repository of initial guidelines for the creation of new hypotheses and ideas in science, or new scientific knowledge. Speculative knowledge also provides science with many research problems, so its power consists not in solving problems (whereby lies the power of science) but in posing them.

VI. Irrational Knowledge

This is knowledge which stands in contradiction to reasonable principles, and in any case, is difficult to capture by the mind. It does not fulfil the condition of intersubjectivity: it is neither intersubjectively communicable, i.e. cannot be verbalized or passed on to others, nor intersubjectively testable, i.e. cannot be submitted to public control.

A question may be posed whether this kind of "knowledge" is at all possible or whether the very idea of "irrational knowledge" is not internally contradictory, and thus whether it is not something of a "square circle" or "childless mother."

Mystic knowledge is an example which illustrates that irrational knowledge is possible. Mystics usually claim that knowledge which they obtain through "direct contact with the supernatural world" (in the mystic way) cannot be expressed in human language and thus passed on to other people (or, in any case, to people who are not mystics). All the more so, it cannot be submitted to any forms of control on the part of those who surround them. It can be admitted that mystical knowledge, connected with the peculiar type of experiences available only to some (or even all) people in definite mental states, does indeed exist.

Another kind of irrational knowledge may be constituted by those fragments of mythological and religious knowledge which do not fulfil the elementary requirements of logic, especially the principle of non-contradictoriness, but are admissible according to principles of blind faith, according to the well known thesis of Tertulian: "credo, quia absurdum." This thesis recommends faith in what is a nonsense, paradox, absurd to reason. According to such an approach, the deepest truth is just that which is impossible from the point of view of logic.

Such components of beliefs can be found in any religion. These components cannot be submitted to rational analysis and can be called, following Thomas Aquinas, "the mysteries of faith" (e.g. "the mystery of the Trinity" in Christian religion). In spite of the popularity of the claim that "faith" means "lack of knowledge" and that where knowledge is complete there is no place for faith, one may agree that faith can also be the source of knowledge. It may also be recognized that knowledge based on blind faith and not verified in any way is irrational knowledge, especially if it does not fulfil the canons of rational thinking (laws of logic).

Yet another source of irrational knowledge, this time philosophical in nature, is associated with a peculiar type of intuition (sometimes called an "irrational" one), which is neither an intellectual intuition or a sensual intuition since reason and senses are excluded as possible sources of irrational knowledge. Intuition, which according to Bergson's meaning, is to be an independent source of disinterested knowledge, and at the same time deeper than reasonable knowledge, may be considered as the one belonging to the kind of irrational intuition.

In my opinion, examples of irrational knowledge supplied here (mystic knowledge, religious knowledge, intuitive knowledge) indeed indicate the existence (possibility of existence) of irrational knowledge in a given sense. However, I think that this is not a "purely irrational" knowledge, i.e. such which would be completely devoid of intersubjectivity values and totally accessible to the reasonable and sensual (experimental) cognition. It seems that these mythological, religious and philosophical systems should simply be included in irrational knowledge, which to a great extent are burdened with irrationalism, i.e. contain as their essential components theses which are nonsensical to reason and absurd from the point of view of logic.

VII. Conclusion

Obviously, the evaluation of a given kind of knowledge depends on the criteria adopted. Yet it can be asserted that the qualifications of the first three kinds of knowledge, i.e. scientific, common, and artistic-literary knowledge, should be positive from the point of view of all reasonable criteria. Especially from the point of view of cognitive criteria, the social rank of these kinds of knowledge is unshaken. Meanwhile, the other two kinds of knowledge – speculative and irrational – may be evaluated in a different way. In their purely cognitive aspect, they are mainly of historical importance in the sense that they served the cognition of the world well when scientific, and artistic-and-literary kinds of knowledge were in their infancy. In contemporary times, irrational knowledge inhibits rather than promotes general cognitive progress. As far as speculative knowledge is concerned, its cognitive role is fairly ambivalent. On the other hand, one cannot deny them, even at present, some other (extra-cognitive) advantages connected with the formation of attitudes towards life, especially ethical and aesthetic attitudes.

Their very presence in human life is undoubtedly a cultural fact of great importance, particularly if it is accepted that the richness of content and variety of the means of expression constitute an autonomous value of human culture.

SCIENTIFIC AND EVERYDAY KNOWLEDGE*

I. Introduction

According to the current opinion, the difference between scientific knowledge and common knowledge consists in the fact that the former is true and certain while the latter (also called general, practical, common-sense knowledge) is either completely devoid of these values or possesses them to a very low degree.

This view does not adequately reflect either what links these two kinds of knowledge, nor what actually separates them. The task of this paper is to outline a more adequate view on the similarities and differences occurring between common knowledge and the scientific knowledge.

II. Truth and Certainty

Since the time of Plato and Aristotle, certainty and truthfulness have been considered by philosophers to be two fundamental values of human knowledge. For the philosophers mentioned above, they were the fundamental characteristics of *episteme*, which differentiated it from the common opinions (*doxa*).

However, we now know that in spite of persistent efforts of both philosophers and scientists to make the kinds of knowledge – the philosophical and the scientific knowledge – true and certain, the above values do not constitute their adequate distinguishing factors, especially if they are compared with common knowledge. This is obvious for philosophical knowledge. As far as scientific knowledge is concerned, it is enough to take any example. Such an example may be a comparison of any law of science with its common-knowledge equivalent. Let us take, for instance, Newton's law of gravitation (a quantitative law) and its common equivalent which had been known long before Newton and is now called the "qualitative law of gravitation" according to which: "All the bodies attract (each other)."

---

* Translation of "Wiedza naukowa a wiedza potoczna," published in B. Kotowa, J. Such (eds.) (1995), *Kulturowe konteksty poznania* [Cultural Contexts of Knowledge]. Poznań: Wydawnictwo Naukowe IF UAM, pp. 23-31.

It is obvious that, in spite of the same level of generality, Newton's law, because of its greater exactness and, consequently, greater (both logical and empirical) informational content, precludes more possible states of things, and is more easily falsifiable than the qualitative law of gravitation. No wonder then that it has been falsified (by counterexamples showing that gravitational attraction of fast moving bodies with respect to each other or having an enormous mass does not come under Newton's formula: $F_{gr} = Gm_1m_2/r^2$, although it is compatible with the equations of the general relativity theory) while the latter is still considered to be true and given its long persistence will probably also defeat other quantitative laws of gravitation, subsequently formulated in physics (e.g. by Einstein in the general relativity theory).

Hence Ajdukiewicz's conviction that scientific knowledge as a whole is not distinguished by any high degree of truthfulness or certainty. If it did, scientists would not have any right to find out new things and formulate new hypotheses, since the latter ones are prevalently true to an insignificant degree and initially are also of little reliability. According to Ajdukiewicz, a principle of rational recognition of convictions holds in science. The principle states that "the degree of conviction with which we claim a given view should correspond to the degree of its justification." According to this principle, a scientist has the right to formulate and proclaim views of any degree of certainty, provided only that s/he does not present his/her new ideas and loose scientific hypotheses as mature and well-confirmed theories (and vice versa). It can be seen that the principle of rational recognition of convictions has two "edges": one directed against dogmatism, the other against undue scepticism.

It can be considered that from the point of view of the certainty of knowledge both kinds of knowledge considered here are comparable with each other. There are some fragments of scientific knowledge which are more certain than various fragments of common knowledge but there are some which are less certain. This shatters the myth of scientific knowledge as that kind of knowledge which is completely true and completely certain. From the point of view of certainty, scientific knowledge appears to be totally inhomogeneous: both well-grounded facts and theories as well as daring concepts that have only recently been proposed enter the treasury of scientific knowledge. If it were otherwise, a scientist as a scientist would not have any possibility at all of proposing new hypotheses, which initially are not very reliable, and sometimes outright false in the light of previous knowledge, as they are simply discordant with it.

Thus, if not truth and not certainty, what constitutes a real discriminant of scientific knowledge, causing its cognitive status to be considerably higher than the status of common knowledge?

The opinions of scientists and philosophers are divided here since there are many candidates claiming the name of the main discriminant of scientific knowledge.

III. High Logical Conciseness

Some theoreticians consider a high degree of logical systematization as the main distinguishing factor of scientific knowledge (and hence also the chief criterion of the scientific character of knowledge). A logical ordering of knowledge is achieved by means of the relation of consequence (inference); it leads to the construction of (strict or approximate) deductive systems in science.

There is no doubt that this is one of the real distinguishing factors of scientific knowledge. According to Einstein, the main aim of the successive theories constructed within the realm of physics is "the internal perfectness," a feature which can be explicated as logical simplicity of a theory. A theory is logically simpler (1) the lesser the number of initial assumptions (postulates or principles) on which it is based and (2) the higher its informational content (i.e. the number of derivative statements). Therefore, the formula is its determinant:

Common knowledge, if it is ordered, its order is not so much a logical

$$\text{logical simplicity} = \frac{\text{informational content}}{\text{number of initial assumptions}}$$

order, as, for instance, an alphabetical order (like in a telephone directory) or a "natural order" in the sense that together is grouped the knowledge concerning either particular classes of objects or particular professions, and the like.

IV. High Explanatory Power

Another basic difference between scientific knowledge and common knowledge is seen by others in the degree of explanatory power: scientific knowledge is capable of explaining much more than the everyday knowledge (see e.g. Nagel 1961, pp. 1-14). The proverbial grandmother or grandfather is a personification of common knowledge. S/he is not well-acquainted with scientific knowledge, but unusually experienced as far as common-sense knowledge and life wisdom is concerned and can (truthfully) communicate a great deal to her or his grandchild how things really are in the world. However, when an inquiring grandchild starts asking them why things are the way they are and not another then the grandmother's or the grandfather's answer start showing their imperfections; as it is, common-knowledge does not usually suffice for the real explanation and reference to scientific knowledge is necessary. At this point, the superiority of scientific knowledge is indisputable.

V. High Degree of Theoretization

The explanatory power of knowledge is undoubtedly connected with the degree of its theoreticality. Theoretical knowledge, in contrast to observational knowledge, does not have to skim the surface of phenomena: it is able to penetrate the nature of things, to capture internal mechanisms of phenomena and describe regularities that govern them.

Indeed, the theoretical character of scientific knowledge makes it more distant from reality since abstraction procedures, modelling and idealization which lead to obtaining such knowledge cause the direct subject of its interest not to be actual (directly observable) objects but their certain idealized models. Such models are simplified and lend themselves to strict description by mathematical means. However, one can think that this departure is only a deviation from the surface of phenomena and it brings science closer to important features of reality.

If this is really so, then common knowledge – as knowledge of a low degree of theoretization – is, as it were, bound to a superficial interpretation of the studied phenomena. The dimension of depth (essence of phenomena), on the other hand, is accessible to scientific knowledge.[1]

VI. High Informational Content

According to Popper the main distinguishing feature of scientific empirical knowledge is a high degree of informational content (especially empirical content) which at the same time marks the degree of testability of knowledge, including negative testability (falsifiability). Knowledge, in turn, is all the more resourceful with empirical content the more general and more strict it is. Common knowledge, due to its low level of generality and usually even lower strictness, is characterized by a low degree of informational content. The statement of the low degree of informational content is also distinguished by an insignificant degree of falsifiability since the class of its potential falsifiers is narrow.

One may think that such are the main differences between scientific knowledge and common knowledge if they are considered in their synchronic aspect. A more complete expression of differences looked for requires taking into consideration the diachronic aspect of human knowledge. This aspect is revealed both when we analyze research

---

[1] An attempt at explicating the notion of the depth of knowledge from the point of view of hypothetism was undertaken by John Watkins in his work *Science and Scepticism* (1984).

procedures – procedures of gaining new knowledge and when we establish the regularities of the development of a given sort of knowledge.

Here I shall restrict myself to the indication of one principal difference between scientific and common knowledge for every of these two dimensions of the diachronic aspect.

VII. Self-Criticism and Self-Control

Scientific knowledge is achieved in a methodical and planned manner, while common knowledge grows in an unrestricted manner, usually unplanned. The fundamental difference in the attitudes of the scholar and the layperson (as of the one who promotes common knowledge): the scientist's attitude is characterized by a high degree of self-criticism and self-control, while the layperson's attitude is one of trust and simple-mindedness toward the common-sense knowledge received.

Whitehead sees this critical attitude as the main feature which distinguishes scientific activity. He states that science was shaped amongst the ancient Greek community when the first European philosophers from the Ionian school (Thales and his disciples) began to relate critically to the knowledge accumulated by their predecessors and teachers, and scientific knowledge also owes to this feature the knowledge gained by their fellow philosophers and by themselves. According to Whitehead, scientific knowledge also owes it its unparalleled success and the terrific pace of development, and also the fact that no result of scientific knowledge is eternal (everlasting), i.e. each one of them enters its treasury only for as long as science achieves a better one. On the other hand, common knowledge (and, for instance, to an even greater extent, religious knowledge) is apologetic in the sense that it is satisfied with the state which it achieves, and this makes its development considerably slower.

This leads us to an important difference concerning the character of development of both kinds of knowledge considered here.

VIII. Correspondence Accumulation of Knowledge

According to Popper and some other philosophers, the main distinguishing factor of scientific knowledge (the main criterion of the scientific character of knowledge) is that it is concerned not with readily accessible knowledge but with the process of the growing of knowledge.

The increase in scientific knowledge undergoes the principle of correspondence formulated *explicitly* by Niels Bohr.[2] The nucleus of this

---

[2] To put it more precisely, according to Popper, the principle of correspondence steers the development of not a whole (mature, well-developed) knowledge but only the

principle (which is very differently formulated) is a view that a new scientific theory, replacing its precedent, in a given field preserves certain important content of the previous theory, however, in a modified (improved) form. The correspondence relation occurring between the subsequent scientific theories precludes, on the one hand, the exact accumulation of knowledge (preservation of empirical generalizations, and in particular, of laws and theories that have already been discovered) and on the other, precludes such revolutionary breaking (to use Bachelard's term, such an "epistemological cut") which does not refer to nor take over anything from previous knowledge.

The supporters of the correspondence principle usually also think that this principle ensures for science the cognitive progress consisting in science penetrating with a growing degree of preciseness the ever more extensive, and at the same time, deeper levels of reality.

Other kinds of knowledge, including common knowledge, develop according to other principles.[3]

IX. Conclusion

On the other hand, such fields of science as applied (practical) sciences and social and humanistic sciences are still much akin to everyday knowledge. Applied sciences are connected with everyday knowledge by their common subject matter and, to a great measure, by mutual problems. It could be said that these sciences aim at extending, deepening and testing common knowledge as well as at making it more precise. The latter function of applied sciences is emphasized by R. Manteuffel, an eminent specialist in the field of agricultural sciences, when he writes: "As far as applied sciences are concerned scientific knowledge serves to verify the common-sense knowledge" (Manteuffel 1985, p. 137).[4]

And social and humanistic sciences are related to the common knowledge also by the subject matter and problems (at least in the degree to which they are practical sciences) and also common language (to the

---

development of theoretical disciplines. If, however, applied sciences are concerned, they develop by means of the growing specialization and differentiation, or in a way which is almost opposite. See Popper (1972, p. 262).

[3] According to J. Kmita, while scientific practice realizes a "functional requirement of referring to its own tradition" in the form of correspondence, then the reference to the tradition of artistic practice "has no features of correspondence but is a selective *adaptation*," therefore it is adaptive in nature (in the definite sense of this word). Cf. Kmita (1976, pp. 208-209).

[4] In this connection, R. Manteuffel even raises the following objection against representatives of science: "Often ... scientists do not want to admit that they do not operate concrete scientific knowledge in a definite problem and simulate this knowledge." Manteuffel (1985, p. 138).

extent in which these sciences still in principle use everyday language). It is obvious that mutual influences which occur between the scientific knowledge and common knowledge bring the two closer to some extent. At the same time, the influence of common knowledge on scientific knowledge occurs in two forms. Firstly, new scientific disciplines often arise in areas which have been so far penetrated by common knowledge. This means that they codify and systematize (at the first, not yet mature stage of their development) the social experience accumulated by this knowledge (cf. Kmita 1976). Secondly, common knowledge exerts noticeable influence also on those disciplines that have been fully formed and that have matured. One well-known and positive instance of such influences occurred in the 19th century biology, when the breeding practice (in England and in other countries) superseded the biological theory as far as the problem of changeability of species is concerned, and as such constituted one of the sources of Darwinism. The influence of this kind may be observed also now, both in the field of social and humanistic sciences and applied sciences connected with the study of nature, such as agricultural sciences or medical sciences.

As far as the influence of science on common knowledge is concerned, it is undoubtedly even greater. Some even think that in the countries which have achieved a high degree of scientific and technological development common knowledge has been transformed so much – when compared with common knowledge of, let us say, ancient Greeks – that it no longer deserves the name of common knowledge. I think, however, that even for the most scientifically advanced countries it is a premature opinion. Common knowledge cannot be replaced by scientific knowledge partly because of the fact that it also concerns areas of reality which have not yet become the subject matter of scientific research. Thus, the kinds of knowledge considered deal with intersecting areas. What makes the scientific knowledge approximate common knowledge, is obviously also the popularization of scientific knowledge, which consists in great measure in transferring the scientific theorems into the everyday language that is understood by everybody.

UNIVERSALITY OF SCIENTIFIC LAWS*

I. Introduction

The opinion that "perceptive sentences *as such* are not scientific statements and *as such* do not belong to science" (Ajdukiewicz 1965, p. 225) is becoming common in present-day methodology of science. It is so because scientific statements must be substantiated in an intersubjective manner, which is as a rule subject to control on the part of any adequately prepared scientist who has the intention of undertaking it and adequate means of cognition at his/her disposal. Perceptive sentences as such do not fulfil this condition of intersubjectivity since – as direct results of observation, tests and experiments – they are always singular and concern singular unique phenomena. Thus, they always require indirect (intersubjective) substantiation. It follows that perceptive sentences cannot be considered scientific statements until they have been substantiated in an intersubjectively attainable and reproducible way, i.e. through methods other than the direct experimental method (Ajdukiewicz 1965, pp. 224-225; also Popper 1959, pp. 64-70).

It is one of the reasons why general statements formulated in science play such a significant role. They are useful for deducing and substantiating (through reasoning) singular statements, e.g. for predicting singular facts.

Popper states in connection with this: "Kant was perhaps the first to realize that the objectivity of scientific statements is closely connected with the construction of theories using hypotheses and universal statements. Only when certain events recur in accordance with rules or regularities, such as in the case of repeatable experiments, can our observations be tested – in principle – by anyone. We do not take even our own observations quite seriously, or accept them as scientific observations until we have repeated and tested them. Only by such repetitions can we convince ourselves that we are not dealing with a mere isolated 'coincidence', – but events which, on account of their regularity and reproducibility, are in principle inter-subjectively testable. Kant realized that from the required objectivity of scientific statements it follows that

---

* Translation of "O uniwersalności praw nauki," *Studia Metodologiczne*, 11, 1974, 35-58.

they must be at any time inter-subjectively testable, and that they must therefore have the form of universal laws or theories" (Popper 1959, p. 45).

Among general statements encountered in sciences the greatest role is attributed to those which are given the status of scientific laws. The value of a scientific law depends in turn to a considerable extent on the degree of its generality – it is generally considered the more important the more it is general. In this respect, the viewpoint of present-day authors is identical with that of former methodologists. J.St. Mill, for instance, when discussing the chief problem of science, states: "What are the fewest assumptions, which being granted, the order of nature as it exists would be the result? What are the fewest general propositions from which all the information existing in nature could be deduced?" (Mill 1947, vol. I, p. 311). Albert Einstein expresses the same idea when he states that the chief task of the physicist is "to search for those highly universal laws from which, the image of the world can be received by means of pure deduction" (Einstein 1934, p. 162). In this context, the problem of conditions to be fulfilled by a statement which could serve as a formulation of a scientific law becomes important.

To start with, it is worth pointing out that for some basic reasons it seems impossible in practice to give a precise reporting definition of a scientific law. One of the reasons is that sentences of a different type are considered laws in different fields of science. Moreover, even among specialists in a given discipline there is no full agreement as to what conditions must be fulfilled by general statements in order to be considered scientific laws. For example, there have been long disputes as to whether statements such as "All planets move on the same plane and in the same direction," "All ravens are black," Kepler's laws, etc., ought to be considered "real" scientific laws or not. On the other hand, an analysis of the very synthetic structure of the general sentence proves evidently insufficient in ruling whether it does or does not represent a law of nature. In this case, a "functional analysis," i.e. a review of functions of the given sentence in the system of knowledge also seems essential, as well as an "epistemological analysis" to establish whether the given sentence fulfils the cognitive conditions attributed to laws. For both these and other reasons, the question of conditions which one imposes or would like to impose on statements which we grant the very honourable but in fact disputable status of a scientific law is – at least to a considerable degree – not a question of proof but of decision (Popper). In this sense, it is not possible to discover what a scientific law is. It is only possible to attempt a formulation of the most serviceable explanation of the notion of a scientific law, one that would be as much as possible in agreement with the intuitions of specialists in various fields of science. Consequently, any explanation assuming a definite demarcation line between the statements suitable for the formulation of a law, i.e. lawlike statements, and statements that are not suitable for this purpose, i.e. non-lawlike statements, is to a certain extent at least, according to Nagel, "bound to be arbitrary" (Nagel 1961, p. 49). However, Nagel does not consider the situation hopeless. It is not impossible to establish reasons for a numerous class of statements to be granted the special status of laws of nature. The "objective" situation and the function of a certain class of statements predestine them, according to most scientists at least, to be called laws of nature (pp. 49-50). The present paper is limited to a discussion of the intuitions connected with the notion of a law which concern the question of universality of scientific laws alone. This concerns those requirements of universality which, according to the majority of competent scientists, ought to be applied to statements considered by them to be scientific laws. In short, what type of universality is represented by a scientific law? Naturally, it is essential to refer to views and intuitions on this subject that are most common in science in order to avoid too much arbitrariness in solving the question. With this in mind, we shall describe various types of general statements in science distinguished by present day methodologists and discuss their suitability in the process of formulating scientific laws.

II. Strict and Numerical Universality

A discussion on the generality of laws usually begins with deliberations concerning the difference between sentences which are strictly general and those which are numerically general. Those deliberations can be brought down to an analogical difference between terms occurring in the subject of general sentences. The terms can be strictly general, or numerically (enumerationally) general. According to K. Ajdukiewicz, "a term is called enumerationally general, if its content gives us a method of enumerating all the designates of the term in a definite time and to ascertain that they have all been enumerated" (Ajdukiewicz 1965, p. 145 (gloss)). This is possible, for example, when the term contains a limitation of its designates to "some space-time region" which is in turn closely connected with the occurrence in the sentence of proper names or supplementary descriptions (e.g. individual descriptions or space-time co-ordinates) which cannot be defined without the aid of proper names (p. 145). Thus, "the lack of proper names or expressions which it is impossible to define without the aid of proper names proves a strictly general character of the statement" (Giedymin 1964, p. 151). This is connected with the fact that the content of a term in the subject of a strictly general statement cannot provide us with a method for an exhaustive enumeration of its designates. For example, such descriptive terms as "apples which ripened in this orchard last summer," or "letters sent from Poland in the year 1966" are numerically general terms, whereas phrases "ripening apples," or "gravitating bodies" are strictly general terms.

Among contemporary methodologists, K.R. Popper introduces an analogical division. He emphasises that it is not a matter of proof but of convention whether laws of nature are strictly or numerically general. However, he does not refrain from considering this difference between strict and numerical generality a fundamental one. According to Popper, its significance results from a basically different role played in science by general statements on one hand, and individual ones on the other (Popper 1959, pp. 62-64). Similarly to Mill, Popper considers it useful to attribute the term general statements only to strictly general statements and to attribute the status of scientific laws to those alone, if they fulfil some further conditions. "The other kind, the numerically universal, are in fact equivalent to certain singular statements, or to conjunctions of singular statements, and they will be classed as singular statements here" (p. 62). As to the former, they can be put in the form: "Of all points in space and time (or in all regions of space and time) it is true that ..." By contrast, the statements which relate only to certain finite regions of space and time can be provided with what is called a historical quantifier: "within the given space-time co-ordinates it is that ..." These are called singular statements (p. 63).

Popper also points out that although the difference between general and singular statements is closely connected with the difference between general and singular notions and terms, it is not sufficient to characterize general sentences as statements in which proper names or their equivalents do not occur (pp. 64, 68). This is because in many statements only general names occur (as in the sentences: "some ravens are black," "there are black crows") and yet these statements are not general. If, however, the logical form of the statement is general (i.e. the statement contains a general quantifier given at the beginning), it does not contain singular names or spatiotemporal co-ordinates (Popper mentions that singular names which are proper names or could be defined only with the aid of proper names often occur "in disguise" of spatiotemporal co-ordinates or vice-versa) the statement is universal, i.e. strictly general. Thus, according to Popper, not every statement which is general in the logical sense of the word (i.e. not every statement which contains at least one general quantifier) can serve to formulate a scientific law. Such a statement is only a lawlike statement, i.e. a strictly general statement. He believes that only a synthetic strictly general statement which is falsifiable has been subjected to a sufficient amount of testing (confirmed) and has definite informative functions (e.g. prognostic ones) can be considered a scientific law.

J. St. Mill represents a similar viewpoint on the universality of laws. He distinguishes strictly general and numerically general statements and points to the difference in methods of obtaining and substantiating each of the types of sentences, as well as to the fact that a numerically general sentence, which he calls "apparently general," is equivalent to a definite number of singular sentences. Mill states: "A general proposition is one in which the predicate is affirmed or denied of an unlimited number of individuals; namely, all, whether few or many, existing or capable of existing, which possess the properties connoted by the subject of the proposition. ... When the signification of the term is limited so as to render it a name not for any and every individual falling under a certain general description but only for each of a number of individuals designated as such, and as it were counted off individually, the proposition, though it may be general in its language, is no general proposition but merely that number of singular propositions, written in an abridged character" (Mill 1947, vol. I, p. 169). Mill definitely rejects as inadequate the concept of a general sentence as a conjunction of a finite class of singular sentences. He emphasizes that "generals are but collections of particulars, definite in kind but indefinite in number" (p. 186). Thus Mill calls a general sentence one which we call here a strictly general sentence.

According to Mill, of all argumentations that are called inductive only induction proper (enumerative or eliminative) leads to statements which are really, i.e. strictly general. The argumentation called complete induction, which is in fact only "an abridged recording of known facts and does not consist in concluding from known facts about unknown facts" is not real inductive argumentation since it does not lead to actual generalizations, i.e. to general propositions (pp. 188-200). Induction proper occurs only when we conclude "about a general sentence from individual cases" (p. 188).

Thus, undoubtedly, when identifying scientific laws with well substantiated general sentences, Mill refers to strictly general sentences and not numerically general ones equivalent to a number of singular sentences such as, for example, historical generalizations concerning past events from a given period of time. His concept of scientific laws as strictly general sentences undoubtedly played a positive part in the times when extremely nominalist and early positivist trends promoted contrary views, i.e. the concept of laws that were conjunctions of a finite number of singular sentences. E. Mach, L. Wittgenstein, M. Schlick, F. Kaufmann, and others are among the authors who do not recognize laws as strictly general statements but consider them pseudostatements.

The view that laws ought to formulate a type of universality that is higher than purely numerical generality (i.e. a strict generality) or "inlimitable generality" (which will be discussed in paragraph four) is at present practically generally accepted both in methodology and in particular sciences. The problem of the character of the openness of class which designates the scope of application of the law, i.e. of its non-empty fulfilment is more disputable.

III. The Problem of the Openness of a Domain of Application of a Law. Ontological and Epistemological Openness

By discussing laws as strictly general statements we assume a certain incompleteness of our knowledge of the number of cases covered by the domain of application of a law; we do not, however, further postulate that the number of cases be unlimited. Some authors do lay down that further condition and require the number of cases within the validity of the law to be not only undefinable to us but also undefinable in the ontological sense of the word. This is because our intuition does not accept as law a statement whose scope of non-empty fulfilment is identical with the scope of empirical evidence at our disposal. This occurs when the examined cases are all the cases embraced by the validity of the given statement, i.e. all the cases for which the law can be applied. The condition that the scope of the law's present applicability and cases examined up to the present moment should not be identical with the law's validity in general is connected with functions generally attributed to laws. In this case, they are the functions of predicting and explaining.

According to a widespread view, shared – among others – also by Mill, a genuine proof of every general truth (i.e. of a true strictly general sentence that is of a scientific law) lies in its containing power of prediction. Such a truth always applies also to further facts that have so far not been implemented (Mill 1947, p. 192). "The sentence 'all men are mortal' does not refer to all now living men but all men present, and to come" (p. 189). Hence it can form a basis for predicting future facts. Thus, the law applies not only to past reality but also to reality which has not been realized yet, i.e. to the future because it covers cases which can be predicted by applying the law and some additional data called initial conditions. In this case, a law always applies to a class which can be called ontologically open for the future, since a part of the range of the law projects beyond the frame of the present and past and covers future facts not yet realized. Many present-day authors are of a similar opinion. According to J. Pelc, for example, a scientific law ceases to be a law when its scope is completely fulfilled, i.e. when it is ontologically, though not necessarily perceptively, exhausted by the cases which have occurred so far. The law then loses all direct power of prediction (prognosis) as it cannot be applied to any phenomenon in the future. Such a law becomes a historical generalization (a historical sentence) concerning only cases realized at a given period of time (Pelc, Przełęcki and Szaniawski 1957, pp. 22-32) and serving only for the purpose of retrodiction (postgnosis).

There is, however, no unanimity with regard to this problem (see e.g. Malewski and Topolski 1960). Some authors, when speaking of the openness of a class designating the scope of validity of a law, mean another type of openness which can be called epistemological openness. A class is open in this sense when the range of proved cases which are its elements is not identical with the whole class, thus with the full scope of application of this law but with its typical sub-set. In this case, too, it is assumed that one of the basic and essential functions of a law is its previdistic power (the law occurs as one of the substantial data in the deductive process of prediction). The term "prediction" is taken to embrace both prognosis and postgnosis (retrodiction) (see e.g. Popper 1959, p. 60). Thus a law is expected to allow deduction "about unknown cases on the basis of known ones" which fall under its scope. It is, however, not required for this deduction to form a way from "past cases to future ones." According to this view, a law need not be characterized by both the previdistic functions: serve both for prognosis and postgnosis. It is sufficient if it performs one of them and forms one of the data of a widely interpreted process of prediction. (Naturally, it becomes evident eventually that in many cases postgnosis is not possible without prognosis in that the process of verifying postgnosis usually necessitates reference to prognosis, predicting results of future observations. This, however, means only that some scientific laws must also perform the prognostic function but does not mean that all laws need perform this function). From this point of view it is inessential for the law whether a class which is epistemologically open is also ontologically open. Naturally, ontological openness results in epistemological openness of a law but not vice versa.

Evidently, ascribing epistemological openness to a law suffices for the statement that the law as a strictly general sentence is not verifiable in an exhaustive way: at the moment of its ultimate verification, i.e. verification of all its cases it would *ex definitione* cease to be a law since the class designating its scope would be epistemologically closed. It would not then be able to perform functions as important for a law as the explanatory function or that of predicting unknown phenomena.

There seems to be no doubt that in sciences that are called theoretical (e.g. in physics) we ought to require from laws that they be ontologically open, and in fact it is generally so. In historical sciences, however, epistemological openness seems sufficient in many cases since postgnosis is perhaps the chief previdistic function which they are expected to perform.

The question arises, however, whether historical generalizations (or speaking more generally – historical sentences) could not also perform previdistic functions including prognostic functions, i.e. concern future cases on the condition that they occur in a given in advance strictly defined (limited) space-time area. Naturally, the answer depends on the explication of the term historical sentence. If we assume that a numerically general sentence containing space-time determination is a historical sentence independent of whether it refers only to historical, i.e. past reality or not, the answer will naturally be positive. For a statement referring to all Poles living in the 20th century will, on the basis of this explication, not to be a law but a historical generalization which could not perform direct prognostic functions, i.e. serve as a datum in predicting future cases within the range of that generalization. Those who object to the above explication on the assumption that a historical statement is one that concerns only past historical reality will naturally also reject the proposition that historical sentences can be applied not only for postgnosis but also for predicting future facts. Thus, the question may be considered open. The problem of the type of openness of a class which designates the validity of a law is, however, not the one that arouses most controversies in the current dispute on the demand for strict generality of laws. The main controversy concerns the occurrence in the definition of a law of proper nouns and is closely connected with the difference between strict universality and what is called unlimited universality.

IV. Strict Generality or Unlimited Generality?

Although it is a general belief nowadays that a numerically general statement does not represent the type of generality (the degree of universality) required from a law of nature, the demand of strict generality seems to many authors too exacting. It is pointed out that the condition that the definition of a law should not contain proper nouns or individual names, singular descriptions or space-time co-ordinates which could not be eliminatory, is too restrictive since it automatically excludes from the family of laws a large class of statements considered to be scientific laws both currently and by the majority of specialists. This concerns such statements as the laws of Kepler, Galileo's law of free fall of bodies in the vicinity of the Earth (according to which a falling body covers the distance $S = 981t^2/2$), etc.

In this way, many statements, such as those referring to phenomena occurring in the vicinity of the Sun ("heliocentric laws") or the Earth ("geocentric laws") whose definitions include, *explicitly* or *implicitly*, proper nouns ("in the vicinity of the Earth"), ("round the Sun") would not be included among scientific laws even though they are granted this status by an overwhelming majority. Two of the proposed solutions require particular attention. We shall present them following E. Nagel (Nagel 1961, pp. 57-60).

The first one is based on classifying predicates which can be used in formulating laws into two categories: purely qualitative predicates and the rest which are often not qualitative. A purely qualitative predicate occurs when the definition of its meaning does not require reference to any particular object or spatiotemporal area. Its content can be explained without reference to proper nouns or spatiotemporal co-ordinates (or supplementary phrases). Under this proposition laws are not required unconditionally to contain purely qualitative predicates (although some sciences, such as theoretical physics, tend to formulate their fundamental statements with purely qualitative predicates alone). It is sufficient for them to fulfil one of the following conditions that are not mutually exclusive:

1) they contain only purely qualitative predicates, or
2) they are derivative statements which can be deduced from fundamental statements alone, without reference to statements of another kind. Moreover, a statement is considered fundamental if it does not contain proper nouns or individual constants and all its predicates are purely qualitative.

It is now possible to answer the question concerning the conditions which must be fulfilled by a statement suitable for a formulation of a law. The answer is as follows: a statement is suitable for a formulation of a law, i.e. it is lawlike, if it is a fundamental or a derivative statement.

However, the authors of the above concept seem to forget that the second condition is in fact of no importance if it is not in turn supplemented by some additional restrictions. Evidently, it is always easy – through a purely formal manipulation – to find for each given statement such a fundamental statement, or even a number of such statements which contain only purely qualitative predicates and from which the given statement can logically be deduced. Thus, a fundamental statement can be transformed into its derivative. In order to exclude this purely formal evasion, the admissible fundamental statements ought to be supplemented with some limitations concerning their veracity or degree of substantiation, e.g. the condition that they should be selected only from among statements considered in their time to be scientific laws. The last restriction, however, seems too exacting, since a derivative statement can – when deduced from some law – be considered not only a formal lawlike statement but a law proper and that substantiated to a degree no lesser than the fundamental law. It is now evident that formulation of adequate conditions to be fulfilled by the fundamental statement for another statement that can be logically derived from it to fulfil the requirements for a lawlike statement is not an easy task. But let us return to the proposal under discussion.

The occurrence in a statement (e.g. in a Kepler's Law) of proper nouns is not – according to the concept under analysis – contrary to qualifying it as a scientific law, if only it can be proved that it is a derivative law which can be derived (explained) on this basis from fundamental statements which contain only purely qualitative predicates. Does it follow that, for example, Kepler's Laws could be qualified as scientific laws before Newton's days, or at least nowadays?

The answer is no for two reasons. Firstly, fundamental laws (Newton's laws) from which Kepler's laws could be derived did not exist. Secondly, it is practically certain that also today it would be impossible to derive Kepler's laws from Newton's laws alone or any other fundamental laws for that matter[1], without referring to what is termed statements concerning the "collocation of causes."[2]

It is natural then that assumptions concerning collocation of causes (which, in the case of Kepler's laws, are statements designating the value and direction of gravitation and inertion of planets) cannot be formulated, at least generally, without the use of proper nouns (which occur in definitions of the mass of the Sun, the mass and velocity of particular planets). Hence, is not possible to derive Kepler's laws from data (fundamental laws) containing only purely qualitative predicates. Thus, if we wish to consider Kepler's laws (and other statements such as "the orbits of all planets are situated on one plain which coincides with that of rotation of the Sun," "all bodies in the proximity of the Earth fall with a velocity proportional to 981 t^2") to be scientific laws, the explanation of the law under discussion which refers to purely qualitative predicates and to the classification of laws into "fundamental" and "derivative," is too restrictive.

This is why another proposal is put forward. Nagel considers it satisfactory. It is based on a classification of general statements different from those hitherto discussed, i.e. on the differentiation between unrestricted universals and restricted universals. Nagel quotes the first of Kepler's laws: "All planets move on elliptic orbits with the Sun at one focus of each ellipse" as an example for an unrestricted universal, whereas the statement "all the screws in Smith's car are rusty" is cited as an example for restricted one. He goes on to say: "Both statements contain names of individuals and predicates that are not purely qualitative. Nevertheless, there is a difference between them. In the accidental

---

[1] Nagel states at this point that "it is far from certain whether such statements as Kepler's are in fact logically derivable even today from fundamental laws alone." Nagel (1961, p. 58).

[2] This fact has long been under discussion. E.g. Mill, when discussing the derivation of Kepler's First Law from Newton's Laws, states: ". . . in this resolution of the law of a complex effect, the laws of which it is compounded are not the only elements. It is resolved into the laws of the separate causes, together with the fact of their co-existence. The one is as essential an ingredient as the other" (Mill 1947, Vol. I, p. 340). Moreover, Mill believes that ". . . the element which is not a law of causation but a collocation of causes, cannot itself be reduced to any law" (p. 340). With regard to Kepler's First Law, the assumption concerning the "collocation of causes" would maintain that planets are subject to both gravitation and inertia taken in appropriate proportions.

universal, the object of which the predicate 'rusty during the time period *a*' is affirmed (let us call the class of such objects the "scope of predication" of the universal) are severely restricted to things that fall into a specific spatiotemporal region. In the lawlike statement, the scope of predication of the somewhat complex predicate 'moving on an elliptic orbit during the time interval *t* and the Sun is at one focus of the ellipse' is not restricted in this way: the planets and their orbits are not required to be located in a fixed volume of space or a given interval of time" (Nagel 1961, pp. 58-59). A general statement whose scope of predication is not restricted to objects present in a given period of time is called by Nagel an unrestricted universal even if it contains proper nouns. He postulates that lawlike statements, i.e. formal candidates to the status of laws which fulfil the conditions of universality laid on laws be unrestricted universals, but he does not postulate that they be strictly general.

All basic difficulties in classifying such statements as Kepler's laws and other statements commonly considered to be scientific laws disappear when this general requirement has been satisfied. What, then, is the difference between Nagel's unrestricted universal and Ajdukiewicz and Popper's (their definitions coincide) strictly general statement? The answer is not simple, since Nagel's explanation of the term unrestricted generality is far from clear. One, however, is doubtless: when introducing the notion, Nagel and the others wished to impose on laws a less restrictive condition than the condition of strict generality, one in which the occurrence of proper names should not be fatal for the law and which would save a large class of statements (e.g. in physics or astronomy), commonly classified as scientific laws from being denied the status of a law.

It is worth noting that Popper who lays the condition of strict generality on laws, believes also that lack of proper nouns or their equivalents is an essential condition but not a sufficient condition of strict generality. Sentences containing only general names are called by him strict or pure sentences, and he emphasizes that strictly general sentences form the most important but not the only kind of statements containing no proper nouns. He says further: "But in many other statements, such as 'many ravens are black' or perhaps 'some ravens are black' or 'there are black ravens', etc., there also occur only universal names; yet we should certainly not describe such statements as universal. Statements in which only universal and no individual names occur will here be called 'strict or pure'" (Popper 1959, p. 68). Undoubtedly, it is a question of a terminological decision and not of deduction or fact whether we adopt Popper's or Nagel's position in this matter. Nevertheless, it seems that Nagel's proposal to classify as scientific laws some statements containing proper names, if they fulfil certain additional conditions, is more in line with a physicist's or astronomer's intuition and language habits and terminological formulations more or less openly accepted in physics (including astronomy). This is probably even more true with regard to other disciplines of science and humanities. This is due to the fact that in physics we can observe a tendency stronger than in the remaining exact sciences for formulation of fundamental statements on the basis of general names only.

V. Law and Necessity. Nomological and Accidental Universality

Apart from the classification of general sentences into 1) strictly general and numerically general sentences and into 2) unrestricted and restricted universals, methodological literature knows a dichotomous classification of general sentences (or strictly general sentences) into 3) nomologically general and accidentally general sentences. J.St. Mill, K.R. Popper and E. Nagel all apply it in their works, though J.St. Mill uses a different terminology (Mill 1947, Vol. I, pp. 221-223; Vol. II, pp. 53-54; Popper 1959, pp. 420-422; Nagel 1961, pp. 49-69). An analysis of Mill's numerous declarations concerning the classification of laws of time sequence into causative and non-causative laws shows that he was not satisfied with classifying general sentences into strictly general ones (he called them "really" general) and numerically general ("seemingly" general) but was to some extent aware of the above mentioned more subtle difference between two types of strictly general sentences: those expressing nomological universality and those representing accidental universality (cf. Such 1967, pp. 32-41). The difference between the Mill's and Popper's opinion and that formulated by Nagel is that Nagel classifies nomologically general sentences and accidentally general sentences within the general framework of general sentences, whereas Mill and Popper do so within the framework of strictly general sentences.

We shall restrict our explanation of the notions of accidental and nomological universality to true sentences. Moreover, just as Mill and Popper, we shall assume that the classification refers to strictly general sentences only and does not include numerically general sentences. True, strictly general sentences which express accidental universality, i.e. true accidentally universal sentences will be called accidentally true, whereas true strictly general sentences expressing nomological universality, i.e. true nomologically universal sentences – necessarily true sentences (laws). An accidentally true general sentence is one for which there are no counterexamples. This means that if we express this type of sentence in the form of a conditional, there has not been realized in this world, either now, or at any time, such a state which would fulfil the conditions formulated in the antecedent but not the conditions formulated in the consequent of this conditional period. Thus, if we establish that a scientific law can also be formulated as an accidentally universal sentence (not only as a nomological sentence), such an accidentally universal law will be considered true, if there has not existed and will not exist a state which would fulfil the antecedent of the law but not its consequent. For example, the sentence: no man lives longer than three hundred years will be considered an accidentally true statement if no man has so far or will in future live for over three hundred years. Hence, the lack of cases contrary to it is sufficient and necessary proof for an accidentally true sentence.

Is it also necessarily true? In order to answer this, it is essential to establish whether laws of nature admit the existence of circumstances in which humans could live for more than three hundred years. If such conditions exist or are possible, even if humans have never experienced them and so have not lived for over three hundred years, the above statement is not necessarily true because an ("accidental") appearance of humans in such circumstances would invalidate the statement. Hence, a necessarily true sentence is a law for which no counterexample exist, moreover, their existence would be impossible (out of question) because of existing regularities of nature. In other words, a necessarily true law is one in which the antecedent supplies conditions which are from the point of view of existing regularities of nature sufficient for the phenomena expressed in the consequent to take place, i.e. conditions after which – according to the very definition of the notion "nomologically sufficient condition" – the consequent is (non-emptily) fulfilled regardless of any other circumstances. An accidentally true law is one which does not formulate the conditions sufficient for the occurrence of the phenomena specified in the consequent. And that it is true in spite of this is due to an accident, due to the fact that certain conditions which could have been fulfilled as far as existing regularities of nature are concerned, have not been fulfilled. It could be said that necessarily true laws are fulfilled not only in the scope of reality (which embraces all states actually realized) but also in the scope of possibility, i.e. in all the worlds governed by the same regularities as this world, or in all the worlds differing from ours only in the initial conditions but not in the laws. It is obvious that the lack of counterexamples alone in our world does not necessarily guarantee that this kind of law is true. Analogically to Leibniz's concept of logically true worlds, i.e. worlds whose logical structure is identical with the logical structure of our world, i.e. where all the laws of logic (of our world) are fulfilled, we could introduce a concept of nomologically possible worlds with a nomological structure identical with that of our worlds, worlds in which all the laws valid for our (empirical) world are fulfilled.

The terms nomological universality and accidental universality may suggest that – as is believed to be true by many – nomologically universal sentences are suitable for formulating laws, whereas accidentally universal sentences – as the very terms seem to suggest – are not. This need not be discussed here but it is worth noting that laws actually formulated in science do not always aspire to be necessarily true in the meaning discussed above, i.e. to express nomological universality (or necessity). Yet it seems to be the ideal or ultimate end of science to formulate nomologically universal laws, i.e. laws of *sui generis* necessities of nature and thus supplying in the antecedent a complete set of nomologically sufficient conditions, i.e. sufficient from the point of view of existing regularities of nature for the consequent to take place. As has been said before, it is expected from such laws not only that there be no counterexamples for them (a confirmation e.g. through enumerational induction can establish an agreement with our empirical world). It is furthermore expected from them that there be no counterexample for them also in the scope of possibility (agreement with each of the nomologically possible worlds) that is the very possibility of their existence is excluded as contrary to the nomological type of laws of nature. This permits us to interpret necessarily true laws as some prohibitions or limitations imposed on the realization of laws of nature; limitations which express what is called nomological necessity of the world.

Hence, we do not share the view common in present day methodology of science that only nomological statements are suitable for formulation of scientific laws while universally accidental statements are not, even if they express strict universality. We are inclined to share Mill's opinion that if some laws of nature, e.g. clauses of reason express an unchangeable and unconditional relationship, i.e. a necessary relationship between the reason and the result and are – according to our terminology – nomologically universal, then the remaining laws of sequence in time (concerning, for example, the sequence of seasons of the year, the sequence of day and night, etc.) express only unchangeable sequences, i.e. they can prove to be accidentally universal. To avoid misunderstandings, I should like to point out that some present day methodologists use the terms "nomic universality" and "accidental universality" in a slightly different sense. For example, among accidentally general statements Nagel classifies also one which, following Popper, Ajdukiewicz and others, we term numerically general statements. When discussing the difference between the two kinds of strictly general statements, we usually refer to possibility. What is understood by the term possible when we say that the relationship in the law must refer not only to all past and present cases but also to all possible cases? Does it refer only to cases which, though not yet realized, will be realized some time in future (possible would then mean something that will actually take place in future), or also cases which – though not eliminated by the existing regularities designating the nomological structure of reality, will never take place because of the lack of conditions essential for their realization? The latter would refer to cases which are eliminated, i.e. impossible not because of existing regularities of nature but because of conditions valid now and in future. This problem can be formulated in the question: does a law refer to phenomena embraced by an appropriate counterfactual conditional (whose formulation contains the functor "if ... then ..." or an analogical one) which states that if the conditions formulated in the antecedent and not eliminated by existing regularities were fulfilled, then given consequences formulated in the consequent would be realized? If the answer is positive, a law refers to all cases (of a given class) possible from the point of view of regularities of nature, regardless of whether the cases are ever realized. If the answer is negative, the scope of the law should be restricted to cases which actually take place (at any time). The difference between the scope of possibility (determined by laws of nature) and the scope of reality (designated by laws and initial conditions) forms the ontological correlate of the difference between nomological universality and accidental universality. The foundation for differentiating two types of strict universality (and hence two types of laws: necessarily true and accidentally true) lies in the dichotomy of laws and conditions. The dichotomy consists in the distribution of fundamental conditions being – partly at least – independent of laws (nomologically universal) and vice versa, laws (nomologically universal) – partly at least – being independent of fundamental conditions realized in the world. If the assumption proved false, the classification of statements presented above would have to be subjected to far reaching modifications.

To return to the above question, it must be said that if (similarly to Mill's causal laws) laws express a natural necessity in the sense that they are not only accidentally true (i.e. there are no counterexamples for them in existence) but true in a necessary manner, i.e. true in all possible conditions or worlds – from the point of view of regularities of nature – (we could also call them nomologically true), they will always take the form of a nomological conditional containing in its scope the scope of the appropriate counterfactual conditional and true in all initial conditions possible from the point of view of regularities of nature.[3] On the other hand, laws whose scope is restricted to fields of reality embracing only actually realized cases cannot be formulated in this way, since they are only accidentally true, true on account of existing circumstances, i.e. due to coincidence. These circumstances render impossible the realization of other cases equally possible from the point of view of regularities of nature which could prove fatal for the accidental cases under discussion (as their counterexamples). Since the validity of some type of laws can be influenced by coincidence, it is also to a certain degree dependent on people, on his action, i.e. in cases where through production or experiment we are able to bring about such situations which would prove fatal for

---

[3] It is assumed in this concept that regularities of nature can be adequately described only in nomological laws, and that accidental laws describe only some "accidental coincidents" which cannot aspire to the status of laws (regularities) of nature.

these laws by being their counterexamples but which would never have occurred without people's assistance, or vice versa: would occur without fail. This is impossible with regard to necessarily true laws since it is impossible to bring about or prevent their counterexamples on account of regularities of nature. On the other hand, to prove that there are no counterexamples for a law of this category and that in this sense it is in accordance with reality is not in itself sufficient to prove its nomological validity, although it is sufficient to state that it is accidentally true. Accordance with reality must also refer to the scope of possibility embracing cases which may never be realized. It follows then that enumerational induction is not sufficient to prove nomological truth of a law, even if it were possible (which it is not, on account of the openness of the domain of application of any law) to apply full induction to the whole scope of the law, i.e. to all the cases it embraces. Mill was fully aware of this when he emphasized many a time that the enumerational method of induction based on observation is insufficient; it is essential to employ the more perfect elimination methods of induction based on experiment, i.e. on artificial production of facts which might never occur without our interference.

The dispute connected with classification of statements into nomologically general and accidentally general in the sense discussed above refers to three important questions: firstly, is this classification at all sensible, some question its practicability by arguing that it requires reference to the concept of worlds nomologically possible which is indefensible, and that it assumes our knowledge of all the regularities of nature; secondly, should scientific laws always be required to express nomological universality, or could they sometimes be accidentally universal statements; thirdly, can the concept of nomological universality be explained in extensional language and without reference to such intuitive and unclear modal notions as necessity of nature or physical impossibility whose adequate explanation still seems too difficult?

The problem of the nature of necessity of nomologically universal statements has interested thinkers since Hume, if not before. Those of the opinion that the concept of nomological universality can be explained without the use of non-reducible modal concepts are considered to be followers of Hume. Those who believe that the use of some modal concepts concerning regularities of nature is unavoidable in any adequate analysis of nomological universality are considered to be his opponents (in this matter). The difficulty in solving this matter lies in the fact that a nomologically universal statement ought to justify the appropriate counterfactual conditional and be formulated as a nomological implication true not only in all actually occurring conditions (at any time and place) but also in all nomologically possible conditions, including those which are never realized. Thus, a nomological implication cannot apply only to phenomena which actually occur. On the other hand, it cannot apply to such situations as: what would happen if the law of gravitation were not fulfilled, or if there were in its place some law of "antigravitation," etc. It can embrace only the cases (the worlds) which assume at the most non-fulfilment of existing conditions but fulfilment of existing regularities. The assumption of the inviolability of nomological laws differentiates the nomological conditionals and the contrary to fact conditionals which do not have to fulfil this condition.

The difficulty of transforming a nomological conditional into the terminology of categorical and ordinary (real) conditionals consists in the formulation of some conditions which must be fulfilled by these categorical statements or real conditional in order to become equivalent with the nomological conditional, i.e. for the transformation to be correct. So far, a formulation of such conditions has not been possible without reference to some unreal conditionals. The difficulty appears to be shifted but not overcome.[4]

VI. Conclusion

Let us conclude by revising the problems discussed in this paper and concerning the universality of laws which seem to be uncontroversial, and then those which are still to a greater or lesser degree subject to controversy.

With regard to laws, the demand for a higher degree of generality than that represented by numerical generality seems indisputable. Hence, it seems reasonable to assume that the domain of the application of a law is epistemologically open, i.e. that it is never possible to verify effectively the full scope of a law; the scope of a law's applicability cannot be identical with the scope within which its exemplification has been applied and verified so far.

This results in the fact that the number of cases embraced by the scope of applicability of a given law always remain to a certain extent indefinite since the cases established so far do not *ex definitione* cover the whole scope of the law.

Which of the discussed problems can be considered controversial (apart from certain suggestions which have been put forward in the discussion above)?

1. With regard to the occurrence of proper nouns in many statements commonly considered to be laws, there is the open question of whether the condition of strict generality is not too restrictive and whether we should not limit ourselves to a more liberal condition in classifying

---

[4] An interesting attempt to explain the contrary-to-fact conditional was undertaken by J. Kmita (Kmita 1967).

statements as laws, e.g. the condition of unlimited universality (which, however, ought to be defined in greater detail).

2. With regard to the condition of strict generality (or unlimited generality) of laws, the problem of the character of the openness of the scope of application of a law (epistemological or ontological openness?).

3. The following problems seem controversial with regard to the classification of general statements into nomological and accidental:

   a) Is this classification at all theoretically substantiated and useful in practice?

   b) Should it be performed within the framework of strictly general statements, or of all general statements?

   c) Should scientific laws express nomological universality alone, or can accidentally universal laws also occur (and do they)?

   d) Hume's problem: is it possible to explain the notion of nomological universality in extensional language and without referring to unreal conditionals and can the notion of natural (nomological) necessity be reduced to non-modal notions with a clearer content which would be easier to perceive (such as the notion of strict universality, exceptionlessness, constant sequences, unchangeable recurrence, etc.)?

The above problems are related; a solution of one may imply a definite attitude with regard to the rest.

Translated by Hanna Grabińska

THE ROLE OF THEORY IN PHYSICAL SCIENCES[*]

I. Law and Theory

The main subject of the logical and methodological analyses which have been carried out up to the present and concerning the results of scientific cognition has undoubtedly been the law of nature. However, it turns out that from the point of view of the epistemological and methodological reflections on science and in view of the functions which science fulfils in society, it is not a law but a scientific theory that can reasonably claim to be the essential autonomous unit of theoretical knowledge. This is especially true with reference to the scientific knowledge that is as mature and developed as the knowledge represented by physics.

The fact that it is not a theory but a law has so far been the subject of the most extensive and thorough theoretical analyses can be partly explained by the fact that the task of analyzing a law is considerably simpler than that of analyzing a theory. Moreover, a law constitutes the basic building block, if one might say so, in the structure of a theory.

Furthermore, the terms "law" and "theory" are occasionally used interchangeably in the considerations on science, or used in such a way that a law represents a specific case of a theory. This results from the fact that both these terms are ambiguous and, in certain specified meanings, their contents overlap partially or completely. Both the term "law" and the term "theory" are used in methodological considerations in three meanings mostly.

A law in science is usually understood as:

1) "a regularity of nature," i.e. a certain objective relation between classes of natural or social phenomena characterized by universality, necessity and internality;

2) "a scientific law," being a strictly universal assertion reflecting, to a certain degree of accuracy, a specific regularity of nature (i.e. a law in the first sense);

3) "a law proclaimed by people" in the sense given to the term by the jurisprudence.

---

[*] First published in T. Buksiński (ed.) (1988). *Laws and Theories in Empirical Sciences*, Poznań: Wydawnictwo Naukowe UAM, pp. 7-23. This text is reprinted here with kind permission of Wydawnictwo Naukowe UAM.

Whereas a theory is usually understood as:

1) "a fragment of theoretical knowledge" as distinct from observational knowledge (also sometimes called experimental or empirical, in a narrower sense of the term), being a result of an ascertainment or a simple inductive generalization of observed facts;

2) "a deductive system of statements (including laws)" ordered logically by relation of consequence;

3) "any fragment of scientific knowledge, systematized in a way," irrespective of the elaboration of its theoretical framework (e.g. a systematized description of a separate phenomenon or of a class of objects; in this sense one, speaks of e.g. the theory of the Moon, the theory of the disturbance of the Sun, the theory of Aurora Borealis, etc.).

Beyond any doubt, the second of the meanings determined above, in the case of both the terms, is fundamental for methodological analyses concerning physics. Therefore, in further considerations, law will be understood as a law of science i.e. a justified strictly general assertion, which, as a conditional statement, is a complex sentence whose consequent describes a given regularity of nature, whereas the antecedent gives, either in the real or counterfactual mood, conditions of occurrence of this regularity. And "theory" will be understood as a system of assertions ordered roughly or completely – by the relation of consequence and thus a deductive system, which – with respect to its logical structure – approximates axiomatic and formalized systems of contemporary logic and mathematics.

Thus, in the light of contemporary logical and methodological analyses, a theory conceived in this way turns out to be a basic structural unit of physical knowledge: a basic unit of description, prediction, heuristics, explaining or verifying in physical sciences. Hence the crucial role of analyses aimed at the logical and methodological reconstruction of a physical theory, and at the examination of a model rendering the structure of such a theory. Besides, it turns out that a theory is the basic unit of theoretical knowledge in the process of its practical use and application. Hence the well-known statement by Gauss that "there is nothing more practical in the world than a good theory" continues to be legitimate.

Let us now ask the following: why does science in general, and physics in particular, not confine itself to formulating individual laws but, wherever possible, aims at building deductive systems of such laws, i.e. scientific theories? A partial answer to this question is not difficult; conversely, it is a banal one. It is not banal, however, to attempt to provide a possibly complete answer to this question, an answer exploiting in full all essential points. The arguments presented further in this paper constitute such an attempt.

The essential reasons for building a theory are apparently contained in two crucial procedures carried out by scientists in their research, namely in explaining and verifying.

II. The Postulate of Verifiability

As is well known, the degree of verification of a given fragment of knowledge depends, among others, on the degree of verifiability. Verifiability, on the other hand, is strictly dependent on the number and diversity of observational consequences, including predictions, which can be derived from that fragment of knowledge and from the so-called initial conditions, i.e. from a description of conditions to which a given fragment of knowledge is applicable. That number quickly increases as separate fragments of knowledge merge into bigger entities that are deductive in nature. Putting it more strictly, the amount of observational consequences of two isolated fragments of knowledge is smaller than the number of observational consequences which can be derived from a uniform deductive system consisting of those fragments of knowledge. In this way, merging laws into a theory raises the degree of the empirical verifiability of laws, increases their possibilities in the domain of testing: a law merged into a theory can be verified – indirectly or directly – with respect to a greater number of cases than a law considered in isolation. Besides, what is important here is not only the aspect of quantity but, naturally, the aspect of quality of the matter as well. A law included in a theory can be verified not only in a more numerous set of cases but also in a more diversified set, a set that is more varied.

Physicists have noticed the fundamental significance of this fact as soon as the first extensive physical theory was formulated, i.e. after the formulation of Newtonian classical mechanics, jointly with the theory of gravitation. The number and especially the diversity of phenomena and often the phenomena which till that moment seemed to have nothing in common with one another, and which Newton and his followers managed to explain on the basis of one theory, aroused astonishment of both scientists and lay people throughout centuries. Those phenomena included the following heavenly and earthly ones: movements of various heavenly bodies (planets, moons, comets, stars, etc.), the falling of bodies, ballistic movements, oscillatory movements, tides of seas and many others. The raising of the degree of verifiability is, naturally, connected here with the increase of the predictive power of knowledge, formulated in the form of a theory.

Merging laws into theories leads not only to the increase of the degree of their confirmation when they are true within a certain domain but also facilitates their falsification when they are false. For instance, the law of the invariability of the speed of light has been refuted by appropriate astronomical observations and physical experiments, based on the general theory of relativity, only due to the fact that it was included – as one of postulates – into the special theory of relativity, which has turned out to be strictly true exclusively for territories with no gravitation.

It seems even that for certain laws of science, especially of physics, a stronger thesis holds true: if not included within a wider context of theoretical knowledge, the verification of laws is impossible at all, i.e. neither their confirmation nor falsification is possible. This refers to, for example, Newton's law of universal gravitation, which, if not included into a system of mechanics, connected with it, seems not verifiable at all. Here we witness a difficulty that was first noticed by Pierre Duhem. The difficulty lies in the fact that those observational consequences which – from the point of view of the procedures of verification – are the most interesting (and sometimes perhaps even only possible) are usually derived not from a single law or even a single theory but from a broader "body of science," as Duhem puts it, from more capacious grounds comprising that law and certain concomitant hypotheses.

Thus, we face a situation which in case of not complying with our predictions does not lead univocally to the falsification of a law being verified. This can be described in the following way:

$$(L \wedge H_1 \wedge \ldots \wedge H_n \rightarrow 0) \wedge \neg 0 \rightarrow \neg(L \wedge H_1 \wedge \ldots \wedge H_n),$$

where L is the verified law, $H_1, \ldots, H_n$ – the concomitant hypotheses and 0 – the observational consequence (I leave out the fact that to derive the observational consequences and predictions from the system of knowledge under the process of verification it is indispensable that the system include also the initial conditions C, i.e. sentences describing conditions in which the theoretical knowledge to be verified is applicable.) It is not the law L being verified but the conjunction whose elements are the law L and the concomitant hypotheses $H_1, \ldots, H_n$, which proves to be falsified by experience.

In terms of one of de Morgan's laws, the consequent of the presented scheme is equal to the following expression: $\neg L \vee \neg H_1 \vee \ldots \vee \neg H_n$, which clearly shows that we are actually not dealing with falsification here but with disconfirmation, a "weakening" of the law L being verified. Hence, only if we assume veracity of all the concomitant hypotheses can we infer that the law being verified is false itself.

A pertinent thesis, formulated in this connection by Duhem (which is usually called the weaker thesis of Duhem-Quine), states that it is not a single statement (e.g. a law) but a broader "body of science" that is submitted to the process of verification. (another argument given by Duhem in support of the thesis under discussion is the fact that verification is a procedure which also refers to technical tools of research, for example experimental devices or measuring instruments, and consequently to theories of their operation, whose veracity ought to be assumed if one wants to accept the results of the experiments carried out as dependable). Besides, this thesis itself is formulated now in a stronger, then in a weaker form. The weaker formulation of the thesis has it that it is not particular laws but particular theories that are liable to individual experimental verification and the stronger formulation that not even particular theories but larger theoretical systems or even the whole "body of science" or simply the whole human knowledge as such is liable to verification.

The fact that, at least in some cases, a more capacious theoretical system and not a single theory can be a fragment of verified knowledge is illustrated by the example of physical geometry which cannot be empirically verified independently of a broader system of theoretical physics (e.g. classical or relativistic mechanics) in which it is contained (as the so-called "geometrical" part G attached to the "physical" part F of the system: $G + F$).

The Duhem-Quine thesis partly explains the fundamental question itself: why systems of laws, i.e. scientific theories, are constructed in science at all and why the main task of science is not confined to formulating only the laws but consists in discovering the laws and constructing theoretical systems that are deductive in nature, composed of the laws. Scientific theories constitute most often natural units liable to an autonomous process of verification. Their size and degree of elaboration are, as it seems, adjusted to the requirements of that independent verification in every discipline and at every stage of its development. The peculiarity of both the theoretical notions and theories, lying in the fact that they do not occur in isolation but constitute ordered systems, becomes at least partly comprehensible in the light of the above. Therefore, the Duhem-Quine thesis, in its weaker formation, appears to be fully convincing. As one may presume, the smallest unit of theoretical knowledge liable to independent verification is most often just a theory (which does not rule out the possibility that individual laws are at times such units and in other cases only sets of theories are). The thesis discussed above could be called the moderate theoretical holism thesis. Besides, it seems that this thesis should be complemented with the moderate experimental holism thesis. It states that we verify our theories referring usually not to individual experiments but to greater experimental entities consisting of several (at least two) experiments characterised by a considerable decidability power, especially experiments that complement one another. Here I mean such experiments whose results taken together lead to the exclusion of some existing hypotheses, or at least ones that are likely to be formulated, in spite of the fact that the results of those experiments taken separately, in isolation, do not lead to such an exclusion. The experiments conducted by Fizeau and Michelson may provide a good example: the experiments taken jointly led to the refutation of the theories of those times assuming the existence of ether, since one of those experiments (Michelson's experiment) was inconsistent with theories that ether is immovable and the other (Fizeau's experiment) with theories stating that ether is movable and entirely "drawn" by moving bodies.

As a result, we arrive at the position of the theoretical-experimental holism which emphasises the significance of a theory in comparison with particular laws and the significance of a set of various (complementary to one another) experiments and which emphasizes even more strongly the importance of building the "mixed" theoretical-experimental structures in science, among which the so-called crucial situation[1] is the most prominent. The crucial situation is that theoretical-experimental structure which leads – as one may suppose – to the definite falsification of some theoretical systems competing with one another at a given time and in a given discipline.

A further fact indicates the raising of the degree of verifiability and consequently of law verification, obtained due to the inclusion of the law into a deductive system. A law included in a theory can often be verified against a more definite empirical material than that comprising the *cases* directly liable to that law. Because it is an obvious thing that not all the observational consequences of a theory can be verified by means of the experimental and measuring observational techniques that are available at a given time, and those which can are not verified, generally speaking, with the same degree of accuracy. In physics of any given period, one distinguishes the domains in which the existing experimental, and measuring techniques enable the carrying out of the most precise experiments, available at that stage. For instance, by the end of the 19th century the most accurate measurements were obtained in physics by means of instruments examining certain qualities of the electromagnetic waves. Their perfect example was Michelson's interferometer. By incorporating certain laws into such theories as Maxwell's electrodynamics, Lorentz's theory of free electrons or the special theory of relativity, the laws have obtained the possibility of being accurately verified by means of instruments of this kind, used for the verification of the above mentioned theories. A theory functions here as if an accuracy transformer from one domain of research to another: it "transforms" the accuracy from one field onto another, "infecting" new territories of exploration with the accuracy.

The fact that theories are usually those smallest units of theoretical knowledge which are liable to independent verifying is significantly testified by the history of science especially of physics. The history of

---

[1] I introduce the notion of crucial situation in Such (1975a). See also my article (Such 1982).

physics shows that progress in physics has been achieved mainly in the refutation of whole theories and not individual laws, and in replacing them with new theories which give a more universal and strict description than the previous ones, and constitute the corresponding generalization of the letter which, on their part, are treated as approximate particular (or contiguous) cases of new theories. In physics, individual laws are rarely refuted since, due to logical and other relations between laws, the refutation of one results usually in deranging a whole sequence of laws. On the other hand, in physics we also rarely witness the refutation in one act, as it were, of a fragment or a system of knowledge consisting of more than one theory.

Even the fact that formerly in science the whole theoretical systems (e.g. the empirically interpreted Euclidean geometry and other systems of applied mathematics, the classical Newtonian mechanics and some other physical theories, some systems of philosophy) had been taken as absolute truths of a theoretical character points to the role of a theory in the process of verifying scientific knowledge.

Also, the analysis of the attainment of new knowledge shows that hypotheses – especially in the sciences that are theoretically applied, i.e. mature, just as physics – are usually proposed not individually but in "clusters" or "bunches," which in a natural way arrange themselves into systems: such systems of hypotheses after verification, as a rule, attain the status of scientific theories. The more distant a given statement is from experience, thus, the more of universal, theoretical and abstract (including idealizational) nature it shows, the more "holistic" and indirect its formulation, use for theoretical purposes and, finally, its verification. Observational statements, exclusively, can "afford" to be put forward or verified individually (in isolation), whereas separate theoretical assertions in isolation do not deserve to be treated as assertions well and comprehensively verified and, in particular, they usually do not deserve the designation of laws of science. This designation can be attributed predominantly to the theoretical assertions incorporated in a theory. Thus, such assertions whose verification either strengthens or weakens the whole system and which themselves receive certain indirect confirmation, i.e. the confirmation by means of other laws of the system. The isolated theoretical assertions are mostly only tentative hypotheses, i.e. hypotheses which become laws only as they are incorporated into elaborate theoretical systems and the latter are verified. Thus, the transforming of hypotheses into laws is determined by incorporating them into extensive systems of knowledge.

Therefore, both the context of discovery and the context of justification confirm the assumption that a theory constitutes the basic unit of scientific knowledge. This is so at least in the theoretically mature sciences.

There is another aspect of the problem, connected with the context of discovery, namely that merging laws into theories increases the heuristic power of knowledge, increases its qualities of, so to speak, creatively extending knowledge. This is connected with the fact that the construction of a theory increases not only the predictive power but also the explanatory power and the heuristic power of knowledge, which will be further discussed.

III. The Postulate of Logical Simplicity

The criterion of inner perfection and the so-called postulate of logical simplicity that is connected with it, both postulated by Einstein, cast some light on the role of a theory in scientific cognition, particularly with respect to physics. Einstein noticed that in spite of the justified emphasis on the decisive role of experience and empirical facts within the process of verification, it can be proved that the criterion of the agreement with experience, contrary to appearances, is not sufficient to defend the empirical standpoint in science. This is evidenced by the so-called paradox of flat empiricism that has been familiar to methodologists for a long time. It was formulated extremely suggestively by two physicists, both well-versed in methodology and, if one may put it this way, displaying great methodological flair, namely by Duhem, the had already been mentioned, and Einstein.

The paradox consists in the fact that any theory can be adjusted to experience, to new facts which contradict it, if only one shows enough invention in the process of theory modification or in the modification of the remainder of knowledge.

In this connection, if the admissible modifications of a theory are not restricted in any way then the empirical criterion (the criterion of the concordance with experience) actually becomes absolutely insignificant. Hence the necessity of supplementing it with some other criterion or criteria of an extraempirical character or, at any rate, not having a directly empirical sense. According to Einstein, such a criterion, supplementing the external criterion (i.e. the criterion of the agreement with experience) is the criterion of "inner perfection" (i.e. logical simplicity of the system). Besides, this criterion was formulated earlier by Leibniz. According to the Leibniz-Einstein criterion, the logically simpler a theory, the more perfect it is. Furthermore, the logically simpler it is, the smaller the number of premises, i.e. reciprocally independent initial postulates on which it is based on the one hand, and the greater its informational content (in other terms, logic power, i.e. a set of logical consequences) on the other. Thus the following formula is the determinant of logical simplicity:

$$\frac{\text{information content}}{\text{number of initial assumptions}} = \text{logical simplicity}$$

It is evident that giving knowledge the shape of deductive systems (theories) makes knowledge logically simpler.

Moreover, as Einstein himself has stated, logical simplicity is negatively correlated with mathematical simplicity. In physics, for instance, it can be achieved only at the expense of introducing a more and more complicated mathematical apparatus (e.g. replacing equations of a lower order with equations of a higher order, and linear equations with non-linear ones, etc.). This is so, according to Einstein, because the logically simpler a theory is, the longer and the more complicated chains of arguments lead from its initial assumptions to its observational consequences.

The flat empiricism paradox demonstrates that the extreme empiricism standpoint is transformed into its opposite: it leads to conventionalist views, under which experience is not a decisive or even significant factor in the process of selecting a theory. This being so, criteria such as simplicity (usually understood subjectively), elegance or convenience are themselves to decide about the fate of theoretical systems. This is an illustration of the old dialectical maxim stating that extremes meet.

The main reason of preference for logically simple theories is that the more logically simple a theory is, the smaller number of postulates (initial statements) it contains, and therefore, the more reliable it is as a whole. To prove this, it suffices to accept the following two assumptions:
– the reliability of a theory is the logical product of the reliability of its individual initial postulates (by analogy, it is assumed in inductive logic that the logical probability of a conjunction of statements is determined by the logical product of the logical probability of individual statements, i.e. elements of the conjunction);
– the reliability of initial postulates of a logically simpler theory does not yield the reliability of postulates of a more complex one concerning the same domain of phenomena.

With regard to the first assumption, it is an analogue of the corresponding assumption concerning the logical probability of a conjunction of statements, whose correctness have already been defended by de Morgan. De Morgan has noticed that there is a widespread illusion that a conclusion derived from several equally probable premises is itself no less probable than every one of them. In fact, the reverse is the case. If, reasoning is composed of some sentences in the form of premises and each of them is uncertain, then the conclusion is of very little probability. To obtain the probability of a conclusion, one has to multiply the fractions expressing the probability of the premises by one another. If, for instance, the probability that A is B equals 1/2 and that B is C also equals 1/2, then the probability of the conclusion that A is C equals – with regard to these premises – 1/2 × 1/2, which is 1/4. To generalise, if the probabilities of premises (necessary for deriving of a conclusion) equal respectively p, q, r, etc., then the probability of the conclusion with regard to those premises equals $p \cdot q \cdot r \cdot \ldots$ The numerical value of this product is small, unless each of the quantities p, q, r, etc., is close to one (1) and the quantities themselves are not many.

Thus, if two theories differ in the number of premises and do not differ in the probability of any of these, then the more probable theory is the one that contains a smaller number of premises; similarly, when the theories do not differ in the informational content, it is the one which is logically simpler. In this case, a more probable theory means a theory whose derivative statements are, generally speaking, more probable or whose conjunction, composed of initial postulates, is more probable, which boils down to the same thing. In this connection, it is worth emphasizing that the consequences of a theory are less probable than its initial postulates, that is, of course, if each of the consequences is derived based on many (i.e. more than one) postulates. The larger the number of the initial postulates from which they were derived, the less probable the consequences. This means that the theory (as a whole) is less probable than any of its postulates. Kazimierz Ajdukiewicz comes to similar conclusions. In his work *On Some Ways of Justifying in the Natural Sciences*, he argues very convincingly that combining laws into theories raises not only the degree of their verifiability but the degree of their verification as well. Ajdukiewicz's conclusion is the following: "the tendency to build theories with the help of a possibly small number of hypotheses explaining possibly much is justified not only in that aiming at the economy of means but also in the greater probability" (Ajdukiewicz 1960, pp. 62-78).

It seems, however, that with regard to the difficulties with determining the logical probability of strictly general statements, including scientific laws and theories, with regard to the inference of inductive logic, stating that with all reasonable premises the initial probability and consequently also the relative probability of all strictly general statements (in the world consisting of an infinite or a large number of events) equals or approximates to zero, in all considerations of that type the notion of logical probability should be replaced with the notion of reliability (which does not satisfy the axioms of the theory of probability). Instead of saying that, for example, a theory is less probable than any one of its initial postulates, one should say that it is less reliable than any one of its postulates, etc.

Hence, the following question arises: is our second assumption right or not, i.e. are the single initial postulates of a theory with a smaller number of premises – generally speaking – not less reliable than the postulates of a theory with a larger number of premises?

It turns out that there are no reasons to assume that the former are, generally speaking, less reliable than the latter. In fact, the reverse seems to be the case. The less numerous the postulates of competing theories of comparable informational content, the more general and strict they are as a rule, and the greater informational content they have, the higher the degree of their verifiability is. As a result of this, as shown by Popper, they can be more rigorously verified and, in the case of positive results, attributed higher reliability.

Thus, it turns out that a theory built with the aid of a smaller number of postulates of a greater informational content is, on the whole:
1) more reliable than a theory of the approximate range but containing more postulates of a smaller informational content,
2) much more reliable than an ordinary conjunction of statements not ordered in a system and describing the same fragment of reality.

The former is more reliable than the latter both with respect to the greater informational content of every one of its premises (the higher degree of their verifiability, enabling more rigorous verifying) and with respect to the smaller number of them, i.e. to the fact that its reliability determined by the logical product of the reliability of its initial postulates agrees with the reliability of the conjunction of a smaller number of postulates. Theoretical knowledge explaining a given class of events proves to be the more reliable, the more informative and less numerous premises it utilises.

One should not wonder then that logical simplicity of knowledge (theory) and, reliability of knowledge (theory) are positively correlated. This justifies the postulate of building a theory of maximum informational content and minimum number of assumptions, i.e. of the knowledge, of the logical simplicity and – at the same time – of the outmost reliability, being also knowledge that is highly systematized and highly theoretical in nature.

Put briefly, the aim of building theories is the logical simplicity and all that what the logical simplicity implies. Moreover, it should be recognized that logical simplicity of knowledge constitutes the basic internal aim (purely cognitive) of science, and a theory is not the only one but the main means of accomplishing this aim.

The role of the logical simplicity postulate is indicated by the following diagram embodying jointly the fundamental internal (cognitive) aims of science:

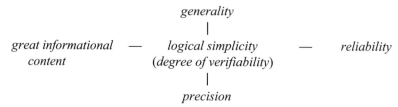

The above diagram, which may be assumed to be consistent with Einstein's standpoint, is a modification of Popper's diagram, according to which it is not the logical simplicity but the great informational content that determines the degree of the verifiability of knowledge and, at the same time, comes into prominence as far as the internal aims of science are concerned. In the diagram, the so-called external aims of science are left out: the theoretical aim (explaining) and the practical aim (predicting), connected with the functions science fulfils in society.

In my opinion, the superiority of the simplicity postulate over the postulate of great informational content and, consequently, the superiority of Einstein's approach over Popper's approach is evidenced, among others, by the fact that although the informational content is determined by the degree of both the precision and generality of knowledge, logical simplicity is nevertheless the function – if one can put it this way – of two arguments, one of which is just the informational content of a theory.

IV. The Postulate of Heuristic Productivity (Dynamic Simplicity)

The significance of the postulate of simplicity (and, indirectly, of the role of a theory in scientific cognition) is evidenced by the fact that it leads – as one may suppose – to the right evaluation of the so-called *ad hoc* procedures, summary procedures, to which scientists sometimes resort in their research, although these procedures are usually severely condemned by methodologists. This refers to such procedures as the making of *ad hoc* hypotheses, whether independent or auxiliary (e.g. auxiliary hypotheses made to save original theories), the procedures of *ad hoc* modification of the existing knowledge for reasons other than verifying, the procedures of *ad hoc* explaining, of *ad hoc* reduction and other summary procedures of this kind.

We witness *ad hoc* procedures when the postulate of logical simplicity is violated in the process of modifying the existing knowledge, e.g. when one passes from a hitherto existing theory on to the building of a more logically complex one. The postulate of logical simplicity, applied to a sequence of theories, is sometimes called the postulate of dynamic simplicity, since it concerns the developmental aspect of science, hence its dynamics. In its essence, however, it is a postulate of the heuristic productivity of knowledge. The characteristic feature of *ad hoc* procedures is the fact that they lead to knowledge of limited heuristic productivity. It is no coincident that the heuristically productive hypotheses are set against the *ad hoc* procedures which only temporarily eliminate obstacles on the way to the development of science and therefore the introduction of one *ad hoc* hypothesis results in the necessity of introducing the whole sequences (chains) of such summary hypotheses which only temporarily succour the existing systems instead of radically improving them.

Therefore, the postulate of logical simplicity, which, when observed, prevents the application of *ad hoc* procedures, reveals another quality of knowledge expressed in the form of a theory: knowledge constructed in this way, as logically simple and without the features of *ad hoc* knowledge, is characterised by great heuristic productivity. And the logical simplicity of knowledge, obtained through a theory, is valuable also because knowledge constitutes not merely the description of perceived, recognised facts but also involves us in particular predictions, heuristics anticipations of new experimental and theoretical results and other procedures or statements concerning facts that are not yet recognised (i.e. future facts). Thus, it is important to see not only the connections of the logical simplicity of knowledge and its concordance with previous experience but also the connection with the heuristic functions of our knowledge.

The heuristic qualities of a theory are also not without impact on to the degree of veracity of a theory. If a theory needs the introduction of still new modifications and *ad hoc* hypotheses for its adjustment to contemporary or future experience, then this fact points not only to its heuristic unproductivity in a given situation but also to the fact that the degree of its veracity does not agree with the level of precision reached by the measuring and experimental technique. By contrast, if a theory, without any modifications and adjustments, is not only concordant with previous experience but also makes it possible to predict new phenomena (i.e. phenomena of a new type), or even to put forward and solve new problems, this means that the degree of its veracity and accuracy fully complies with the level of development of the experimental techniques and is even higher. Owing to this, as new problems are being solved and experiments are being made more accurate, etc., a heuristically productive theory begins to gain an ever greater superiority over less productive theories, even if they initially seemed observationally equivalent. One might even suppose that since eventually theories turn out – strictly speaking – to be false, the heuristic productivity is most important, for it indicates that a theory is true not only within the range of the hitherto existing applicability but also a little further in a broader domain of reality.

In view of the connection of the logical simplicity of knowledge with its veracity outlined earlier it is clear why the saving of a theory, not concordant with experience, through accepting still new additional *ad hoc* hypotheses, leading to its complication, is an indication of its falsity, and the falsity against the facts already recognised and not only the future ones. It must be then required that the auxiliary hypotheses that are introduced should not diminish the informational content, the logical simplicity, and consequently the degree of verifiability and the heuristic productivity of the original theory.

It seems that physics and the methodology of physics have worked out a significant and practically unfailing (or, in all events, efficient) tool for determining whether the transition from a given theory on to a new one (being possibly a modification of the former) complies with the criterion of dynamic simplicity or not. The tool is the principle of correspondence formulated by Bohr in the early 1920s. The fulfilment of this principle by two subsequent theories in a given field means that – simply speaking – the earlier theory, limited only to the range of its actual (rather than to the whole presumed) empirical applicability constitutes only an approximate, particular (contiguous) case of the later theory. The later theory describes reality not only more precisely but also in a more general way; thus, it is both more informative and logically simpler than the former. Besides, the compliance of a new theory with the postulate of correspondence does not only mean the fulfilment of the dynamic simplicity criterion which requires only that a new theory should represent at least the same degree of logical simplicity as a previous one. Instead, it also means a significant progress in this respect, which is desirable in view of the fact that with the development of a given discipline the requirements toward such qualities of a theory as a high degree of verifiability (and degree of verifying), of informational content, of logical simplicity, of predictive power, increase all the time.

Great heuristic productivity of a theory is also connected with the fact that the degree of the heuristicality of knowledge is intrinsically dependent on the degree of the theoreticality of knowledge. However, the issue will not be discussed here. I will confine myself just to saying that a theory turns out to be the most productive way of formulating knowledge since, among others, by providing ordered information it enables one to approach problems in a field and at a time in which scientists are interested as a sequence showing that the solving of one problem leads to putting forward (and solving) further ones. "Only within a theory," writes Mario Bunge, "problems ripen in whole bunches, so that the solving of one throws some light on similar issues and, in turn, puts forward new problems in the field or in other related fields. And only within the limits of a theory the confirmation of one thesis results in confirming or refuting several others" (Bunge 1967, p. 149).

A uniform theoretical system is something irreplaceable also due to the fact that only such systems enable us to solve the most complex problems in science. In addition, a new theory includes, as a rule, much more information than that which had been indispensable for the solving of an issue on which the authors of the theory worked.

The heuristic productivity of knowledge is partly dependent on the mathematical and logical apparatus employed. The formulating of knowledge in the shape of a theory offers a possibility of employing the mathematical and logical apparatus on a wide scale in the empirical sciences – as it actually happens in physical sciences and some others – in the process of both formulating statements and combining them into deductive sequences, which increases the predictive, explanatory and heuristic power of knowledge.

Theoretical holism is also recommended by the fact that great scientific discoveries and scientific revolutions are always, at least partly, theoretical in nature. Among others, they always consist in building new theories. The essential progress in scientific cognition is made not through discoveries that are experimental in character and not through formulating or modifying individual laws but through essentially modifying or constructing anew of whole theoretical systems.

Apart from that, the theoretical holism emphasizes not only the significance of a theory in comparison to individual statements but also the significance of larger theoretical "units," i.e. of the whole systems of theories in comparison with individual theories. According to some authors, nothing confirms the notion of scientific theories as of the reflection of reality able to reach the objective truth, more than the fact that independent theories merge into larger coherent systems which prove to be concordant with experience. There has been quite a number of mergers of this type in the history of physics (e.g. the combination of optics, electricity and magnetism into the electrodynamics made by Maxwell) (Laue 1957, p. 20).

This is manifested, among others, in the fact that as science develops, the newly-built theories become more and more extensive and they more often cross the borders of the traditional divisions of a given science into particular disciplines. This requires and, at the same time, enables their verification against a more diverse material than that on which the previously existing theories were based. For instance, the general theory of relativity is not only the theory of mechanics (i.e. theoretical mechanics) but also the general theory of gravitation, then the theory of the metrical structure of space-time (i.e. physical geometry), and also, to a certain degree, the theory of optics or, more generally, of electrodynamics. This refers, even to a larger extent, to quantum mechanics. Quantum mechanics has abolished the traditional borders separating mechanics from other fields of physics, and even chemistry from physics, and has been successfully applied to both these disciplines, combining and, at the same time, containing some fields of both theoretical physics and theoretical chemistry. From quantum mechanics, it has been possible to explain the periodical law of Mendeleyev and also to discover the principally electron nature of chemical bonds.[2] Apart from that, the construction of quantum mechanics meant that not only a new physical theory had been created but that it was a theory of a new type representing a completely new model of a theory not complying with the patterns worked out up to that time by methodologists.

The changes which are under way in contemporary science, in its structure, and especially in the structure of contemporary physics, head towards the logical and methodological integration of science (physics). Also, they head towards the strengthening of the links between individual statements and theoretical systems, between experiments and observations carried out, finally, between theories and their applications and experiments confirming them. Therefore, the holistic outlook on science and on the scientific procedure of acquiring and justifying of knowledge gains more significance as sciences develop, and becomes more and more productive.

As far as the justification context is concerned, possibly the most important result of such an overall approach is contained in the thesis stating that with regard to the non-existence of absolutely certain, irrevocable, basic theorems of science (and, on the whole, of the cognitive results, both theoretical and empirical), general knowledge that is best justified is acquired with the aid of the alliance between the theory and the experience, with the aid of the mutual correction of the theoretical components of knowledge, on the one hand, and of its experimental components, on the other. Due to this mutual check-up, the verification of knowledge takes place on all its levels and on all its planes, at all the territories and not only at the purely experimental level, i.e. at the point where observational statements meet experience and possibly at the point where all other theorems meet observational statements At any rate, the above abolishes the position of the traditional atomistic empiricism, according to which it is the individual atoms of theoretical knowledge, i.e. isolated statements (e.g. laws of science), that are liable to empirical verification and, even more so, that they are liable to it due to being based on the atoms of experience, i.e. individual results of observations and experiments.

It should be assumed that the fundamental structural unit of a developed science is a theory. New scientific conceptions usually assume

---

[2] Feynman, in his well-known *Lectures in Physics*, presents a bold supposition that quantum mechanics perhaps contains, in its equations whose full content has not been yet determined, the solution to the mystery of life and even of consciousness. In such a case it would become a general theory referring to all sciences, going beyond the borders of the natural science.

the shape of a theory and as theories (or aspiring theories) they are verifiable. Thus for methodology as well especially for methodology practised through basing on the logic of formal and semantic systems, including the theory of models – a theory constitutes the most important subject of analysis: it is usually the smallest theoretical whole, liable to independent verification (the smallest "verifiable" unit).

Undoubtedly, a theory constitutes the most perfect form of the systematization of knowledge. One should only remember that this "form" itself is liable to gradation, comprising both loose theoretical systems, deserving rather to be treated as views or conceptions, on the one hand, and highly formalized axiomatic theories, on the other hand.

V. The Postulate of Explanatory Power

Should present some reasons which concern the explanation procedure and which would explain why knowledge is formulated in the shape of a theory in a mature science. We already have one of these reasons: combining laws into theories enables us to increase the explanatory power of knowledge with regard to the number and diversity of the phenomena explained. Apart from that, at least the most complicated scientific explanations, if not all, require referring not to individual laws but to whole theories.

However, there exists a more profound aspect of the matter. The explanation of phenomena, in accordance with intuitions fostered by scientists, must not simply consist in the fact that the phenomena are included under certain general principles and shown to be particular cases or examplifications of certain empirical generalisations. The procedure of genuine explanation always explores certain deeper layers of reality than those which can be directly observed, certain internal relations, "inner" mechanisms of the phenomena being explained, or the essence of the phenomena, as it is sometimes called.

Theories are more capable of reaching into those deeper layers of reality and arriving at substantial descriptions of those internal mechanisms or essence of phenomena than laws taken in isolation. The further away those layers and mechanisms lie from the observable surface of phenomena, the longer chains of logical and mathematical reasoning lead from their description to the description of the observable layer and, thereby, the more complex and systematized is the character of our knowledge. Therefore (and for some other reasons not mentioned), some authors simply put forward the postulate stating that there is no correct explanation without a theory and that even a law or a set of isolated laws is unable to provide duly for the explanatory function. Besides, we sometimes meet an opposite approach, as it were, demanding that the ability to explain should be accepted to be one of the characteristic features of a theory, which should be included in the definition of a theory.

VI. Conclusion

In conclusion, the pragmatic (especially didactic and utilitarian) aspects of the issue must not be ignored. Such pragmatic advantages, obtained from formulating knowledge in the shape of a theory, as the economy of thinking, the economy of gathering and imparting information (including remembering) and the economy of processing and transferring information should not be disregarded. If a theory were to be treated as a deductive system of laws $T = C_n(A)$, containing all logical consequences of a given set of initial postulates A (axioms) of a theory, i.e. consisting of a theoretically infinite set of statements, then the stock of knowledge contained in a theory is twice (as if two-dimensionally) unlimited. On the one hand, each law entering the composition of a theory as a strictly general statement contains an unlimited stock of knowledge (information) involving (the unlimited number of) cases of a given class, contained within a law. On the other hand, a theory as a deductive system embodying all statements and laws which are potentially contained in it and which can be, if need be, derived (irrespectively of the number of those actually derived by scientists in this time or other), contains an unlimited stock of information concerning the (infinite number of) regularities described by those laws. If, then, an individual law provides information that is unlimited in one dimension (referring to an unlimited number of cases), then a theory does so in two dimensions (where an unlimited number of regularities, each of unlimited number of cases is concerned). Thus, we witness the double inexhaustibility ("information unrestrictedness") of a theoretical system:

a) the information inexhaustibility of a law – being an element of a system – against an individual fact (an exemplification, a case of applicability of a law),

b) the information inexhaustibility of a theory against a law (a strictly general statement, being a basic element of a theory).

Recording in our memory the initial postulates of a theory (the axioms) and being competent in logical and mathematical ways and means (e.g. the principles of deduction) of deriving all possible derivative statements from the initial postulates of a theory, we gain enormous knowledge that is unrestricted in two dimensions and concerns a vast domain of reality. Thus, it is unquestionable that a theory provides the most economical description of the world and is, consequently, an extremely powerful and efficient tool of mind.

The logical systematization of knowledge expressed in the constructing of theoretical systems of a deductive character, is the basic structural feature of scientific knowledge; it is, so to speak, its discriminant among the remaining types of human knowledge.

On the other hand, a theory can be treated as the basic structural unit (basic element) of scientific knowledge only due to the functions which science fulfils in society: the functions of a theory turn out to be strictly connected with the social functions (theoretical and practical) enacted by science. A theory is the best way of formulating knowledge, the most efficient means of handling knowledge and using it for heuristic and practical aims. Hence the correctness of Gauss's statement cited earlier that "there is nothing more practical in the world than a good theory." The saying is unerringly directed against the narrow pragmatism of many practitioners and the non-theoretical positions of some scientists and some science researchers burdened with a positivist tradition.

THE LEIBNIZ-EINSTEIN PRINCIPLE
OF THE MINIMIZATION OF PREMISES*

1. In Book III of G. W. Leibniz's *New Essays on Human Understanding* (cf. Leibniz 1981, p. 285) there is a statement on the subject of the so-called goropism, which provided the basis for the principle that is referred to by some scientists and scholars as the Leibniz-Einstein's principle (or postulate) (cf. Merkulow 1971, p. 86; Such 1975a, p. 179).

Leibniz uses the term "goropism" to refer to the procedure of correlating a false theory with any feasible experiment by means of reacting to empirical results which contradict the theory with the continual introduction of newer and newer assumptions (Leibniz 1981, p. 285).

2. Since an additional assumption which "saves" a theory from the danger of empirical experiment is usually an *ad hoc* hypothesis, it diminishes the modified theory's logical coherence and inner perfectness. Thus, it violates Leibniz-Einstein's postulate of minimization of a theory's premises, as well as the so-called postulate of a theory's dynamic simplicity.

The latter is based on the notion of a theory's "logical simplicity," which Einstein used interchangeably with the notion of a theory's "inner perfectness." In Einstein's view, the fewer independent premises (axioms, postulates, principles, input theorems) a theory contains and the higher its volume of information (logical and empirical), the larger its logical simplicity (inner perfectness). A theory's volume of information, in turn, is measured by the number of its logical (and also empirical) consequences and by such features as their generality and exactitude.

3. The relation between a theory's logical simplicity and its credibility explains why the saving of a theory which is contradicted by empirical experience by means of introducing additional initial *ad hoc* premises and thus complicating the theory is a quite certain indication of its falsity (cf. Such 1975a, pp. 155-180).

This, in turn, explains the necessity – postulated by Einstein and others (including Leibniz and Duhem before him) – of combining the criterion of conformity with experience with that of logical simplicity (inner perfectness). The rejection of the latter would bring about a conservatism

---

* First published in *Leibniz und Europe*. VI Internationaler Leibniz-Kongres. Vorträge I. Teil. Hannover: 1994, pp. 765-769.

of reasoning: the modified theory may be made to conform with experience more closely if its input system is made more complicated and if additional *ad hoc* hypotheses are introduced into it. Thence, if one rejected the criterion of simplicity, one would still support the old theories and introduce into them, whenever the need arose, new *ad hoc* modifications and particularizations. Consequently, one should strive for a theory's conformity with experience, although not at any price: we have already demonstrated that the criterion of conformity with experience need not be – contrary to appearances – the only criterion of the correctness of knowledge.

4. The above conclusion introduces the notion of a theory's dynamic simplicity. Unlike logical simplicity, which is of static nature, since it constitutes a theory's quality at its certain stage (in time), dynamic simplicity is a feature of the long process of the development of a scientific theory, and not of certain settled state of a theory (cf. Schlesinger 1963, pp. 32-38).

The notion of dynamic simplicity may be presented in the following way.[1] Let us consider two theories T_1 and T_2, each of which provides a different explanation of the group of facts E_1. Let us also assume that the theory T_2 is correct, even though at this moment there are no special reasons to favour it, as T_1 is equally confirmed and simple (or complex) as T_2.

Let us now assume that further research discovers the group of facts E_2, which confirm T_2 but not T_1. Still, the proponents of T_1 try to save their theory by introducing a new hypothesis H, in such way that $T_1 \wedge H$ are evidenced no worse than T_2. Practically, this process may be continued infinitely, by introducing all possible hypotheses in order to remove the contradictions between the continually modified theory T_1 and experience. If the continual introduction of new hypotheses H_1, H_2 etc. increases the logical complication of the initial theory T_1, then the theory may be said not to meet the criterion of dynamic simplicity. A theory satisfies this criterion either when its logical simplicity increases as the theory develops, or when its level of logical simplicity at least remains constant.

If the theory T_2 meets this criterion, and the alternative theory T_1 does not, then dynamic simplicity becomes the indication both of the (relative) correctness of T_2, and of the falsity of T_1. Thus, dynamic simplicity becomes also the criterion of choice between T_1 and T_2 and an indication that the hypotheses H_1, H_2 ..., which complicate the initial system of T_1, are *ad hoc* hypotheses, and thence – as more and more of them appear –

---

[1] Cf. Merkulov (1971, p. 86). This method of presentation is more interesting than the others because it demonstrates the relation between a theory's dynamic simplicity and its (relative) correctness.

they convey more and more of their principal limitation (which is their being *ad hoc*) to the continually modified system, i.e. to the theory T_1.

5. The general situation which we presented above may be exemplified by the story of the Ptolemaic system, which could not be "falsified," as its proponents benefited from the right which they enjoyed at least until Copernicus' time, i.e. the right of introducing more and more (increasingly whimsical) eccentricities and epicycles, or additional *ad hoc* hypotheses, which basically made the system consistent with any results of observations. In the case of the debate about the heliocentric system, the issue was more complex, as Copernicus' theory, which constituted the alternative to the Ptolemaic system, was modified at least twice, by Kepler and by Newton (that is, if we limit our account to classical physics only; as we know, the third important modification in the Copernican system stems from the general relativity theory). However, in each case the nature of the modifications was different. While the modifications in the Ptolemaic system made it more and more complicated, until eventually it lost all of its initial logical simplicity, the modifications in Copernicus' theory always presented our knowledge of the Solar system in a simpler way. Kepler's modification, in particular, simplified Copernicus' original theory considerably when it replaced the numerous differentials and epicycles with the movement of planets along ellipses.

6. The mere fact that a theory, which experience has made less plausible may often be saved by being made more complex (as in the case of the Ptolemaic system or the theories which supposed the existence of ether) or by the introduction of *ad hoc* hypotheses but not by simplification, proves that simplicity may imply correctness, or, more precisely that relative (comparative) simplicity may constitute an indication of relative (comparative) correctness: simpler theory → more correct theory (where, however, the sign "→" designates not material but stochastic implication). As newer and newer facts emerge, theory T_1 (in our case Ptolemy's theory) becomes more and more extensively falsified, and although each new modification in it is congruent with the experience accumulated up to a certain moment in time, the theory itself becomes more complex and requires more and more modifications as new empirical facts appear. Can there be more conclusive proof of the falsity of both the initial theory and of its *ad hoc* modifications?

7. In order to perceive the relation between dynamic simplicity and heuristic productivity on the one hand and correctness on the other, it is enough to realize that a theory's heuristic productivity means, in terms of experience that the theory will be consistent with future experience, and in terms of theory – consistent with any future theory based on the same corpus of experience. An *ad hoc* hypothesis complicates a theory and

makes it more congruent with a certain amount of experience (given or existing) but does not increase its heuristic productivity or capacity of prediction (i.e. does not make it congruent with future experience). Thence, an *ad hoc* hypothesis may provide an apology for the accomplished facts, while a heuristically productive (simplifying) hypothesis may be compared to a lantern which lights the way for the facts of the future, for tomorrow's experience. If the criterion of practice is to be complied with unconditionally, theories ought to be judged not only according to their previous practical applications but also to the prospect of their future applications, which means that the postulate of the heuristic nature of knowledge must be followed.

8. As we have demonstrated, the introduction of *ad hoc* modifications into a theory destroys its initial volume of information and its heuristic value. The conventional strategy of adding more and more auxiliary hypotheses as new facts, contradictory to the theory, appear, may deprive it of its heuristic strength and even of its empirical nature. In such a case, the theory forfeits its logical simplicity and consistency, and through an incremental process of changes in the public attitude to its criteria of falsity (liberalization or even sheer omission) loses its empirical nature: it remains congruent with newer and newer facts with which it may be confronted. Such a theory is no longer controllable by experience.

9. The idea of goropism, which was first expressed by Leibniz, may be considered to introduce, albeit in an implicit manner, the criterion of a theory's dynamic simplicity. Thus, the German philosopher contributed considerably towards the formulation of one of the most important methodological postulates of the modern philosophy of science – the postulate of dynamic simplicity (also called the principle of the minimization of a theory's premises), the necessity of which was especially stressed by Duhem and Einstein. Thence, the proposition is justly referred to as Leibniz-Einstein's postulate. The former scholar formulated the postulate's main idea, while the latter pointed out its significance in the modern study of science and its development.

10. To conclude this brief presentation, I would like to emphasize the fact that Leibniz's contribution to the formulation of the postulate of dynamic simplicity must by no means be considered accidental. The German philosopher paid much heed to the issue of the simplicity of the Universe, and especially of the simplicity of the laws that govern it. Thus, for example, in the *Theodicy* he tried to make the idea of the simplicity of laws compatible with the idea of the complexity of phenomena. In Leibniz's view, out of an infinite number of possible worlds, God chose and created the one that combines the maximum simplicity of laws with the maximum abundance of content.

ON KINDS OF INTERPRETATION PROCEDURES IN SCIENCE*

I. An Introduction

While browsing through some more recent works on methodology of natural and humanistic sciences, I have found that the word "interpretation" appears in them very frequently. It is found more often than many popular methodological terms such as "concluding," "reasoning," "generalizing," "abstracting," "analyzing," and only few methodological terms such as, for example, "explanation," "prediction" or "testing" appear more often on the pages of works on methodology than the word "interpretation."

In older works on the methodology of science, e.g. in the 19th century publications by, say, Mill or Jevons, it appeared very rarely. It can be said that the word has had a successful methodological career and now belongs to the fundamental methodological terminology, to the basic vocabulary of the study of science. Undoubtedly, its currency is connected with the more and more vivid criticism of the traditionally understood inductionism and the so-called atomistic empiricism, and also with the dissemination among methodologists of the standpoint of hypothetism and the so-called holistic empiricism.

At the same time, our attention is drawn to the shortage of writings dedicated to the analysis of interpretative procedures. For example, in handbooks and monographs in the field of methodology and practical (pragmatic) logic we often find chapters and paragraphs dedicated especially to various procedures and cognitive operations. However, there are no appropriate fragments devoted to interpretation. Such a state of affairs is evidence of the current interest in these problems.

This situation is not conducive to a responsible application of the term interpretation – the term that is not only one of the most often used ones but is also one of the most differently explicated ones in contemporary methodology. Both in common use and in scientific practice it is usually so indefinite and ambiguous as only few terms used in science. In addition, it is also often abused. If in the same discipline a given author uses the

---

* Translation of "O rodzajach procedur interpretacyjnych w nauce," *Studia Metodologiczne* **6**, 1969, 103-130.

term "interpretation" on different occasions in senses that have nothing to do with one another (i.e. its various uses do not reveal any semantic relations), then s/he abuses the term.

The evidence of the ambiguity of the term "interpretation" is the fact that in various disciplines the words that are synonymous, sometimes at least similar or close in meaning include: "translation," "explanation," "explication," "understanding," "reading" (metaphorically, reading someone's thoughts or feelings), "making something explicit," "making the meaning of something unambiguous," "making the meaning of something obvious," "giving meaning," "giving sense," "giving an additional meaning," "determining denotation," "indicating a designation," "modelling or construction of a model or reference to a model," "formulation, expression (from some definite point of view)" and the like.

Due to this, the subordination of interpretative procedures that occur in natural and human sciences to the general notion of interpretation seems impossible at first. In any case, it is obvious that in order to give a reporting definition of this term, if only an approximate one, i.e. embrace within this concept possibly all the procedures termed interpreting, this term should be understood very broadly.

Not so broadly, however, as to consider interpretation as an indispensable component of all the cognitive operations. Interpreting is not a universal component of research operations; that is, not every cognitive operation is at the same time an interpretation to some extent. I also think that not every cognitive result is obtained, among other things, thanks to the application of interpretation operation although this question may be disputable.

On the other hand, it seems that an optional cognitive result, i.e. every result of a cognitive operation may be submitted to some endeavour at interpretation. Thus, we can speak of an interpretation of sensual data, of facts and of notions, and also of sentences of all kinds (ranging from introspective sentences and observation reports which describe results of experiments to axioms of theoretical systems), we can speak about interpretation of laws, principles, postulates, rules, norms, questions, about interpretation of theories, models, axiomatic systems, language, knowledge, works of art, and the like. What is submitted to interpretation or "a raw material" of interpretation (obviously, raw in the relative sense, relativized to a given attempt at interpretation, and not in the sense that it is always some initial element of cognition in general, something of a kind of "raw fact" or "immediate data") may therefore be everything which is a result of a cognitive process. Each cognitive result and any cognitive operation may be submitted to interpretation.

We often speak of an interpretation of subjects which are not results of cognitive processes. Thus, we can speak of an interpretation of phenomena, facts interpreted not as certain cognitive constructs, of historical sources, various extra-cognitive situations, about an interpretation of the subject of cognition, interpretation of nature and the like. It seems that in all these cases the term "interpretation" is also used in its literal rather than merely metaphorical meaning, and that the use is fully justified; thus, what is meant is not an interpretation of some cognitive results concerning phenomena, facts, historical sources or nature but also submitting to interpretation the very phenomena, sources, and the like. I mention this at this point because I have encountered a view that interpretation concerns exclusively the cognitive results, and that we should not speak of interpretation of phenomena as we cannot speak of generalization of phenomena but only about the generalization of knowledge, e.g. concepts or judgements concerning phenomena. Therefore, we assume that the initial material for interpretation, i.e. what may be submitted to any interpretation, is either a definite cognitive result, or a definite cognitive operation or a certain phenomenon or an extra-cognitive situation. However, even though everything which exists may be submitted to certain interpretation, certainly not everything may be submitted to interpretation of a given kind since it requires, e.g. a prior reconstruction of what is to be submitted to interpretation. For instance, a semantic interpretation (in the sense of constructing a model, a pseudo-field) of a natural language or of a language of any empirical science is impossible without a previous logical reconstruction of such a language (and this should be a reconstruction made by means of rather advanced means of contemporary logic).

Let us now leave aside the question of the scope of the term "interpretation" and the consideration of what may be an initial material of interpretation, and let us pass on to the problem of the components of the procedure of interpretation and the kinds of interpretative procedures.

II. Components of the Interpretative Procedure

Five components are distinguished in the interpretative procedure. These are the following:

1) *the initial material* of interpretation, i.e. that which is submitted to interpretation;

2) *the system of reference*, i.e. the knowledge in the light of which we interpret;

3) *rules* of interpretation, or certain principles, according to which we interpret, the principles which as if provide "a key" for interpretation (the terms – rules of interpretation – is understood very broadly: what is meant here are not only the so-called methodical rules which depend on the initial material of interpretation, on the reference system of interpretation, and the like; thus, they are different for various procedures but also more fundamental methodological assumptions of

interpretation are meant, principles which may concern all interpretative procedures);

4) the fourth element is an *operation* of interpretation itself, as if "the process of interpreting";

5) finally, the *final result* of interpretation, or that into which "the initial material of interpretation" was transformed or a certain derivative material as a result of submitting it to the process of interpretation.

It is worth mentioning that "interpretation" is commonly understood as either the act of interpretation, or the result of this activity, thus either the fourth or the fifth element of those mentioned above.

As an example, we shall consider the semantic interpretation of primary terms of some axiomatic system. 1. The initial material here is constituted by the primary terms of the system. 2. The reference system is a certain body of knowledge in the field of semantics, particularly, knowledge concerning semantic models. 3. Rules of interpretation are appropriate semantic rules or, let us say, denotation rules. 4. Generally speaking, interpretation itself consists in finding or constructing an appropriate class of models of the system. 5. As a final result, we obtain a semantic system whose axioms, and theses in general, may be ascribed semantic truthfulness in the sense of the correspondence theory of truth.

An interpretation process seems to be an operation which is unequivocally determined by the first three elements (thus, it can be defined as an operation with three arguments): initial material, a reference system and rules of interpretation. However, it should be emphasized that in the interpretation practice we are usually only partly aware of the content of these three components of interpretative procedure, which results in the process of interpretation being partly (and sometimes even wholly) spontaneous, not fully aware of the factors which determine it, in the overwhelming number of cases.

I shall now pass on to the discussion of certain kinds of interpretational procedures.

III. Kinds of Interpretative Procedures

Theoretically, the criterion of classification of interpretative procedures may be looked for in any of the above-mentioned components of the procedures. When searching for principles of division, one could also refer to the aims of interpretation and functions which interpretation is to fulfil in the system of human knowledge and action.

The most important role in the classification of interpretative procedures should undoubtedly be played by the first three elements. And, indeed, in scientific practice we usually find divisions of interpretation according to (1) that which is interpreted, i.e. the initial material of interpretation, (2) by means of which it is interpreted, i.e. the reference system of interpretation or (3) by which rules one interprets, i.e. the "key" of interpretation. When we speak, for instance, about the psychological or sociological interpretation of historical facts, it is obvious that interpretation of historical facts is first distinguished as a separate type; therefore, the kind of interpreted material is used – as a principle of division – and then the reference system, i.e. a definite kind of knowledge serving interpretation is used.

I am going to give neither a definition of interpretative procedure in general nor an exhaustive classification of various interpretative procedures encountered in science, as I am not capable of that. Furthermore, I shall discuss only some kinds of these procedures occurring in science, hoping to have selected those that are of most importance.

The most fundamental division, both theoretically and practically, seems to be the dichotomous division of interpretation into the following kinds: (1) *interpretation of theoretical systems* (including *axiomatic ones*) or *their parts* and (2) *"theoretical" interpretation of knowledge from the observational (experimental) level.* Like Kuhn, we could speak of two "worlds" of science, i.e. about two "worlds" which a scientist submits to interpretation: the world of facts and the world of theories (including mathematical formulas). There is one common fundamental goal of both kinds of interpretation: to establish a relationship between these two fields, these two "worlds," which surely is one of the most difficult tasks that a scientist faces. However, each of the two interpretation procedures runs in an opposite direction: one in the direction of empiria and specification, and the other in the direction of theory and generalization. Apart from that, they seem to differ from each other so much that when we first encounter both and confront them with each other, it seems that they have nothing in common and that a separate name should be introduced for each one of them. I have noticed that students, for one, when talking about interpretation have only one of the two kinds in mind: e.g. a student of mathematics usually thinks about the interpretation of axiomatic systems, while a student of physics or a student of chemistry usually thinks about the theoretical interpretation of experimental results.

It seems significant to find out whether these two procedures should be considered as fundamentally different and whether to contrast one with the other or whether, because of the possible profound structural similarities, to combine them and subsume them under a common name of interpretation. I assume provisionally that both these procedures are sufficiently similar to be considered as interpretation procedures.

A. Interpretation of Theoretical Systems

As far as the first kind of interpretative procedures, i.e. the interpretation of theoretical systems, is concerned, it seems advisable to distinguish three basic types thereof:

1. *Semantic interpretations* of theoretical systems (in particular, axiomatic ones) or of their parts;
2. *Empirical interpretations* of theoretical systems or of their parts;
3. *Ontological interpretations* of theoretical systems.

Let us now discuss these three types.

1. The initial material for *semantic* interpretation is usually some formalized axiomatic system or its part. An interpretation of a system consists in constructing a class of models for it. Every model marks one of the possible interpretations of a system. To give an example, let us take an axiomatic system, representing an empirical theory. An interpretation of systems of this kind is usually understood as constructing models for those systems, which is connected with establishing of denotations of specific (descriptive) terms which occur in those systems. It is usually assumed that such a model of theory T, being a semantic interpretation of the axiomatic system consists in the non-empty set of individuals U, being a universum of a theory, or a set of objects run by its variables, and of relations $R_1, \ldots, R_n$, occurring between elements of the U set, relations being denotations of predicates of a theory and constituting the so-called characteristics of the model.

In semantic interpretation of a system, the word interpretation has a strictly defined sense (or rather several strictly defined senses. For instance, we can speak of an interpretation of language in two senses: in the first one, interpretation is a model in the meaning of a field about which we can speak in a given language; in the second interpretation is a pseudo-field, i.e. a model in which the statements of logic and linguistic postulates are true). Instead of an interpretation of a system we usually say: the construction of a model in which the system is fulfilled; instead of an interpretation of a term, we usually speak of designating its denotation. I think that the case of semantic interpretation is one of those few (not very numerous) cases in which the word interpretation is given a strictly defined meaning. The theory of models (it could also be called a theory of semantic interpretation) of axiomatic systems constitutes, as we know, one of the younger fields of mathematical logic (we also have model theories in the fields of cybernetics and of the general theory of systems but what is of concern in them are models, generally speaking, which are non-semantic, thus, they are not directly concerned with semantic interpretation).

It is often emphasized that there is such a way of interpreting axiomatic systems (the so-called syntactic interpretation, or the purely formal one, about which Ajdukiewicz and Kokoszyńska, among others, wrote) in which the problem of (semantic) interpretation of the axiomatic system does not arise at all. A given system remains a pure syntactic system or an uninterpreted formal (logical or mathematical) calculus. Obviously, such a system is not a semantic system and to an even lesser degree it is a theory of an empirical nature.

The problem of semantic interpretation is of great epistemological significance since it is closely intertwined with the problem of truth (as is well known, the concept of truth is relativized not only with respect to the language but also to the model). As it seems, it is also of great importance as far as transmitting achievements and methods from one mathematical discipline to another is concerned. Therefore, thanks to one or the other interpretation, the theory of groups is applied in various fields of mathematics, e.g. in geometry and in the theory of numbers, Boolean algebra in the set theory, mathematical logic (e.g. sentence calculus) and many other kinds of pure and applied mathematics and the like. Interpretation of axiomatic system is also a powerful research tool of metalogic, particularly as far as the study of such properties of theoretical systems as non-contradictoriness and independence of axioms is concerned.

I am not going to deal at length with semantic interpretation, since logicians are much more familiar with those issues than I am. I shall only say that as far as semantic interpretation of theoretical systems representing empirical theories is concerned, what is methodologically important is the differentiation of the standpoint of realism in this question from the standpoint of instrumentalism. This is a trend which was started already by Berkeley, who supplied an instrumental interpretation of Newtonian mechanics. According to realism, all specific (descriptive) terms of such a system undergo interpretation, independently of whether these are theoretical or observational terms. According to instrumentalism, only those observational terms are interpreted in the field whose universum are observable objects while theoretical terms – as the ones which apparently concern fictional objects – do not have a semantic interpretation. Such a division into terms concerning real objects and terms concerning fictional objects was conducted for the first time just by Berkeley with reference to Newtonian mechanics. The lack of semantic interpretation of theoretical terms involves a purely instrumental interpretation of theses in which these terms occur: these theses are a convenient tool of operating sentences constructed using exclusively observational terms, and they constitute an apparatus for prediction which does not perform any semantic function. Obviously, it is in particular assumed that they have no logical value (truthfulness or falseness).

The standpoint of instrumentalism is based on an assumption that the sense of observational terms is independent of the sense of theoretical terms, thus that the first ones are interpreted independent of the second ones. That is why the adherents of this trend sometimes interpret the language which reconstructs the actual language of a given discipline as consisting of two various languages: the semantically interpreted observational language and uninterpreted semantically theoretical language. Both these languages are to be combined into one by means of rules of correspondence, which to some extent also allows us to speak of empirical interpretation of theoretical terms as well. Perhaps one can agree with J. Kmita that the viewpoint of instrumentalism in the above version collapses together with the recognition of the so-called "bilateral dependence" of the sense of theoretical terms and the sense of observational terms. Thus, when it is recognized that also the sense of observational terms is not independent on theoretical terms (Kmita 1966, pp. 187-188, 203-205).

I shall now move on to the question of empirical interpretation of theoretical terms.

2. Speaking of *empirical* interpretation one usually keeps in mind also axiomatic systems but those axiomatic systems, which represent or can represent empirical theories (physical, biological, psychological and the like). A fundamental and extremely complex problem arises at this point: in what manner are specific extralogical (elementary and theoretical) terms of the theories, and particularly primary terms, given a definite empirical meaning which can be used on the basis of experience?

M. Przełęcki, who can be said to represent a traditional point of view of logical empiricism on this question, distinguishes two ways of interpreting specific terms: *an intrasystemic interpretation* and *an extrasystemic interpretation* (Przełęcki 1966, p. 210ff). The first one is determined by the theses of the system itself (namely, by its axioms), the second is supplied "from the outside," i.e. by "semantic rules" of a definite kind, thus by the statements of the metasystem. The intrasystemic interpretation corresponds to this which is usually called definition (or pseudo-definition, definition in entanglement) by postulates: it is recognized that axioms constitute the meaning of primary terms. This "constitution of meaning" is resolved to the adoption of a terminological agreement, which demands that the specific terms be understood as names of such a meaning where axioms of a theory become true sentences. If there is only one model of a given theory (i.e. a field in which axioms are true) then terms of a theory have a univocal interpretation; otherwise, introduction of a definite model, and with it, of a definite interpretation of specific terms, must occur "from the outside."

It results from some statement on isomorphism that no theory can have only one model since each field isomorphic with a field being a model of a theory T is also a model of this theory. One can expect at the most from a theory that all its models are isomorphic. As is well known, a theory which has exclusively mutually isomorphic models is called a *categorical* theory (one can also speak of *univocally categorical* theories, in the case when, for the two optional models of the theory T, there is one and only one perfect relation establishing their isomorphism, and also of theories *categorical in power* if only models of a theory with an absolute notion of identity of a universum of a definite power are isomorphic). However, even in the case of categorical theory, the interpretation of its specific terms is very indefinite. It also appears that a categorical theory may only be a theory which has only finite models, and because empirical theories are assumed not to fulfil this condition as a rule since they are universal, i.e. strictly general, they are therefore not categorical theories.

The conclusion is that a system of axioms of an empirical theory not only does not determine one model of this theory but usually determines such a family of models the composition of which is also entered by non-isomorphic models. The system of axioms does not lead to an interpretation of an empirical theory; in particular, it does not allow unequivocally to determine that which a given theory is about (e.g. the denotation of its predicates). That is why a specific term, equipped only with an intrasystemic interpretation should be considered as devoid of empirical sense (as "totally fuzzy" as Przełęcki says since one cannot judge about any object whether it falls under a term or not) (Przełęcki 1966, p. 215). The above demonstrates that an intrasystemic interpretation is not an empirical but a semantic interpretation.

An empirical interpretation may only be provided to a theory by an interpretation "from the outside." Distinction from amongst a family of models of a theory of its *proper model* (in a definite sense, namely a model being an empirical interpretation of a theory, "an empirical model" of a theory) may be performed by means of semantic rules (statements of a metatheory) which subordinate definite denotations to particular specific terms of a theory.

As far as mutual relations in which the specific terms of a theory have entered, it generally suffices to interpret only some of them using semantic rules, and the remaining ones will obtain an indirect interpretation. The terms obtaining a direct interpretation may be certain perceptional terms, introduced by means of ostensive definitions, i.e. by the procedure including direct indication of definite observable objects. In this way, we obtain a division of specific terms of a theory into theoretical terms, which have only a indirect interpretation by means of the so-called rules of correspondence, and observational (elementary) terms obtaining a direct interpretation by means of semantic rules which are ostensive definitions.

It appears that a direct interpretation of perceptional predicates is something unusually complicated and does not allow to univocally subordinate denotations to these terms, hence the notorious blurring of these terms. It is even worse in the case of a univocal empirical interpretation of theoretical terms. Only in the case of the definability of theoretical terms by means of elementary terms does the interpretation of the latter guarantee an unequivocal interpretation of theoretical terms. But theoretical terms are never, or almost never, definable on the grounds of empirical theories by means of elementary terms (it is emphasized that it is just the opposite). Mutual relationships which occur between these two classes of terms there are looser – they can usually be captured by means of the so-called partial (conditional) definition or generalization (an "ordinary" or a probabilistic one).

Therefore, a direct interpretation of elementary terms does not involve a univocal indirect interpretation of theoretical terms (Przełęcki 1966, p. 220). But through detailing and making empirical theories more precise, and in particular, through adding ever newer partial definitions, the range of possibilities of interpretation is becoming narrower as a theory develops. In practice, however, an empirical theory never achieves an ideal state of a unequivocal empirical interpretation: all its empirical terms, and in any case theoretical terms, remain blurred. It can be said that empirical interpretation is, unlike at least some kinds of semantic interpretation whose characteristic feature is adequacy, always inadequate, approximate, inaccurate, and ambiguous to some extent.

In many points, the question of empirical interpretation of axiomatic systems is explicated in a completely different way by the opponents of logical empiricism, who call themselves hypothetists, adherents of holistic empiricism or in some other ways. But I shall speak about this viewpoint when discussing the theoretical interpretation of results of experimental knowledge.

3. A few words now about *ontological* ("metaphysical," philosophical) interpretation. What is concerned here are non-empirical interpretations of formal sciences and their systems by the currents which assume that those systems constitute a structural description of a certain non-linguistic and non-empirical reality. They consider axiomatic systems from some transcendental point of view, assuming that the subject-matter of mathematics (logic including) is either a non-empirical material world, or a world of ideal beings of the type of Plato's ideas, or some third (besides a material and an ideal) "neutral" world of logical beings and mathematical structures (concepts, numbers, classes, judgements, functions and the like), at the same time non-physical and non-psychic, thus also non-empirical ones. The latter are sometimes defined by a common name of meanings (*Bedeutungen*). These standpoints interpret at

least certain systems of formal sciences (e.g. set theory) as certain general ontologies (the so-called formal ontologies) of a kind of Leśniewski's or Wittgenstein's ontologies. The standpoint of neoplatonism in the philosophy of mathematics as represented by Frege, Meinong, Husserl or Russell at the end of the 19th century, and which still has its advocates, may be considered a classical position of this kind.

The question of ontological interpretations is closely connected to a problem that seems to remain so far unresolved, namely the problem of the nature of formal sciences: the problem of whether they are sciences describing the structure of some actually existing (or only possible) non-linguistic reality, or whether they are rather only an instrument of science, i.e., let us say, a language serving the scientific description of reality.

Now let us move on to the discussion of interpretations of the second kind, interpretation of "the world" of facts.

B. Theoretical Interpretation of Experimental Results

What is concerned here is a theoretical interpretation of all kinds of results of cognition on the experimental, sensual level: interpretation of observational data, raw and scientific facts, results of observation and experiments, interpretation of historical sources, observational terms and sentences, empirical statements and empirical laws, and the like. Here *a theoretical interpretation of observational terms* and *sentences* comes to the fore, and will be given some attention.

Generally, at present no one believes in the existence of uninterpreted "direct data," "bare facts," or of an observational language that would be "pure" and uncontaminated by, say, hypothetical interpolations of an experience or theoretically "neutral." It is assumed that theoretical terms that are specific to empirical sciences have an empirical sense, and certainly even an observational sense, while observational terms have a theoretical sense and some of them are interpreted by means of the other. Thus, both kinds – if they occur in theoretical systems of empirical sciences – are inevitably interpreted; however, their interpretation runs, as it were, in two opposite directions: in the first case, it consists in supplying an empirical sense, and in the second in supplying a theoretical sense.

In connection with a thesis of the reciprocal dependence between elementary expressions and theoretical expressions, observational terms and sentences are defined so that it results from the definition itself that they have a certain theoretical sense, extending beyond the framework of what is given directly in a currently occurring experience. An example of such a definition of an observational term may be the following one (given by Giedymin): Term O belongs to an observational vocabulary if under proper conditions a definitive solution based exclusively direct experience is possible, or term O is applied for a given situation, or it is not. This definition emphasizes the nature of any observational term that is in some measure dispositional thus also theoretical to some extent. This can be seen in the phrase: "under proper conditions," occurring in the definition quoted here (see Giedymin 1966, pp. 91-92).

This leads us to the conclusion that elementary terms do not have a sense, regardless of the sense of theoretical terms occurring in scientific theories. Therefore, not only is there no a sharp division into theoretical and observational terms but in general there are no observational terms which would be devoid of theoretical sense.

The recognition of the non-observational, and thus to some extent of the theoretical and hypothetical nature of all terms and sentences of the language of empirical sciences leads to the conclusion that there is no empirical basis of science consisting of statements which are irrevocable, e.g. statements the rejection of which must involve shaking of empirical rules of language. Sensual impressions received by an observer do not constitute either criteria of sense or criteria of truthfulness of observational expressions of the language of empirical sciences (Giedymin 1966, p. 94). Examples are given of typical observational sentences which were recognized in some period and then were rejected because new hypotheses and theories were adopted (and the old ones rejected) which were joined as far as meaning is concerned with the terms of these observational sentences. The rejection of an observational sentence, even of the observational report, being a description of a direct experimental result, i.e. the statement of a raw fact, results from an improvement of methods and tools of observation, claiming the unreliability of observers, revealing of systematic errors of observation, or observing a phenomenon which so far could not be captured due to its insignificance.

It can be seen from the above that perceptional sentences are not recognized exclusively on the basis of perception that is currently experienced; that recognizing a sentence founded on the basis of perceptions we adopt them due to some ordinary, tacit assumptions, which are not justified in this perception but do not provoke any doubts at a given moment (p. 107). The effect is just that when some of these assumptions appears to be false we revoke an appropriate perceptive sentence. Here we have the famous Popperian examples showing that we recognize a common observational sentence about a presently perceived object (e.g. about a glass of water) only as long as the object observed behaves according to our knowledge about this object, and thus according to our expectations, it behaves typically, as if "regularly" from the point of view of what we know about objects of a given class.

From this point of view, observational sentences may be described as sentences which are recognized or rejected on the basis of direct experience but only when at the same time it is held that there is no reason to question the non-observational assumptions which are involved.

Either the recognition or rejection of an observational sentence occurs always within some theoretical framework and is a result of making a decision which has not been determined unequivocally by observational data, thus a decision which is to some extent a theoretical one. Such a decision is at the same time a theoretical interpretation of experimental data, and constitutes as if a transformation of a raw fact (expressed, e.g. in a sentence "the arrow of this instrument deflects") into a scientific fact (expressed, e.g. in a sentence "current flows in a conductor"). To put it differently, this is a decision as to whether a given result should be considered as an indicator of an objective state or a certain scientific fact which is definite and inaccessible to the present observation, or whether not to recognize it. This leads to a conclusion that every recognized observational sentence may be then rejected for theoretical reasons, i.e. on the basis of some theory or law (even without obtaining new observational results concerning a given object).

The rejection of Campbell's and Carnap's theses on independence of the sense of observational terms and sentences of the sense of theoretical terms and sentences leads to the rejection of a traditional scheme of interpretation of axiomatic systems of empirical sciences, presented, for example, in Przełęcki's works. It is a scheme according to which an empirical theory is a formal calculus enriched with rules of correspondence which subordinate it at least one domain of observational objects as a model. It also leads to the rejection of a view that there is a direct and definitive (established once and for all) observational interpretation of elementary expressions of the language of empirical sciences.

We have two distinctly different conceptions as to the status of observational sentences and the structure of scientific theory; both found their expression, for instance, in a collected work *Teoria i doświadczenie* [Theory and Experience] (the first one is represented in the book by Wójcicki's and Mejbaum's works, the other, by Giedymin's and Kmita's works) (Eilstein, Przełęcki 1966). Mejbaum can be said to recognize irrevocable introspective sentences (the so-called absolutely basic sentences) since he thinks that without accepting them we fall into an infinite regress or conventionalism (sometimes called a "from below" conventionalism) (see Mejbaum 1966, 111-128). Giedymin, on the other hand, claims that any observational sentences (including observational reports) are equipped with a theoretical sense and are revocable, hypothetical, since the conditions of truthfulness of every observational sentence include assumptions which have not been tested by a given perceptive act, and that the content of each such sentence goes beyond the perceptive content. Therefore, stating any sentence engages us into an acceptance of some theory. Giedymin remarks that for some term of a given language to be recognized as totally and definitely interpreted ostensively it should certainly be considered as a term totally isolated in a given language from other terms. Such a term could not occur, for example, in any sentence with another term because if it appeared in any sentence with another term then this sentence might appear to be a hypothesis, which we will tend to preserve even at the price of modifying the initial ostensive interpretation of the term (Eilstein, Przełęcki 1966, p. 162). According to Giedymin, a typical example of an observational term dependent, as far as meaning is concerned, on the theoretical terms is the term (stellar) "parallax" understood as "apparent shift" (of a closer star with respect to the further one). It depends on theoretical terms occurring in the heliocentric hypothesis and in the assumption that the closer star is static with respect to the further star. The adjective "apparent" indicates this dependence: after all, the fact that the movement of a given body is apparent is not given to us in perception but constitutes a theoretical interpretation of perception; interpretation in the light of heliocentric theory from which it results that it is just the observer who is on the Earth that moves in relation to stars, and not the stars in relation to one another. The sense of the term "stellar parallax" depends on the sense of such predicates as "revolves around" or "immovable in relation to" which appear to be important for the formulation of a heliocentric hypothesis and for the assumption of immovability of both the stars (p. 163).

In this connection, Giedymin has formulated a thesis on the systematic ambiguity of results of observation, saying that there is no unequivocal logical relationship between the result of observation, represented by an isolated observation report, and any isolated observation sentence and any isolated hypothesis or theory (p. 165). To put it differently, there is no unequivocal relationship between a raw fact and a scientific fact. Observational terms are not "interpreted wholly ostensively," and observational sentences are neither exclusively or definitely justified by perception.

As Kmita shows, this conception leads to a conclusion that may seem strange at first sight, namely that what occurs as an object of description in a given science (even if it is described as an observation object) is actually a result of abstraction and, one may add, also of interpretation, based on definite knowledge. That is why when describing the facts observed a scientist at the same time describes very complicated "theoretical states of affairs," which he "sees" (in the non-phenomenalist meaning of the word) in these facts. It follows that even if two people use the same term in reference to the same observable object, experiencing the same direct data (under the same conditions of observation) it still is not a proof that they "see" the same thing (i.e. that this term has the same meaning for both). For instance, an adherent of heliocentrism and the adherent of geocentrism while looking at the Sun "saw" two different things. According to Kmita, it can be expressed in the following way: the predicate "the Sun" had the same rules of direct partial interpretation with respect to both astronomers, i.e. in analogical observational situations they made judgements about objects (the temporal phases of the Sun) that belonged to the same class. However, the meanings of this predicate were different for both of them, as for the first one the sentence "the Sun is a planet" was analytical, while for the other the sentence "the Sun is not a planet" was just of such a character (Kmita 1966, p. 186).

The standpoint presented here leads to the rejection of the conception dividing scientific knowledge into two separate sets of sentences: sentences dependent on experience but such which have no theoretical sense, and the other sentences. Thus, the position leads to the so-called holistic empiricism, according to which all sentences, and therefore the whole of knowledge (excluding the so-called formal sciences) is determined by experience, although none of the sentences are determined unequivocally (p. 196). Thus, every sentence requires a certain interpretation of the material on the basis of which it was formulated and accepted. When such an attitude is adopted, the viewpoint of instrumentalism collapses together with atomistic empiricism. It is a viewpoint which also assumes the fundamental dichotomous division of sentences into two classes of a distinctly separate cognitive status that had been mentioned above. From the point of view of holistic empiricism, it is not important whether each theoretical term is connected semantically with observation terms by means of correspondence rules or the so-called partial definitions, in particular, operational definitions. It is essential that a theory as a whole have an empirical sense, so that it could serve e.g. for predicting, and that theoretical terms which enter into its composition had such predicting power on the grounds of this theory. This assumes a "double" combination of theoretical terms with experience, a combination which just allows to interpret theoretical sentences as essential premises of prognostic reasonings.

The standpoint on this question of holistic empiricism seems to be in line as that represented by many physicists-theorists who, just as Einstein, while rejecting operationism, claim that they do not worry about some theoretical term of their theory not being connected with observation terms by means of correspondence rules but with experimental situations by means of operation definitions; it is important that a theory as a whole have an empirical sense and could perform experimental functions. On the other hand, the viewpoint of atomistic empiricism seems to better correspond to the traditional standpoint of experimental physicists.

And in general naturalists – as distinct from methodologists and presumably humanists – are not very much interested in the problems of relationships of theoretical terms and sentences with observations, when these problems are considered at length. They are interested mainly in a certain section of these problems, namely, in the question of interpretation of particular, concrete results of experiences and experiments, results expressed as certain initial facts. What comes to the fore here is the question of a raw fact stated in the observational report, describing as if only what is directly given (coincidences, belonging to the objects of certain observable features and the like) to the scientific fact which is already a certain theoretical construct, a fact stated in sentences such as "a given piece of metal is a magnet," "current flows by a given conductor," and the like. The above problem, posed by Duhem and Poincaré has not been satisfactorily solved.

It seems that naturalists, as opposed to methodologists, have always been hypothetists (with few exceptions) rather than inductionists, and even when, like Newton, they thought that the basic method of obtaining law and theories was the method of inductive generalization in the traditional narrow sense of the term but in practice, they rather used a method of posing hypotheses, the method of idealization, modelling and analogy, thus methods which do not fit in the strictly (narrowly) conceived induction. They as if also realized since the beginning that experiments may be variously interpreted, so that one never arrives at the same cognitive results on the basis of observation itself. In this connection, what we call *experimentum crucis* occurs extremely rarely, rather only in exceptional situations.

This standpoint is expressed in remarks which amaze by their paradoxicality but which are very accurate. These remarks circulate among physicists and concern the changeability of interpretation of the same experimental results, remarks of a kind that Armin Teske offered in relation to Michelson's experiment. Teske states that if the famous Michelson's experiment had been made three hundred years earlier, at the time when existence of ether was assumed (Descartes and others), and the motion of the Earth became a problem, the result of the experiment would have been considered as a proof that the Earth does not move. By contrast, in our times the motion of the Earth is considered to be a doubtless assumption, and the existence of ether is a problem; that is why the same result speaks in favour of the non-existence of ether (and not in favour of the claim that the Earth is at rest) (see Łubnicki 1967, p. 288). The conclusion is obvious: the interpretation of an experimental result depends on the whole knowledge which is the scientists has acquired in a given field at a given period of time. No less instructive is an interpretation of results of experiments concerning the nature of light: a conclusion was drawn from Joung's and Fresnel's experiments, which clearly indicated the wave nature of light, i.e. the fact that light is not corpuscular in nature. This was based on a seemingly obvious assumption that the sentence "light is either of wave or corpuscular nature" is a disjunction. But in the 20th century the disjunctiveness of the component parts of the sentence quoted here was questioned; hence the thesis that light has a corpuscular nature was no longer rejected and the wave-corpuscular dualism of light (and of matter) was acknowledged (Łubnicki 1967, p. 288). Again, the theory that was accepted appeared to be dependent on interpretative assumptions, and not only on the result of experiments.

Apart from problems with the interpretation of experimental results, the physicist-theorist encounters another serious problem with interpretation in the course of his/her research practice. The problem actually concerns the first kind of interpretative procedures, i.e. how to interpret the mathematical apparatus which has been applied or, as it is expected, should be applied in physics in a specific physical theory. Can Maxwell's equations receive a mechanical interpretation? What is a proper interpretation of Lorentz's transformation: kinematic, which leads to special relativity theory, or dynamic, the Lorentzian, based on the rejection of the principle of relativity? Should the equations of the general relativity theory be interpreted according to the principle of general covariance and relativity in the spirit of Einstein, or according to the principle of Fock's harmonic systems? How to interpret the calculus of operators of quantum mechanics? Should the psi (φ) function be given a principally probabilistic interpretation, or a different one, which is in agreement with the hypothesis of hidden parameters? Interpretative questions of this kind are a constant concern of a physicist-theorist.

Every field of science puts forward its own problems. For instance, a historian (and to some extent every humanist) is also interested, as far as I can think mainly in two questions: the problem of the interpretation of historical sources and the problem of the interpretation of historical facts (I have mentioned earlier that usually a psychological interpretation of historical sources is distinguished as well as the sociological one). Another important problem, and the one that is specific to normative sciences such as law or ethics, is the problem of the interpretation of norms; a specific problem of axiological sciences is the problem of the interpretation of value evaluations, value judgements, and the like.

It seems that particularly in historical sciences the problem of interpretation is always unusually topical and highly present. Historians are often ready to consider the diametrically different solutions as a result of simply various interpretations of the same sources and facts. When I was in the last grade of grammar school I studied a history textbook which stated that Suvorov brought carnage to the citizens of the Warsaw district of Praga after he had seized Warsaw. One year later, when I was a university student, I used a history textbook which reported that after capturing Warsaw Suvorov treated its population in a very humane way. (the exact Russian sentence was "zanyav Varshavu Suvorov otchen gumanno otnyossa k yego zhytyelyam," i.e. "having captured Warsaw, Suvorov treated its population very humanely"). This case is somewhat shocking but what was even more curious was the explanation that I received from my professor. When asked which version was true, he answered that it was a matter of interpretation. Some methodologists of history (e.g. Moszczeńska 1960), while observing a notorious abuse of the term "interpretation," in particular when its application for historical sources is concerned, seem to avoid the term wherever possible, and in any case recommend using it with utmost caution.

Apart from the two basic kinds of interpretation which I distinguished, discussing only some types which belong to them, I think that there are many interpretative procedures that are often applied freely and commonly without defined rules or principles which cannot fit easily into any of the basic kinds. These are, for example, interpretations in which the initial material is of the same level of generality, abstraction and of empirical character (or non-empirical character), and so forth, as the final result of interpretation; they are interpretations in the course of which e.g. one sentence goes into another, one text into another, one explanation into another and the like. It seems that this is also the case when the law which is initially considered to be a dynamic law receives a statistical interpretation (e.g. this was also the case of the second law of thermodynamics, and the same applies to the many statements of social sciences), and also when the law is interpreted as a principle restricting the possible states of affairs (e.g. laws of thermodynamics were interpreted by Planck as principles of the impossibility of the existence of a *perpetuum mobile* of the first and second kind); the case when rules of the system are expressed as statements of a metasystem or when we give two different interpretations of the same saying, e.g. Newton's famous words "Hypotheses non fingo!," of which one emphasizes the word "non fingo" and the other the first word of the saying.

My claim that I was not going to make an attempt at an exhaustive classification of interpretative procedures occurring in science allows me to disregard the question of classifying those as well as many other interpretative procedures which I have not even mentioned.

THE PROBLEM OF THE RATIONALITY OF SCIENCE*

I. The Rationality of Means

Just as any other spheres of human activity, the problem of the rationality of science may be approached in different ways. It can be studied either in narrowly or broadly. In a narrow interpretation, rationality is a concept applied during the evaluation of means leading to given goals but not in the evaluation of goals themselves.

In this interpretation, rationality of science depends exclusively on the efficiency of scientific activities in achieving the goals of science. This kind of rationality is sometimes termed instrumental rationality of science, whereby the rationality of science is interpreted as the function of the efficiency of research procedures and methodological principles of science in achieving its goals; having the goals of science given, science is rational (just instrumentally) in the sense that its research methods ensure achieving them.

In the discussions that have been conducted so far on the problem of the rationality of science, the above narrow instrumental interpretation clearly achieved the dominant position. The rationality of science is conceived of in such a way by the majority of philosophers of science who deal with this question, among them Hempel, Kuhn, Laudan, Salmon and others. For instance, Hempel states that ways of proceeding or rules which refer to this proceeding may be rational or irrational only in reference to goals, which it is assumed they are to achieve (Hempel 1979, p. 51). On the other hand, Hempel does not mention the rationality of goals themselves.

Thus, it is clearly seen that we have to do with an instrumental interpretation, in particular treating science as rational, if its procedures and methodological directives are instrumentally efficient in achieving its goals.

The above narrow (instrumental) interpretation of the rationality of science is in line with decision theory, and may even be supported just by that theory. It is just this theory (and to a certain extent also praxiology, understood as the theory of efficient activity, as a peculiar kind of "the

* Translation of "Problem racjonalności nauki," *Człowiek i Społeczeństwo* **2**, 1986, 13-18.

methodology of practice") which interprets the question of rationality exclusively in categories of goals and means, while viewing only fitting the (efficient) means to goals that have been given beforehand.

II. The Rationality of Goals

A wider interpretation of rationality of a given field of activity will be obtained when we ask a question about the rationality of goals, and not only of the means that lead to them. It is right to say that setting some or other goals is not an activity that is justified in itself; defining goals that we will endeavour to achieve is not tantamount to establishing that they are proper, valuable or rational. On the other hand, one should remember that the greatest disputes between people concern goals, and not means; more specifically, they concern ultimate, highest goals. We should also admit that the conception of rationality which would preclude the possibility of rational choice of goals or, put differently, a rational solving of disputes over goals, would be severely curtailed.

It seems that, in spite of suggestions from the decision theory, the concept of rationality may referred equally to goals and to the means that lead to them. If so, then rationality is a broader concept than the one which is used by the decision theory.

Therefore, goals of scientific cognition must also be rational and selected in a rational manner if science as a whole is to be rational. And if the rationality of science depends on the rationality of its goals in equal measure as on the efficiency of the means that lead to them, then this rationality cannot be an exclusively instrumental question. If we act efficiently with reference to certain irrational goals, then our activity, although instrumentally efficient, is irrational. Thus, in the process of defining the rationality of actions we should also take into consideration the rationality of goals. Otherwise, we will come to the wrong conclusion that if scientific procedures and methodological guidelines are instrumentally efficient with reference to the goals of science which are not rational themselves, then in spite of that scientific activity is a fully rational activity.

Thus, scientific rationality, as any rationality of human actions, extends beyond considerations connected with instrumental efficiency.

Taking into account the two aspects of the problem of rationality of science mentioned above, i.e. (1) the rationality of means and (2) the rationality of goals, let us ask the following: what does the rationality of science consist in?

III. What Does the Rationality of Science Consist in?

What does it mean to say that science is or is not rational? Science is a rational undertaking if the goals of scientific research are rational and the means leading to their realization are efficient. If we admit that the methodological awareness of researchers consists of methodological norms that determine goals of scientific research realized at a given stage (cognitive values) and of the methodological directives that determine the means of carrying out of these goals (manners of their realization) then we can resolve the rationality of science into rationality of norms and guidelines, i.e. into the methodological rationality of scientists.

It is well known that there exist many competing views as to the goals of scientific cognition. And various goals lead to different views on the rationality of scientific procedures and methodological perspectives, as efficient ones in achieving the set goals of scientific cognition.

In this essay, Einstein's view of the goals and ideals of scientific cognition as an adequate expression of the research tasks of science is adopted. According to Einstein, the historic task of science is an ever more adequate reflection of the world, which complies with the norm of objectivism, i.e. with an ideal of the fully objective cognition of the world. Thus, we can say that truth (true knowledge) is a regulative idea which guides scientific research. However, this is not a banal truth (knowledge), which is easily accessible but of little interest from the theoretical point of view and often useless from the point of view of practice. According to Einstein, scientific knowledge, as a non-banal knowledge, should be characterized by certain values, among which he foregrounded the so-called inner perfection or, in other words, logical simplicity.

Scientific theory is logically the simpler the smaller the number of premises (postulates, axioms, principles) on which it is based on the one hand, and the higher their informational content on the other. The concept of logical simplicity may be considered to be an explication of the criterion of "inner perfection." According to Einstein, this criterion, next to the criterion of "external compatibility" (agreement with experience), plays a fundamental role in the process of selecting a theory. Since the informational content of knowledge (of a law, of a theory) depends on the degree of its exactness and generality, then according to Einstein, the goals of scientific cognition are as follows: logical simplicity, high informational content, a high degree of generality and exactness, and naturally also the maximum reliability of knowledge.

However, it appears that the above-mentioned goals of science are its "internal" goals that are purely cognitive in nature. Their rationality is not obvious by itself but should stem from the correlation with the functions which science performs (or should perform) in the society. It can be agreed that these functions be termed the "external" goals of science.

It is quite common to admit that these external goals are both the theoretical and practical goal. The theoretical goal is to explain phenomena, and the practical goal is to predict, as without predicting there is no efficient action.

In this interpretation, internal goals of science (purely cognitive goals) are not its final goals but constitute the means to achieving the external goals (social functions) of science. In this manner, the former appear to be functionally subordinated to the latter, which means that "the structure of means and goals" determining the rationality of science is a three-element structure: it consists of methodological guidelines which determine the means (procedures and methods) of achieving the internal (cognitive) goals of science (i.e. attaining knowledge of maximal logical simplicity that is maximally general, strict and certain), and these constitute (or should constitute) the means of realizing functions that are fulfilled by science in the society (the external goals of science).

Thus, ultimately, social practice and its requirements as well as other social needs, including the need to understand the world ever more thoroughly, that are assigned to science define the conditions of the rationality of science in the sense that science, as a cognitive undertaking of social nature, is the more rational the greater the degree to which it satisfies the requirements and expectations that the society assigns it.

However, when speaking of the rationality of science we cannot disregard certain principles which are known as postulates of rationality. What are they and what is their role in the above mentioned three-element structure determining the rationality of science?

IV. Postulates of Scientific Rationality

Among methodological principles controlling scientific research, the principle of intersubjectivity and the principle of (rational) recognition of convictions can generally be considered to be postulates of rationality.

The principle of intersubjectivity, which is sometimes referred to as a weak principle of rationality, states that human knowledge should be intersubjectively communicable and intersubjectively testable. This means that it should be accessible to every normal, adequately prepared human being (the recognizing subject) and accessible to public control. This principle concerns all human knowledge (possibly apart from the so-called irrational knowledge), which is why it is called a weak (or weaker) principle of rationality: all knowledge should be rational, i.e. discursive and available to public control.

In turn, the principle of (rational) recognition of convictions, also called a strong (stronger) principle of rationality, states that the degree of conviction which we proclaim for a given view should correspond to the degree of its justification.[1] The above principle is very difficult to fulfil even within science, and outside of science it is notoriously violated, especially in the areas of such kinds of knowledge as religious knowledge, knowledge in the field of political ideology or occult knowledge.

It is obvious that postulates of rationality that which constitute methodological principles guiding scientific studies, are directly referred to the element of rationality of science which we have called the instrumental rationality (rationality of means). What is meant here are two important methodological principles (directives) that make appropriate demands on the ways of recognizing reality that is carried out within science, which is to yield (intersubjective and properly justified) knowledge realizing internal goals and then also external goals of scientific cognition.

Thus, the three-element structure outlined here is, I believe, a full structure determining the rationality of science. The structure should be taken into consideration in the process of providing answers both to question (1) about what the rationality of science consists in as well as to question (2) on whether real science (both historical and contemporary) is actually rational, and if so, then to what degree.

---

[1] K. Ajdukiewicz usually formulated this principle in a somewhat different form: the degree of conviction should not exceed the degree of the justification of knowledge. The above Ajdukiewicz's formulation is directed against dogmatism in science. However, I think that the formulation supplied in this text is better since it is directed both against dogmatism (the degree of conviction should not be greater than the degree of justification) and against excessive scepticism (the degree of conviction should not be smaller than the degree of justification).

PRINCIPLES AND KINDS
OF SCIENTIFIC RATIONALITY*

I. Introduction

Science is considered to be the most rational kind of human knowledge. This conviction arises from the fact that, apart from those principles of rationality that are generally fulfilled by various kinds of knowledge, science also recognizes as obligatory certain other principles of rationality that are not fulfilled by non-scientific knowledge.

For instance, the principle of intersubjectivity, sometimes termed the weak (or weaker) principle of rationality is holding in all kinds of knowledge, considered to be rational, i.e. obligatory in science, common knowledge, artistic and literary knowledge and speculative knowledge, and it is not observed only in the so-called irrational knowledge. On the other hand, the principle of rational recognition, also termed the strong (stronger) principle of rationality, is actually fulfilled only in science. Let us look more closely at these principles and the kinds of rationality that are valid in science, especially in contemporary science.

II. The Principle of Intersubjectivity – the Weak Principle of Rationality

The first of such principles is the principle of intersubjectivity, initially formulated within psychology, in connection with introspective methods. As is well known, the principle has two components: it is a principle of intersubjective communicability and intersubjective testability. In its first component, it requires that knowledge be accessible to each normal, appropriately prepared (e.g. adequately educated) person, and – in its second component – that it be submitted to public control.

As has already been mentioned, this principle, fulfilled by all kinds of knowledge with the exception of irrational knowledge, is observed to a different degree in various disciplines of science, whereby the degree is higher in all disciplines of science than in the other kinds of knowledge.

---

* Translation of "Zasady i rodzaje racjonalności naukowej," published in J. Bańka (ed.) (1999), *Filozofia jako sposób odczuwania i konceptualizowania ludzkiego losu* [Philosophy as a Way of Feeling and Conceptualizing of Human Fate]. Katowice: Wydawnictwo Uniwersytetu Śląskiego, pp. 19-26.

As K. Ajdukiewicz rightly pointed out, it achieves the highest degree of fulfilment in logic and mathematics where the rules of "good work" are established so precisely that there are generally no disputes among specialists on the subject on whether a given set of proofs is a correct proof, or whether it is only an apparent proof. Controversies that are encountered in that field concern either extremely complicated proofs or are disputes between representatives of various schools in the field of philosophy and the foundations of mathematics, e.g. between adherents of classical mathematics and intuitionists.

According to Ajdukiewicz, a lower position in that respect is occupied by the other strict sciences, i.e. exact natural sciences and exact technical sciences, then the other natural sciences, various applied sciences, and finally social sciences and humanities. Especially in the latter case the rules of research procedures are often not lucid enough or so much relativized to philosophical standpoints and particular research schools that disputes concerning competence are frequent, and the results obtained by particular groups of researchers are by no means commonly accepted by specialists. Another problem is that of observing this principle in psychology due to the fact that the only method that allows "reaching" the mental phenomena directly is the method of introspection which, even when "treated most kindly," cannot be recognized as a fully intersubjective one, especially in the light of requirements concerning intersubjective testability.

III. The Principle of the Rational Recognition of Beliefs

In turn, the principle of rational recognition of beliefs demands so much on the part the researchers that in practice it is observed exclusively in science, and even there not without serious problems.

In one of the formulations that are usually supplied by Ajdukiewicz, this principle states that the degree of conviction with which we proclaim a given view ought not to be higher than the degree of its justification. The above formulation reveals the antidogmatic thrust of the principle. However, it can be expressed in a way that will show that thrust to be directed against excessive scepticism as well. This is the second formulation of the principle which is also sometimes given by Ajdukiewicz, and it reads as follows: the degree of conviction with which we proclaim a given viewpoint should correspond to the degree of its justification (i.e. it should neither be higher nor lower, since in the first case we become dogmatic, and in the second excessively sceptic).

In the light of the above principle, the scientist has the right to formulate even the most improbable and "crazy" ideas but s/he must not present ideas of this kind as mature theories (s/he must not present well-confirmed theories as initial working hypotheses either). If the scientist did not enjoy that right, science would not be able to develop since new scientific ideas are initially not very reliable and there is almost no exception to that. It can be seen also that the treasury of scientific knowledge contains theories of various degree of reliability and that, from the point of view of certainty, scientific knowledge is greatly inhomogeneous.

The principle of the rational recognition of convictions can be applied in science (in any case in empirical sciences) only in "rough approximation." In spite of great efforts made by various researchers, especially by the advocates of the so-called logic of induction, with Carnap and Hintikka at the forefront of this group, no one has managed to develop a theory of confirmation that would ascribe a particular scientific theory a numerically defined degree of confirmation or a definite logical probability.

In social and humanist sciences in which results of research are often strongly relativized not only to the research methods used but also to scientific schools, adhering to this principle encounters further serious problems.

Within the kinds of knowledge that are different from science this principle is obviously violated. This also quite evidently concerns philosophy, in which every orientation and almost every school has, let us say, its own "conviction priorities" too, which incline one to recognize as reliable only those theses which are in agreement with its own doctrine.

IV. The Principle of Accessibility

I shall take the liberty to use the above term to describe the principle according to which science addresses and treats seriously only those problems for the solution of which and for the scientific refinement of which the conditions have already matured or are at the stage of realization. For instance, as is well known, the issue of the atom structure of matter appeared two and a half thousand years ago when the so-called speculative atomistics was founded thanks to Leukippos and Democritus. However, science undertook that problem as late as in the 19th century when a possibility appeared – first in chemistry (thanks to Dalton and his law of simple and multiple relations) and then in physics (in connection with Boltzmann's works and the kinetic theory of matter) – of establishing the magnitude of atoms and developing, first theoretically and then also experimentally, a relevant method based on the atomistic hypothesis, and thus of transforming speculative atomistics into scientific atomistics.

Similarly, for thousands of years the kinds of knowledge such as mythology, religion, speculative cosmology and philosophy have confronted the problem of the origin of the Universe, whereas this problem was addressed in science only 30 years ago when the quantum-relativist cosmology "felt" the possibility of its scientific expression and development within the Big Bang theory and the Standard Model of Cosmological Evolution.

Another reason why the principle is the proof of the rationality of science is because it allows scientists to concentrate on the problems the solving of which is "within reach" rather than waste time tackling questions which cannot yet be scientifically resolved with the aid of existing means and methods.

The principle of accessibility allows to explain a number of facts in the field of the history of science, and also some amazing peculiarities of contemporary science. One of such peculiarities is a great and still growing "hiatus" between theoretical physics and experimental physics. While experimental physics "reaches" the level of the structure of matter measured by means of such magnitudes as 10^{-17}cm for space, 10^{-26}sec for time, and 10^3GeV (100 billion electronovolts) for energy, the theoretical physics (and cosmology too) already "reached" Planck magnitudes: Planck length which is 10^{-33}cm, Planck time which is 10^{-43}sec, and Planck energy which is 10^{19}GeV (10^{19} billion electronovolts).

At first sight, it might seem that such a great hiatus (indicated by an exponent of about 16-17) is an evidence that experimental physics lags behind, and thus of a great mismatch between both fields, and thus constitutes some "irrational" trait in the development of physics.

Clearly, this fact may be interpreted in terms of experimental physics "lagging behind" the theoretical physics and it can be said that further development of physics would be much more successful if the former developed experimental ways and methods of penetrating the world on the level of Planck magnitudes.

However, if this situation were viewed through the prism of the principle of accessibility, it would appear that the mutual situation of the two fields of physics in relation to each other is, if not fully, then in any case rather rational. The thing is that the possibilities of penetrating the ever deeper levels of matter by experimental physics on the one hand, and by theoretical physics on the other, are determined by principally different factors. For experimental physics, the level of development of experimental and measuring techniques of fundamental importance in this respect, while for theoretical physics it is the development of mathematics.

In experimental physics, reaching ever smaller objects and areas is connected with the necessity of giving elementary particles ever greater energies. Therefore, it is connected with the necessity of building ever more powerful accelerators. These are enormous, extremely complex and very expensive devices, which is why obstacles of two kinds exist here, namely scientific-technological and financial obstacles. Even if sciences and technology developed so that they were capable of constructing a given equipment in the field of high energy physics it may then appear that there are no funds for such construction.

One of the more recent examples may be starting the construction of a Superconducting Super Collider in the USA in 1990. The device was to collide protons with one another with an energy 20 times greater than the greatest energy that had so far been obtained in the existing accelerators (i.e. with the energy 2×10^4GeV). However, this construction was discontinued in 1993 because the US Congress suspended subsidies (SSC was to cost 10 billion dollars to build, ca. 2 billion had been spent until work was stopped, i.e. 20 per cent of the cost).

As can be seen from this example, interference of a non-scientific factor disturbed the process of approximation between experimental physics and theoretical physics. The first one "stopped" at the level of leptons, quarks and gluons, while the other "reached" the strings. The theory which is at present considered to be the best candidate for "the Theory of Everything" is the theory of superstrings. When it appeared, all the fundamental theories of contemporary physics could be derived from this theory: quantum mechanics, general relativity theory, quantum electrodynamics and the standard model of elementary particles.

According to the newer version of the theory of strings, the elementary objects of physics of the dimension of Planck length (10^{-33}cm) are one-dimensional strings which vibrate simultaneously as if in two space-times of which one has 26 dimensions (clockwise vibration) and the other has 10 dimensions (counterclockwise vibration). This latter space-time arose from the first one through compactification, i.e. coiling of 16 dimensions of the first one to small sizes. After all, also the 10-dimension Universe appeared to be unstable (it was in the state of the so-called false vacuum) and "tunnelled" into the four- and six-dimensional world.

According to the scenario of the theory of superstrings, our four-dimensional Universe, starting from Planck era (10^{-43}sec), expands rapidly while the six-dimensional one decreased into the 10^{-32}cm size.

The superstring theory is the first one that unequivocally determines the number of dimensions of space-time: only strings of the 26 and 10 dimension vibrate in a quantum coherent way, i.e. in a manner that is in agreement with quantum mechanics. The theory is said to be a 21st-century theory which only incidentally emerged in the 20th century. The main restraint of its further development is considered to be not so much the impossibility of verifying it experimentally given the present state of technology but the fact that mathematics, which has not achieved the level of the 21st-century mathematics, is "lagging behind." This proves a thesis that it is not technology but mathematics that is the main determinant of research potential of theoretical physics.

The occurrence in science of two principles of rationality (the principle of the rational recognition of convictions and that of accessibility) which do not function in the other kinds of knowledge, and the circumstance that the principle of intersubjectivity in science is fulfilled to a higher degree than in the other (rational) kinds of knowledge clearly indicate that scientific knowledge is considerably distinct with respect to the degree of rationality.

To conclude, let us look at two kinds of rationality of human actions and their place in scientific research.

V. Kinds of the Rationality of Human Actions

Depending on whether human activity is directed towards setting proper goals, on the one hand, or the means which serve the accomplishment of goals on the other, two kinds of rationality are distinguished: axiological rationality and instrumental rationality. In contradistinction to such fields of symbolic culture as religion, ethics, philosophy or ideology, science has little to say on the question of adopting goals which, for some reason or other, deserve being pursued and which are worth accomplishing. In this respect, the possibilities of science will be restricted to such rather ancillary and formal activities as demonstrating the non-contradictoriness (or co-realizability) of goals or the non-utopian nature of aims (their realizability under given conditions).

On the other hand, science, like technology, has much to say about the efficient means and ways of attaining goals. Various kinds of engineering serve those which have their theoretical foundation in science. That is why the type of rationality that is typical of it is the instrumental rationality. But are social sciences and the humanities a certain exception, and are they engaged also – and to an ever greater extent – in determining goals of human actions in various fields penetrated by science?

There is no doubt that since the degree of rationality of the procedures which set goals is considerably lower than the degree of rationality of the procedures which determine the means of attaining goals; therefore, the fact of science's involvement in the instrumental rather than the axiological rationality is an evidence on its part that the rationality of science exceeds the rationality of all the other kinds of knowledge.

THE RATIONALITY OF SCIENCE AND LIMITATIONS OF SCIENTIFIC METHODS*

I. Paradoxes of the Epistemic Possibilities of Science

One of the paradoxes of scientific knowledge is the circumstance that as science achieves more and more success in research, doubts are also growing as to the real cognitive possibilities of people in general, and of science in particular.

As M. Heller writes that, on the one hand, "the 20th century gave mankind a great new experience: an experience of limitations which are inherent in the scientific method" (Heller 1994, p. 156). What Heller means is the so-called limitation statements in logic and mathematics, which were started by Gödel's famous "three": statements on (1) incompleteness, (2) unsolvability of richer formal systems (that contain arithmetics) and on (3) non-existence of absolute ("internal") proofs of non-contradictoriness.

Limitation statements (that now already form quite a long list of statements connected with the names of Tarski, Church, Turing, Churchland, Lorenzen and others) indicate an incomplete realizability of Hilbert's programme of the formalization of mathematics and logic, resulting from the existence of certain relative (relative since they are moveable), thresholds of methods applied in formal sciences.

On the other hand, we can think that this does not prevent contemporary science from penetrating ever newer areas that were so far virgin territories, and this applies not only to the areas of the macroworld but also to areas that encompass phenomena whose scale deviates more and more "in depth" and "in width" from the scale of phenomena of the macroworld.

What is more, science does not fail to extend its cognitive horizons both at the cost of other fields of knowledge, e.g. taking up problems that before were reserved to philosophy or religion. One of such problems is the origin of the Universe.

---

* Translation of "Racjonalność nauki a granice metody naukowej" by Jan Such and Małgorzata Szcześniak, published in B. Kotowa, J. Wiśniewski (eds.) (1988), *Racjonalność a nauka* [Rationality and Science]. Poznań: Wydawnictwo Naukowe IF UAM, pp. 13-26.

Science, while sensitising other fields of knowledge and human to the question of non-utopian character of goals that one should try to accomplish, itself observes proper procedure. One of the indicators of its serious and responsible approach to knowledge is the circumstance that science generally undertakes only such problems which it is in a position to solve give at the level it has achieved. Among other things, it is characteristic of scientific rationality that it does not take up questions that have not yet "matured" to be scientifically studied. (obviously, this does not mean that these questions are irrational by their nature. This is indicated to by the very possibility of their future scientific consideration.)

In this context, we may adduce Marx's well-known saying that mankind takes up only such tasks that it is developed enough to solve, or which are in the making. The validity of this statement, which is sometimes accused of being a tautology, would mean that the rationality we are speaking about presently is extended to all the domains of human activity, including cognitive activity. In our opinion, that it is not a tautology is shown by the fact that some kinds of cognition other than science do not respect this statement. For example, the question of the atomistic structure of matter or the problem of the origin of the world had for centuries or even millennia been the subject of philosophy or religion, while science took them up only then when the theoretical (e.g. mathematical) or empirical (e.g. technical) means of their scientific analysis appeared. In this way, scientific atomistics appeared only in the 19th century (thus two and a half thousand years later than speculative atomistics), first in chemistry (thanks to Dalton in connection with the law of simple and multiple relations), and then in physics (mainly due to Boltzman and the kinetic theory of matter). It appeared then when it was possible to determine (theoretically) the magnitude of atoms and make calculations providing a provisional theoretical justification of their existence. Empirical, observational grounding of the atomistic hypothesis came half a century later thanks to theoretical works of Einstein, Smoluchowski, Perrin and others, concerning, among other things, Brownian movements and at last due to the construction, in the 1960s, of a microscope which allows the observation of particular atoms distributed on an iridium blade in the form of "dark spots." A similar situation is represented by the history of the origin (eternity) of the Universe. For thousands of years, this problem has put to work the mythological, religious and philosophical thought but was taken up by science only in the recent decades of this century since only now the relativist cosmology, better equipped with quantum ideas, begins to see the possibility of a proper interpretation and solution of the question of the eternity of the world. It seems that two fundamental theories of modern physics came closer to its solution: the general theory of relativity (GTR) and quantum mechanics (QM). In this context, a great difficulty lies in the fact that these theories do not agree with each other, and in this connection a new great synthesis is necessary which will surely be "a corresponding generalization" (to use N. Bohr's expression) of both theories, which are called (allowing for the further development) "the quantum theory of gravitation." Without such a theory, scientific considerations of what was happening before the so-called Planck threshold (which is distant from the Big Bang by $t = 10^{-43}$sec) become very provisional and uncertain. This is because both relativist physics and quantum physics (and together with them surely all the laws of contemporary physics), cannot be applied for densities larger than $q = 10^{94}$g/cm^3, and this is just the density that characterizes Planck threshold. An assumption that the future "quantum theory of gravitation" (sometimes also called the "theory of quantum gravitation") will manage to explain what happened since the Big Bang ($t = 0$) till Planck threshold ($t = 10^{-43}$sec), results in this extremely short period of existence of the Universe being called "an era of quantum gravitation" (or "era of quantum cosmology"). It is assumed that in this era an important role was played by quantum effects of gravitation, which surely "come into prominence" whenever the intensity of the gravitation field assumes enormous values (e.g. in the proximity of black holes).

The necessity of using both the above mentioned fundamental theories (GRT and QM) for the early stages of the evolution of the Universe results from the fact that in its initial stages (before the period of inflation) the Universe was at the same time a cosmic object (due to the magnitude of mass and energy) and a microobject (with respect to spatial and, perhaps, temporal dimensions). It can be seen from calculations that at the time of inflation, i.e. in the period of the accelerated expansion of the Universe, lasting from approx. $t = 10^{-35}$sec to approx. $t = 10^{-32}$sec, the size of the Universe increased from 10^{30} to 10^{50} which caused the Universe to "grow" from an object smaller than a proton to one of a diameter of 10cm, i.e. from a "microobject" it became a "macroobject" (the size of a grapefruit).

It is not clear yet how the restrictions resulting from limitation statements in logic and mathematics may influence cognitive possibilities of empirical sciences. However, these sciences themselves encounter their own cognitive limitations.

The limitations can perhaps be perceived most clearly in cosmology. It was not incidental that Michał Heller, who has already been quoted in this contribution, entitled of the chapters of his book *Nauka i wyobraźnia* [Science and Imagination] "Graniczny charakter kosmologii" [The Boundary Character of Cosmology] (Heller 1995). He perceives at least three restrictions of the epistemic possibilities of cosmology. First, the field of applicability of physical theories used in contemporary cosmology breaks, especially the GRT. Consequently, such questions become boundary ones as the problem of "initial or final singularity, borders of space-time, final stages of gravitational collapse" (Heller 1995, p. 106). Second, boundary problems are

> those problems which the empirical method cannot manage fully but which in some way arise within empirical sciences. In cosmology, such problems are, among other things, the origin of laws of nature, roots of time and of its irreversibility, whether the Universe itself determines its boundary conditions, or what is the ultimate stuff of the Universe (Heller 1995, p. 106).

> Third, "the threshold problems" may be understood in the sense given to this expression by existential philosophy. Boundary problems would be those which elucidate the final fate of people. Sources of cosmology themselves may be perceived in this type of motives of investigation. Such problems as the search for human being's place in the Universe, stronger versions of the anthropic principle (discerning in human beings and their cognition the raison d'être of the Universe), are often inspired by cosmology, and considered in the philosophical context, they have distinctly an undertone of boundary problems (Heller 1995, p. 107).

Further on, Heller deals with the epistemic restriction, connected with the empirical method of searching, and concerning the very concept of "the Universe." Heller writes that theoretical opinions in which the term "Universe" appears

> are always very distant from sentences giving results of some observations. Sometimes they are so "distant" that we begin to fear that the method, characteristic of empirical sciences, is already collapsing. In such cases one should say that "the Universe" becomes a boundary concept (term). . . . Such situations cannot be avoided by means of a simple prohibition in the style of positivist methodology. It can be proven by common scientific practice. What is more, just in this kind of boundary situations at present take place the most current studies of fundamental physics, such as, for instance, studies concerning fundamental interactions and unification of physics (Heller 1995, pp. 110-111).

When emphasizing the boundary character of the concept of Universe, Heller adduces the following opinion of E. Hubble, contained in the final part of his book *The Realm of the Nebulae*. In this way, studies of space end with an accent of uncertainty. And this is unavoidable. By definition, we are in the very centre of the observable area. We know our direct neighbourhood rather well. With increasing distance, our knowledge rapidly gets poorer and poorer, until we reach the nebulous borders, i.e. till the boundary possibilities of our telescopes. There, we only measure shadows, among the hazy measurement errors we seek the more real points of reference (Hubble 1958, pp. 201-202). This opinion commented by Heller as follows:

> one could risk a statement that the Universe of every epoch is determined by such areas in which "we only measure shadows." But we measure not only by means of our telescopes, we can also measure by means of our theory. A property of the Universe is that within its boundaries it is – as far as both measurement and theory are concerned – always blurred, emerging from uncertainty (Heller 1995, p. 114).

The emphasis of the above commentary on the fact that what is meant is not only a "measuremental" but also "theoretical" blurredness is not strange if we consider that the author (in contradistinction to the author of the opinion quoted here) is a cosmologist-theoretician and not an astronomer-observer.

Another paradox concerning cognitive possibilities of science, which is after all closely connected with the paradox discussed at the beginning is that the power of scientific cognition in empirical sciences consists just in the mathematical method of study, i.e. the method whose limitations have been clearly outlined by limitation statements.

II. The Mathematical Nature of the Rationality of the World and of the Rationality of Scientific Knowledge

The belief in the "mathematical" nature of reality, which was cherished in the antiquity by the Pitagorean, Platonian and the neo-Platonian school, has been present in our culture to this day. What is more, mathematics and mathematical (i.e. strict) natural sciences thoroughly reinforced this conviction among modern scholars.

Accordingly, attempts were made as early as in the antiquity at giving a mathematical form to the laws of nature. Greatest success in this field was achieved by Archimedes, an most outstanding mathematician and physicist of ancient Greece. Thanks to the use of Euclidean geometry and arithmetic for the description of the laws of the lever, his statics became the first discipline of exact natural sciences and thus the beginning of modern physics. Significant achievements in this respect were also made by Eratosthenes to whose studies and calculations we owe the first precise quantitative data concerning the size of the Earth.

A certain important exception was the school of Aristotle which followed its founder's considerations on the "qualitative" nature of reality and was deeply convinced that the phenomena of nature cannot be submitted to description by means of mathematics, which determined the development of science in the Middle Ages.

However, Copernicus, Kepler, Galileo and Descartes, following in the footsteps of Archimedes, Eratosthenes, Apolonius and Ptolemy, reestablished mathematics, especially geometry, as the main discipline providing natural sciences with the means of an exact description of the studied phenomena and the laws which rule them. In this situation, Galileo's conviction that "the book of nature is written in the language of mathematics" was treated earnestly by Newton and other founders of the scientific revolution of the 17th century.

The differential and integral calculus, created by Newton and Leibniz, became the second powerful means of cognitive penetration of physical phenomena, next to Euclidean geometry. The success of mathematical natural sciences encouraged Leibniz so much that he considered mathematical methods of studies as the only ones which are worthy of fully scientific methods. This was manifested in his programme of *mathesis universalis*. The 18th and 19th centuries brought an extension of mathematical research methods onto all the disciplines of physics and chemistry, or also onto "the Baconian sciences" (Kuhn's expression), dealing with heat, electricity, magnetism and chemical processes.[1] Still greater power was revealed by mathematical research methods in the 19th century, when Maxwell – thanks to the application of the so-called method of mathematical hypothesis – managed to associate the "displacement current" which was predicted theoretically with the "conduction current," discovered experimentally by Faraday. As a result, a full form of equations of classical electrodynamics could be provided. For the first time, the fact that "equations are wiser than their founders" became noticeable; later, the fact has often been demonstrated in the field of relativist and quantum physics.

It is a well-known fact that Einstein did not appreciate the role of mathematics in physics in his youth. Referring to an elegant mathematical "four-dimensional" interpretation of the special relativity theory (SRT) provided by Minkowski, he said that when his theory was in the hands of mathematicians, he ceased to understand it. The huge edifice of the general relativity theory (GRT), which only amazes by its elegance, based on a powerful apparatus of differential geometry and tensor calculus, made Einstein abandon his critical attitude towards mathematics. Einstein began to understand that without a "geometrical" interpretation of the SRT supplied by his former teacher he would not have been able to create the GRT. Since then, he was always astonished at the effectiveness of mathematics in the modern and contemporary physics (cf. Heller 1995, p. 171).

There is a well-known Einstein's saying that "the most incomprehensible thing in the world is that the world is comprehensible." Einstein interpreted "the comprehensibility" of the world, i.e. the fact that it submits to human understanding, as its mathematicity or in any case mathematizability (i.e. a certain "feature thanks to which we can project our own mathematicity onto it") (see Heller 1995, pp. 168-185). In his work *The Fundamentals of Theoretical Physics*, Einstein writes that physics encompasses the group of natural sciences which define its concepts by means of measurements and whose concepts as well as statements may be submitted to mathematical interpretations. Therefore, the field of their interest may be defined as this part of our whole knowledge which can be expressed by means of mathematics. Together with the progress of science, the area of physics expanded to the degree

---

[1] Kuhn distinguished in older physical sciences two "clusters" of disciplines: (1) "classical physical sciences" (astronomy, geometrical optics, statics, mathematics and harmony, or the theory of music) which had not required a laboratory experiment for their development but only observation and mathematical (mainly geometric) methods of processing of its results, and (2) "Baconian sciences," which were based on experiment and had not applied mathematical methods (see Kuhn 1985, pp. 67-112).

that it seems to be limited only by the restrictions of the method itself (Einstein 1978b, p. 316).

As is well known, Einstein, like Planck, tried persistently to provide a unified interpretation of the whole physics. That is why after creating the GRT he channelled all his efforts into constructing a unified field theory which would be capable of overcoming the duality of corpuscular and field matter and, along with it, the duality of matter and space-time. He called an attempt of this kind "a rational unification" and wrote that whoever experienced a strong emotion, connected with achieving progress in this field, feels a deep reverence toward rationality which is revealed in existence (Einstein 1978a, p. 57).

"Science's more and more decisive move beyond the imaginary possibilities" (Heller 1995, p. 178), and in particular the increasingly undemonstrative nature of contemporary relativist and especially quantum physics result in the fact that, as M. Heller writes, "the achievements of modern science are an incessant victory of mathematics over imagination" (p. 169). The same applies even to a greater extent to the most recent hypotheses of contemporary physics and cosmology, concerning quarks and quasars, superstrings, supergravitation, supersymmetry and superunification, containing "in themselves an unimaginable content, which we would have never found without mathematics" (p. 170).

Let us take as an example S. Hawking's conception of the Universe without borders (singularities). The conception describes the Universe by means of a closed time curve, taking into account the existence of the so-called imaginary time, i.e. time measured by means of imaginary numbers whose magnitude is always a negative number: $t = \sqrt{-1} \cdot t$. In this conception the most probable histories of the Universe may be compared with the surface of the Earth in which the distance from the North Pole is represented by imaginary time, and the size of the space is pictured by a radius of a circle that is equidistant from the pole. On the North Pole begins the history of the Universe, the Universe being a single point. With the increase in the imaginary time, the Universe expands, parallels of latitude as they move south become larger, and the Universe achieves its greatest size in the equator and then shrinks until some point, i.e. to the South Pole. What is interesting in this conception is that in spite of the fact that the Universe has a zero radius on the poles, they are still not nothing peculiar, just as nothing peculiar occurs on the earthly poles in which common laws of science hold (see Hawking 1988, pp. 137-138).

Hawking remarks that the history of the Universe in real time would be described differently. A dozen or so billion of years ago the real minimal magnitude of the radius would correspond to the maximum magnitude of the radius in the history of the Universe in the imaginary time. The next stage would be the expansion of the Universe (like in

Linde's chaotic models but without assuming that the Universe has been created in a state that makes inflation possible) to a considerable size, and in a final stage it would shrink to the "singularity" in real time. Just in real time does the Universe begin and end its existence with "singularities," i.e. boundaries of space-time in which the well-known laws of physics break down. On the other hand, in the imaginary time, indeed also the Universe ends but it is devoid of boundaries and singularities (see Hawking 1988, p. 138).

Hawking emphasizes that the term imaginary time sounds as if it were taken out of a science-fiction novel but in reality it is a well-defined mathematical notion. In order to avoid technical problems in the Feynmanian sum after histories, one should use imaginary time. That means that, in the calculation, time should be measured by means of imaginary numbers rather than real ones. It has an interesting impact on space-time: then all differences between time and space disappear. Space-time in which events have imaginary time coordinates is called the Euclidean space-time to honour Euclid, the Greek mathematician who was the founder of the geometry of two-dimensional surfaces. Euclidean space-time has very similar properties, the difference being that it pertains to four rather than two dimensions. In the Euclidean space-time there is no difference between the direction in time and the direction in space. In real space-time in which events have real-time co-ordinates it is easy to show the difference: in every point the time directions are inside the light cone, and the spatial ones on the outside. In each case in the ordinary quantum mechanics one can consider using imaginary time as a mathematical means (or trick) allowing to calculate, what happens in real time-space (Hawking 1988, pp. 134-135).

Further on, Hawking writes that it would be a mistake to assume beforehand that imaginary time, in contradistinction to "real time," is a fictitious time which in reality does not correspond to anything. In fact, certain considerations suggest that the so-called imaginary time is actually real time, and what we consider today to be real time is only the product of our imagination. In real time, the Universe begins and ends by singularities being the boundaries of space-time in which all the laws of physics break down. On the other hand, there are no singularities or boundaries in the imaginary time. Therefore, imaginary time may be more fundamental, and what we call real time is only a conception invented for the description of the Universe (Hawking 1988, p. 139).[2]

---

[2] Then Hawking emphasizes that such an interpretation of reality of time in his conception corresponds to his view of what a scientific theory is: according to an approach described in Chapter I, scientific theory is only a mathematical model serving the description of our observations and existing exclusively in our minds. So there is no point in asking about what is real, "real" or "imaginary" time? The problem is only reduced to the

Although the conception was presented in as visually as possible, it is obvious that even the most daring imagination would not be able to lead to Hawking's conception of "the world without boundaries" without taking advantage of a proper mathematical apparatus.

M. Heller tries to answer the question of what the efficiency of mathematics in contemporary physics is based on when he outlines his conception of "interpretative structuralism." According to this conception, "mathematicized scientific theories ascribe to the world a structure that is similar to the mathematical structures which they use themselves" (Heller 1995, p. 11). It is at least partly justified since

the human being applies mathematics (which he discovers or creates) for the Universe, and the Universe reacts to it by revealing at least part of the information which constitutes its structure. What is more, history shows that the Universe remained dumb to all attempts at dialogue in a language other than that of mathematics. Aristotle's qualitative physics and "architectonics of the Universe," developed in the Middle Ages, ultimately appeared to be impressive creations of human imagination. The same concerns "physics" and "cosmologies" of other civilizations that we know of. But if only in any of these systems some mathematically formulated problem were found, the Universe sooner or later responded by means of results of measurements which constituted a reasonable whole (Heller 1995, p. 181).

What the mathematical structure "fits into," and what the "fitting in" consists in is explained by Heller in the following fragment:

The structure of the world is also quite a hackneyed phrase, originating from popular scientific books. In addition, the phrase is to some extent misleading. As it is, contemporary theoretical physics suggests that the world *does not have* a structure but that the world *is* a structure. The structure contains certain information, coded in itself (or, in other words, is certain information). Science decodes its fragments by fitting mathematical structures in the structure which is the Universe. The procedure of fitting in consists in approximations. So far, it has been possible to correct mathematical structures so that they were better fitted to the world, and nothing indicates that this process will soon end. The decoded fragments of information are called scientific theories or models of the world (Heller 1995, p. 170).

To answer the second question, Heller later adds the following:

The same may be expressed somewhat bit differently. Between mathematical structures of which our theories are built and the structure, which is the world, there is no identity but there occurs a strange resonance. Thanks to that resonance, although they are much simplified structures when compared with the structure of the world, they "harmonize" with the world, recreating some of its structural features (modelling them). The existence of the resonance is something astounding. All the more so, if we consider that mathematical structures which "harmonize" with the world are the work of human minds, and therefore are cast to their mould and *a priori* no reason for which the potential of our minds would be in any proportion to the richness and extent of complication of the structure, which is the Universe. The existence of the resonance between mathematics and the world is sometimes called the question or, if one prefers, the mystery of the mathematicity of the world (Heller 1995, pp. 170-171).

---

following: which of them is more useful for the description of phenomena? (Hawking 1988, p. 139).

III. Conclusion

So far, we have been discussing the rationality of the world, identifying it with its mathematicity, and we also have being speaking about the rationality (mathematicity) of scientific knowledge as influenced by the rationality (mathematicity) of the world.

However, the rationality of science does not obviously end here. According to Ajdukiewicz a strong principle of rationality is valid in science, which is one called the principle of the rational recognition of convictions. It states that the degree of conviction with which we proclaim a given view should correspond to the degree of its justification. The former should not be greater than the latter because a scientist then becomes dogmatic but it should not be lesser either as then there is the danger that s/he would become excessively sceptical. Thus, the principle requires from a scientist that s/he justify all his/her views and proclaim them according to the degree of justification.

Apart from justification, scientific cognition may be ascribed such kinds (or aspects) of cognitive rationality as: methodicity, intersubjectivity,[3] objectivity, sensibleness and wisdom. The latter seems to express one of the most important aspects of rationality, understood as a feature of scientific cognition. However, wisdom may be understood in various ways. In our article, the wisdom of scientific knowledge has been expressed as a mathematical quality of knowledge.

Postmodernism seems to shake most of the given kinds (or aspects) of rationality. To some extent, it tends, as it seems, in the direction in which the rationality of science is developed. For instance, it tends in the direction of theoretical pluralism (which was outlined already in the works of Popper, Lakatos and Feyerabend), and also in the direction of giving up the modern ideal of "a fully objective description" which quantum mechanics gave up earlier.

However, the scientists' conviction, especially that of physicists and mathematicians, that the rationality of the world (and at the same time of scientific cognition) consists in its mathematicity, seems to pass the time test. And this happens in spite of limitations that are imposed on methods and formal systems by limitation statements. All the more so that the

---

[3] It is noteworthy that while the principle of rational recognition of convictions is called the strong principle of rationality, the principle of intersubjectivity is sometimes termed the weak principle of rationality. The former is obligatory exclusively in science since other kinds of knowledge are not in a position to match it while the latter holds in all kinds of knowledge that are included within rational knowledge. This is also fulfilled by common knowledge, artistic/literary knowledge and speculative knowledge, but not fulfilled by the so-called irrational knowledge (i.e knowledge that refers to mysticism or irrational intuition).

consequences of these statements are so far felt merely within mathematical sciences, and not empirical ones.

Is it so that the mathematicity of knowledge and of the world that was so deeply experienced by Albert Einstein in connection with the effectiveness of mathematics in the study of the world would become the only and reliable support of scientific rationality?

WHAT HAS CHANGED IN THE PHILOSOPHICAL VIEW OF SCIENCE?*

In the philosophical view of science in this century, fundamental changes have occurred that are connected with profound revaluation of previous solutions and viewpoints. They concern such fundamental questions as the subject and aims of philosophy of science and various interpretations of science, both in the structural aspect and in the developmental aspect. Below, I shall briefly discuss the most important of those changes.

I. Static and Dynamic Interpretations of Science

The first of the changes is the transition from the static interpretation of science to the dynamic, developmental one. As is well known, this transition was initiated by Popper and continued both by his adherents (Lakatos, Watkins, Worrall, Zahar) and representatives of other trends in the contemporary philosophy of science.

Popper was, for sure, the first philosopher of science who understood that adequate criteria of the scientific nature of knowledge which set a demarcation line between science and metaphysics and other kinds of knowledge should take into consideration, apart from the structural (synchronic) dimension, also the temporal (diachronic, developmental) dimension of science. That is why his the considerations of the concept of the scientific nature of knowledge, jointly with the principle of falsifiability, included the principle of correspondence, whereby Popper admitted that only those fragments of later theoretical knowledge enter the treasury of scientific knowledge that stand in the relation of correspondence to some earlier theories included in science.

II. The Constancy and Changeability of Models of Scientific Rationality

The second important change that made the dynamic (developmental) view of science more profound, consisted in admitting that the methodological

---

* Translation of "Co się zmieniło w filozoficznym spojrzeniu na naukę?," published in H. Korpikiewicz, E. Piotrowska (eds.) (1997), *Alternatywy i przewartościowania we współczesnej filozofii nauk* [Alternatives and Revaluations in the Contemporary Philosophy of Sciences]. Poznań: Wydawnictwo Naukowe IF UAM, pp. 13-22.

principles (the so-called norms and directives) – despite the earlier views – are historically variable, i.e. hold generally only at some definite stages of the development of science.

Both inductionists (neopositivists including) and hypothetists thought that they were suprahistorical and thus that models of rationality that hold in science are also supratemporal. It was only Feyerabend, Kuhn and other research workers, who were sensitive to historical changes in science, noticed that historicity in science has a deeper dimension reaching the very "rules of scientific play," and that scientific revolutions consist not only in the change of fundamental theories but also in the change of methodological principles, aims and values that have so far controlled scientific development.

III. Science as a Process and as a Result

The third important change is the shift of the main centre of methodological studies from the ready-made scientific knowledge to processes of acquiring this knowledge, and in particular the explanatory and justifying procedures.

At the same time, this meant shifting the emphasis of analyses from questions of a logical and linguistic nature to strictly methodological questions, concerning ways of describing, explaining, predicting and justifying scientific knowledge.

IV. The Context of Discovery and the Context of Justification

This is connected with the fourth important change, which is the fact that the rigid separation of "the context of discovery" from "the context of justification," as suggested by Reichenbach and, in a somewhat different sense, by Popper, designed to focus the attention of methodologists exclusively on the context of justification, has been rejected by "the radical trend" and most of the other trends in the philosophy of science. This led to the enlargement of the subject of methodological analyses, i.e. to the joint interpretation of both contexts.

V. Autonomic Models and Heteronomic Models of the Development of Science

The fifth important change is the ever more serious treatment of heteronomic models of the development of science, i.e. models which perceive the main determinants of the development of science (which significantly influence even the content of scientific knowledge) in the external, non-cognitive factors. Here the main role was that of the above-mentioned radical trend in the philosophy of science, distinguished by the names of Kuhn, Feyerabend, Toulmin and Hanson since both neo-positivists as well as falsificationists (Popper and Lakatos) considered as more adequate autonomic models which emphasize the importance of internal (cognitive) determinants, and thus of the "internal logic" in the development of science.

VI. The Personal and the Institutional Dimension of Science

The sixth change, which is closely related with the fifth one, also consists in the significant extension of the subject of studies of the philosophy of science. This extension results from reaching beyond issues concerning the structure and development of scientific knowledge and methods and procedures applied in science and taking into account in the studies of the personal (subjective) factors such as collectives and styles of thinking or scientific schools and institutional factors of science, which means a certain acknowledgement of the sociology of knowledge, the sociology of science and the psychology of scientific creation.

VII. The Method of the Rational Reconstruction or Sociological Methods

The next, seventh change, is connected with the one discussed above and concerning conceptual means and methods used in the process analysing scientific knowledge and its development. The initial rather limited repertoire of logical and linguistic means used in neopositivism in the process of analysis, which were restricted mainly to the logic and language of science and those means that serve the so-called rational reconstruction of science, has been gradually extended to include means allowing historical comparative studies on science and placing science within the social and cultural context. The philosophy of science partly adopted those means from the history of science, the sociology of knowledge, the psychology of scientific creativity and other disciplines belonging to the studies of science, and in part it began developing them by itself to be used for its own needs.

VIII. The Analysis of Ever Greater Structural Units of Knowledge

The eighth change consists in focusing attention in methodological analyses on ever greater structural units of scientific knowledge. If the neopositivist logical and linguistic analyses were initially concentrated mainly on scientific statements and their components, and then on scientific theories, then at present (i.e. in the second half of this century) larger units of knowledge, such as theoretical systems (consisting of many theories which are submitted to a common checking procedure), paradigms, scientific research programmes and crucial situations are analyzed.

IX. The Problem of Theoretical Proliferation

The next change concerns the form of competition between scientific theories as confronted with experience. Neopositivism inherited the view that was cherished for centuries that every theory is confronted with experience separately, in isolation from its competitors. Indeed, F. Bacon's concept of *experimentum crucis*, of experiments which predetermine in favour of one theory and to the detriment of another one, contains an idea of competition between theories. However, the idea did not play any significant role until the 20th century.

It was only the conceptions of development which represented hypothetism (conceptions of Einstein, Popper, Lakatos, Watkins) and conceptions of evolutionary epistemology referring to hypothetism (K. Lorenz's and G. Vollmer's) which began – not without Darwinism's influence – to approach the problem of competition between theories seriously. The issue grew into its mature form in Lakatos's conception of scientific research programmes, which states that the development of science occurs in the competition between opposite research programmes.

The conception of "theoretical pluralism" (also termed theoretical proliferism) is even more strongly (and at the same time, even earlier) defended by Feyerabend, who thinks that the monopoly which science has achieved in contemporary societies as well as the monopoly towards which particular theories tend, and which some of them actually achieve are very harmful (in his opinion, quantum theory and its Copenhagen interpretation provides a relevant example).

The construction of an alternative theory in a given discipline is not only an efficient means of fighting the monopoly but also allows establishing facts which serve as crude tests (in Popper's sense) for the existing theoretical systems. According to Feyerabend, the strategy of theoretical proliferism doubly contributes to overcoming dogmatism in science in this way.

X. Cumulativism and Anticumulativism

The next change concerns the transition from the extremely cumulativist interpretations of the development of science, through various correspondence interpretations, to interpretations that are more and more revolutionary and which sometimes even assume the ontological, linguistic, logical and empirical incommensurability between fundamental theories which follow one another (and are separated by a scientific revolution).

The extremely cumulativist theory of scientific knowledge was recognized through centuries by both rationalists and empiricists. Neopositivists were also its adherents. Early interpretations of the relation of correspondence in the philosophy of science, subsumed under the heading of the implicational (generalizing) correspondence, also represented extreme cumulativism. Only later, more correct conceptions of correspondence which account for the fact that at least in certain and, importantly, most significant and most interesting transitions the so-called explanatory correspondence (sometimes also referred to as the significantly correcting one) comes into play, which is reminiscent of the Hegelian "removal" (*die Aufhebung*) being at the same time a continuation and breakthrough, were situated in the position of moderate cumulativism (also called moderate anticumulativism). As is well known, a further step in the direction of anticumulativism was made by Kuhn and Feyerabend, who negated the universal value of the principle of correspondence, expressed as a principle monitoring the development of science, especially physics. They did it surely by exposing, in an exaggerated manner, the elements of incommensurability between theories separated by the scientific revolution.

XI. The Problem of Epistemic Progress

The eleventh change is closely connected with the previous one. This change consists in a gradual abandonment of the initially very optimistic idea of scientific progress (born as early as in the 18th century). Neopositivism, and even Poincaré's and Duhem's conventionalism, assumed permanent cognitive progress in science or in the very least progress that was halted only for short periods. Popper's hypothetism considerably weakened the belief in scientific progress because it logically justified only the falsifying procedures and in spite of great efforts of Popper and of some of his followers, the aim of which was to rescue the conception of approaching the truth (verisimilitude). Kuhn's theory, in turn, allows only very limited cognitive progress during the periods of normal science, and in the whole course of development of science takes into consideration merely the "heuristic" progress whereby the new paradigm usually allows the formulation of (and possibly solving) more problems-puzzles than the preceding one. During the period of normal science, progress consists in the accumulation of the puzzles solved, and during the periods of crisis and revolution the place of scientific progress is actually taken, according to Kuhn, by scientific change which does not have any distinct features of cognitive progress.

This is connected with Kuhn's view that a transition to a new paradigm has both advantages and disadvantages. This concerns primarily the whole repository of problems which can be formulated within a given paradigm as well as the resources of explanation which can be obtained within it. The change of the paradigm consists first of all in the revolutionary transition to a new theory which is logically and empirically incommensurable with the previous one. Among other things, it signifies the change in the set of notions which reaches so deep that at least some previous problems (and their solutions) lose their sense and in their place, and new ones appear which cannot be formulated within the old paradigm. The new paradigm also changes perceptions on what is a good explanation, and provides new ways and models of explanation, which sometimes invalidate certain explanations that have thus far been considered correct.

XII. Syntax, Semantics, Pragmatics – Extension of the Semiotic Means of Research

The next change consisted in extending the means of the study of science, which have gradually been provided by the developing logical semiotics. In the first three decades of the 20th century syntax was the only well developed semiotic discipline. It is small wonder then that scholars connected with the Vienna Circle and others who were bent on applying exclusively logical and linguistic means in the study of science (Carnap, Russell, Popper, Ajdukiewicz) initially restricted themselves to syntactic studies of science, i.e. of the "ready-made" scientific knowledge. As a result, they did not, for instance, take up serious studies of the notion of truth, since they were feared the antinomies which were lurching everywhere. Tarski's semantic studies on the notion of fulfilment and truth, and the development of the theory of models on its basis radically changed this state of affairs and led to a large-scale application of the means provided by semantics in methodological studies. In the modern philosophy of science, research tools developed by pragmatics are also applied. Thus, all the three disciplines of semiotics have been implemented in the studies of science.

XIII. Criteria and Values

The thirteenth significant change concerns a certain "blurring" of standards ascribed to scientific studies (research workers). Rigid criteria of evaluating and testing theory (verification, confirmation, falsification, or disconfirmation) that are typical both of neopositivism (inductionism) and Popper's hypothetism have begun to be replaced in the philosophy of science by values which cannot be precisely defined, such as logical simplicity ("inner perfection") of a theory, elegance, explanatory power, heuristic power, predictive power or profundity. Lakatos already paid attention to that when making the criteria of dogmatic and naive falsificationism more subtle while at the same time loosening them up. In particular, works by Kuhn, Feyerabend or Laudan contributed to establishing in the modern philosophy of science a view that, generally, it is not the strict criteria but merely values that are difficult to be made more precise, and even feelings and other rather vague standards of evaluation, determine the fate of theoretical systems in science.

XIV. Science and Metaphysics

Finally, the fourteenth important change concerns the attempts that were made in the 19th and early 20th century to radically oppose science to metaphysics as well as other kinds of knowledge that were previously considered to be mediocre kinds of knowledge. At the time, it was thought that metaphysics exerted a negative influence on the development of science and that it "contaminated" experience and scientific knowledge because it added speculative themes, which brought in pseudoproblems at every step.

A revaluation of the role of metaphysical hypotheses in science occurred partly through Popper, and further on under the influence of the radical current and other trends which perceived certain "environs" of science in metaphysics, and thus viewed it as a kind of parascience (but not pseudoscience),[1] or a "repository" from which science extracts its ideas, next submitting them to precise and reliable scientific treatment.

XV. Conclusion

To conclude, let us consider the following question: what do the changes in and the revaluation of the philosophical view of science outlined above lead to? They seem to mark a certain constant trend in the development of the contemporary (and perhaps future) studies of science.

The first three decades (and, to some extent, the entire first half of the 20th century) of the studies of science interpreted the subject matter (i.e. science) in isolation from the cultural, social and economic environment from which science grows, and tended towards a lucid portrayal of the peculiarity of science, of all the distinctive features which make it a separate and very special kind of knowledge and kind of cognition.

The changes discussed above have fundamentally altered both the object and the goals of the philosophy of science. What is pursued now is ways of "placing science in culture" most firmly and profoundly, the discovery of contemporary traits which connect science with other kinds

---

[1] Parascience and pseudoscience are frequently confused in the philosophical literature. Hence, I would like to make a clear distinction between the two. I view pseudoscience as the fragment of irrational knowledge which simulates (feigns) or "impersonates" science because of its high epistemic status and social prestige.

of knowledge and spheres of culture, especially with symbolic culture, and expressing all the significant interactions occurring between science and the other areas of social life.

This trend is outlined in an exaggerated manner in the modern reflections on science which seem to aim at obliterating all the important differences between scientific practice and the other types of practice, and even at reducing a cognitive relation to (certain special) relations of the causal type.

Changing the name of the philosophical discipline in which we are interested from "the methodology of sciences" (the former name) to the "philosophy of science" (the name that is popular at present) should be considered as an adequate reflection of the changes which have occurred in the subject area at hand and in aims of the discipline.

Part III
The Development of Science

TYPES OF DETERMINATION VS. THE DEVELOPMENT OF SCIENCE IN HISTORICAL EPISTEMOLOGY*

I. Introduction

A full elaboration of the question of the determinants of the development of science requires an answer to three questions: first, what are the main factors (determinants) influencing the development of science (this question has so far been discussed relatively fully); second, what are the mechanisms which help to realize the impact of various determinants on the development of science (this question has not been analyzed as widely as the previous one); third, what kinds of determination are involved in the process of the development of science (this question is still in its infancy).

I think that one of the advantages of historical epistemology is that it has considered, more than the other conceptions, the role of particular types of determination in the development of science. According to J. Kmita, the following types of determination take part in the development of science: functional determination, genetic determination (these two may be treated together as functional-genetic determination) and subjectively rational determination which is a type of causal determination.

The aim of this article is to outline the role of these types of determination in the development of science in J. Kmita's formulation and to present a proposal considering the relation between functional-genetic and subjectively rational determination.

II. Functional Determination

Functional determination is ascribed the main role in the development of science by historical epistemology. Thus, it is no wonder that in the works of the representative of this trend, especially in J. Kmita's works, this type of determination has been described quite precisely.

According to J. Kmita, the term "functional determinant" (the functional equivalent of the "cause") stands for

---

* First published in A. Zeidler-Janiszewska (ed.), *Epistemology and History* (*Poznań Studies in the Philosophy of the Sciences and the Humanities*, vol. **47**). Amsterdam-Atlanta, GA: Rodopi, 1996, pp. 157-168.

The phenomenon Z, in which a certain system has such a property that ... a necessary condition for the system to keep this property is another phenomenon Z', in which a specific element of the system is given a different property. With some further conditions the phenomenon Z will be the functional determinant of the phenomenon Z'. For example, the presence of arterial blood (that is "fresh" blood containing oxygen) in an animal's organism (the system) represents a functional determinant of the so-called beating of the heart. More commonly we can say that the presence of arterial blood is the function of the beating of the heart or that the beating is the reply to the organism's demand for the steady presence of arterial blood (Kmita 1987a, p. 604).

Kmita records that the relation of functional determination is not only different from causal determination (the presence of arterial blood in the organism is not the cause of the heart-beat) but it is inversely correlated with an appropriate relation of causal determination: the functional determinant Z of the phenomenon Z' is the result of the phenomenon Z' and of the fact of the system's maintaining the functional balance. For example, the presence of arterial blood in an animal's organism is the result of the fact that (1) the heart-beat lasts and that (2) the organism functions normally (it exists) (Kmita 1987a, p. 604).

Functional determination meant in this way is applied by Kmita in his model of the development of science. According to that model, the development of science is determined first of all by the "objective function of social scientific practice," which consists in "the coding and deductive systematizing of the predictive elements of the directly subjective social context of the basic practice, and also – the subjective contexts of the remaining types of social practice, functionally subordinate to the basic practice" (Kmita 1976, p. 95). According to this, "the functional determinant of the development of scientific cognition is the supplying of subjective descriptive premises to the specific realms of social practice by social practice" (Kmita 1987a, p. 604).

The claim that the function of scientific cognition is to supply subjective descriptive premises to the specific realms of social practice by social practice, that is that the functional determinant of the development of scientific cognizance is supplying those premises by this practice, says (among other things) that supplying those premises is the necessary condition for the specific realms of social practice to subsist and for this practice, as a unity, to be able to steadily reproduce the objective conditions of social existence (Kmita 1987a, pp. 604-5).

In Kmita's opinion, such a functional dependence of the state of scientific knowledge of the objective needs of social practice, as a further consequence, on the objective conditions of social existence derives from the basic thesis of historical materialism concerning the (functional) determination of social consciousness by the objective conditions of social existence.

The consequence of the above statements is that the historical changeability of the need of the particular realms of social practice for the products of scientific practice leads to a different type of scientific knowledge being the object of demand at different times. Consequently, the development of science proceeds in parallel with the process of general social development, being functionally determined by it. In the last instance it is determined by social existence and the basic practice which has "a specifying character in relation to all other types of practice and, what is more, its development determines (in a functional course) the development of these other practices" (Kmita 1976, p. 94).

Kmita strongly emphasizes that the objective relation of science to non-scientific practice is not steady:

the specification of the way of functioning of social practice in relation to (basic) social practice must be general in nature because this way is historically changeable: different for different evolutionary stages of research practice and consequently generally different for different realms of practice in the same chronological period, because particular realms of the given period (in the chronological sense) may, and usually do, represent different stages of development. However, at different stages in the development of social practice, which functions differently at each of these stages, some invariants of this functioning must recur. Otherwise, we would not be able to consider science as a separate type of social practice. These invariants should be taken into consideration by the general characteristics of the way in which the research practice functions (Kmita 1976, p. 94).

After all, the dissimilarity in the way of objective functioning of different realms of social practice may derive not only from their different stages of development but also from the fact that "they happen to be functionally connected with the basic social practice in a more or less direct way" (Kmita 1976, p. 94). For example, there is no doubt that natural sciences, let alone technical sciences, are more functionally dependent on productive practice than the humanities, which are more directly subordinate to legal-political practice and only through its mediation to the basic practice.

Apart from functional determination which is to have the main role, historical epistemology discerns two other types of determination in the development of science, i.e. genetic determination and subjectively rational determination (the regulation of people's behaviour, including cognitive activities by consciousness).

III. Genetic Determination

According to the assumption of historical epistemology, the notion of genetic determination in the realm of the theory of cognizance is associated with the fact that the social acceptance of a given, new complex of beliefs – also scientific beliefs – is only partially explained by a "functional moment," the fact that this complex represents a more adequate answer to the existing or apparent requirements of the relative types of practice than a complex excluded by it. The full functional explanation needs to take "a genetic moment" into consideration: the fact that this complex of beliefs is the transformation of the existing intellectual material (at the present level of knowledge) (Kmita 1979, p. 102).

The development of science, just like that of any other social phenomenon, is genetically determined by the hitherto existing state of a developing system (the system of knowledge). A new idea never appears in an intellectual vacuum; it refers in a way to the existing intellectual output, that is to the theories socially accepted before (Drat-Ruszczak 1987, p. 738).

In this way, it turns out that, on the one hand, the development of all forms of social consciousness is determined not in a purely functional but in a functional-genetic way: by their hitherto existing state and by objective social needs, which, directing the processes of selection, decide what and in which way the existing intellectual material will be used and adopted. On the other hand however, we can see that the intellectual material gathered, in the course of genetic determination, never univocally settles the shape of the future state of knowledge: the whole class of systems of knowledge may be derived from it, whereas the "functional determinant" will decide which of them will play a "historical role." The "functional determinant" is a kind of social need for knowledge of a specific level and type (Pałubicka 1977, p. 49). That is why functional rather than genetic determination is ascribed the basic role in the development of science by historical epistemology.

This explains why the principle of correspondence, which presents the relation between a given theory and its predecessor, and at the same time expresses the "relation of states" (genetic determination of knowledge at subsequent stages in the development of science) is not ascribed an independent importance by this epistemology but is subordinated to the principle of the objective functioning of science (which expresses the functional determination) (Pałubicka 1977, p. 50).

Let us consider in detail what the mechanism of genetic determination is according to historical epistemology. We should make it clear from the very start that historical epistemology, as a conception representing methodological anti-individualism, is interested primarily in the social rather than individual mechanism of that determination.

This means that it studies and explains not the facts of the individual creation of specific new ideas, that is the individual output of scientists (we need to include various individual factors of a psychological character for that factor which – from the point of view of social processes – are accidental) but the process of social "formation" of new ideas, the effect of which is that these ideas begin to spread socially, possibly forcing out specific old ideas from the consciousness of some groups of people (Kmita 1973, p. 247). The main point here is the mechanism of explaining the effects of forming social, not individual, consciousness. Because these ideas, which get social acceptance, spread, limiting the problem to science, we enter the problems of social acceptance and the legal validity of scientific knowledge at every stage.

The necessary condition for the social acceptance (and legal validity as well) of new scientific knowledge (theory) at a specified stage is that this knowledge (theory) be in a relation of correspondence with the knowledge (theory) existing in a given realm. Thus, the expression of the genetic determination of scientific knowledge is the regularity of the "steady occurrence of the relation of correspondence – either generalizing or essentially rectifying – between the succeeding stages in the development of scientific knowledge. The generalizing correspondence is based on the following system of knowledge being a generalization of the preceding one, insignificantly rectified by limiting the degree of the generality of some of its elements. The essentially rectifying correspondence consists in the following system of knowledge explaining why the hitherto existing knowledge has been able to organize subjective practical activities (including experimental ones) of appropriate effects, though it has given a picture of reality which is logically incomparable with the picture constructed by the new system" (Kmita 1987a, p. 606).

It should be strongly emphasized at this point that J. Kmita's conception of correspondence assumes the logico-"empirical" incomparability of the succeeding theories separated by a "theoretical breakthrough" (scientific revolution) – between which a relation of exactly rectifying correspondence occurs. On the other hand, it assumes comparability because of the so-called practically-objective reference (Kmita 1976, pp. 121-35). This means that the essentially (exactly) rectifying relation of correspondence is not "intellectual," logical, internal but is made relative to the objective conditions of the social practice making its demand on research practice. This moment indicates that genetic determination is not a fully independent kind of determination but is closely connected with functional determination.

In Kmita's view, the role of genetic determination and the principle of correspondence in the development of science are connected with the fact that functional determination "has more than one meaning – the function of one link may be performed indirectly by many alternative states of this link. Especially the socially-subjective context of a practice of a given type may be composed of different alternative systems of opinions. For example, Lutheranism or Calvinism or other religious doctrines could perform the function of the subjective context connected with people's outlook on life in relation to the practice of the commercial beginnings of capitalism created in the sixteenth century. This ambiguity is reduced only by considering a specific historical situation, taking into consideration the circumstances that every socially-subjective context of the social practice of the given time respects in a specific way the 'intellectual material' hitherto gathered, according to F. Engels' – its hitherto existing state"

(Kmita 1987b, p. 243). The way of respecting the hitherto existing "intellectual material" in science is marked by the principle of correspondence.

The links between functional and genetic determination lead J. Kmita to treat them jointly under the name of functional-genetic determination.

IV. Functional-Genetic Determination

According to J. Kmita, the general functional-genetic explanation (called in brief the historical explanation) must include (1) the functional law saying "what function the specific feature *C* of a fragment of the structure performs, in other words to what it is necessary" (Kmita 1973a, p. 214) and (2) "the characteristics of the state which precedes the explained phenomenon; the presentation of the origin of dissemination, the general permanence or disappearance of the given feature" (Kmita 1973b, p. 252).

With reference to science and in translation from the "language of explaining" to the "language of determination" this means that the "mechanism of social self-forming or social acceptance" of another scientific theory *T'* following the theory *T* is marked

> not only by the functional determinant . . . but also by the genetic one. And so, the full determination considered here is functionally-genetic. In particular, the theory *T'* does not appear in an intellectual vacuum; in a way it refers (genetic moment) to the "intellectual material" it guides: to the theory *T* (and the theories created in the range of the same methodologically-theoretical factor) (Kmita 1976, p. 124).

J. Kmita undoubtedly admits that the main component of the determinant considered here (functionally-genetic) is the functional, and not the genetic, determinant. In the case of the functionally-genetic determination of opinions (e.g. scientific), the genetic determinant (the inherited intellectual material) is subjective (or partially subjective) and the functional determinant (the need of the socially-economic sphere) objective.

The consequence of the thesis that the development of science is marked (determined) in a functionally-genetic way is the so-called principle of the development of science formulated by J. Kmita (Kmita 1976, p. 128). It consists of (1) the principle of correspondence, expressing the "genetic moment" of the functionally-genetic determination of the development of research practice, and (2) the principle referring to the "general objective function of every research practice – with reference to the following evolutional phase of the non-scientific practice considered"; expressing the functional component of functional-genetic determination (p. 129).

However, full explanation of the development of science – just as any other type of practice – is, according to historical epistemology, impossible without taking into consideration another type of determination, i.e. subjectively rational determination.

V. Subjectively Rational Determination

According to historical epistemology

each type of social practice, including scientific practice, is controlled in the socially-subjective course by the specific collection of generally respective normative opinions (establishing the values, i.e. aims to be reached) and instructional opinions (establishing the ways to reaching the aims that is, defining the kind of activities which should be undertaken in given circumstances to reach the right aim) (Kmita 1987a, p. 605).

A given collection of opinions, determining a given type of social practice in a subjectively rational way is called the subjective social context of such a practice or its socially subjective regulator (Kmita 1976, pp. 84-85). It is also considered that this regulator is an element of the practice subjectively determined by it, being one of the forms of social consciousness. For scientific practice, it is the social methodological consciousness, which consists of the norms establishing the research aims and the instructions defining the ways and means to achieve these aims.

Why is the type of determination considered called subjectively rational? Among other reasons, because every practice is conscious and intentional. It is the undertaking of rational activities, realizing specific aims on the grounds of specific knowledge. The explanation of such activities (and explanation is – as we can remember – an epistemological equivalent of determination) is called a humanistic interpretation in historical epistemology. It requires reference not only to the axiological order and the knowledge of the acting subject but also to the so-called rationality principle. The subject's knowledge and the system of values (in other words, the beliefs specifying the aims and the ways of achieving them) only with the assumption of his rationality lead to undertaking specific (rational) activities, and so in consequence, they help to explain undertaking those activities. Thus, without referring to subjective motivation and assuming rationality, an explanation of rational activities is impossible.

This is connected with the fact that there is no objective (natural) relation established by certain psychological laws or other laws of natural history between the knowledge or aims of the agent (e.g. the researcher) and the his/her activities (or their products). The dependence which takes place here is subjective and, being rationally motivated, it is termed subjectively rational dependence (or determination) (Kmita 1971, p. 149).

And so, according to Kmita,

The full subjectively-rational determinant of the undertaking of a given activity . . . consists of . . . the following elements: (1) the knowledge of the subject of the given activity, the knowledge which separates the set . . . of activities it is possible to undertake and which can foresee their effects, (2) the preferential order which creates hierarchies from the axiological point of view, represented by the subject, the collection of effects foreseen by him. The description of elements (1) and (2) in conjunction with the assumption of rationality, that is with the assumption generally prejudging the consistent respect by the subjects of activities of their own knowledge and axiology, implies the logical undertaking of the activity leading (according to the subjects knowledge) to the desired effect; I call it the aim or sense of an activity. Pointing to this aim is pointing to the fragment of the subjectively rational determinant of the corresponding activity (Kmita 1985, p. 16).

Because real subjects never fully perform the assumption of rationality (the principle of rationality), it is an idealizational and general law. In the scheme of explaining (humanistic interpretation) it performs the function of an initial premise that is, the nomological part of an explanation, while the subject's knowledge and system of values perform the role of the starting conditions.

According to historical epistemology, the social methodological consciousness of scientists in a given historical period of the development of science is marked in a functional course by the requirements of the social practice of that period for knowledge of a specific kind and level of adequacy. This causes the scientific practice controlled by the social methodological consciousness (which is its socially subjective context and element at the same time) to meet the requirements and, in consequence, the development of knowledge is parallel to the development of the whole of social practice and life.

VI. Functional-Genetic Determination vs. Subjectively Rational Determination

I think that the presented conception of the role of three types of determination in the development of science leads not only to the conclusion (derived in historical epistemology) that functional determination is closely connected with genetic determination (so that they stand for one unity: functional-genetic determination) but also to the conclusion that subjectively rational determination is in fact an indispensable element in the mechanism of functional-genetic determination wherever human activity (or its products) is concerned. It is an indispensable element, i.e. without it the functional determination of a social practice of any kind by other kinds of social practice of the objective requirements of socio-economic conditions would be impossible. In essence, other types of practice (including productive practice) functionally determine science (scientific practice) through the social methodological consciousness rather than not directly. This consciousness directly controls scientific practice in a subjectively rational course, which makes subjectively rational determination a necessary link, without which the functional-genetic determination of human activities (or their products) would not succeed.

However, a question arises whether such a formulation of the relations between subjectively rational and functional determination (in the case of explaining practice or human activities and their products) is not contradicted by the fact that, according to historical epistemology the order of the functional determination of the norms and methodological directions of social practice, i.e. of the social methodological consciousness of science, is opposed to the "ideological" order. Research results specified from the "ideological" point of view should be appreciated, because they correspond to the appropriate methodological norms and directions. On the other hand, from the point of view of objective determination, a given type of social methodological consciousness becomes obligatory, because there is a need for sanctioned scientific knowledge (Kmita 1987a, p. 60S).

The reverse of both orders of determination – the objective (functional) one and the subjective one (subjectively-rational) – is not an obstacle here: it only proves that the requirements of the objective conditions of social existence, or of the basic social practice directed towards scientific practice, are not able to mark its specific shape (functional in relation to the conditions) in a direct way that is, without the mediation of the social methodological consciousness. Without realizing to some extent the requirements concerned (and "realizing" not only in its individual sense but also in the social sphere concerned – the "reflection" of these requirements in the social consciousness in the shape of proper social cognitive values, proper cognitive ideals), they would not be able to stimulate people to appropriate actions as elements of research practice. This is the reason for the inevitability of this subjective link in the process of the functional explanation of the historically given shape of research practice. Naturally, it concerns explaining all human activities (and their products) rather than merely cognitive actions.

It follows from the above that the functional (or functional-genetic) explanation in the humanities fundamentally differs from such an explanation in biological sciences, where the indirect subjective link is certainly not considered. Only the functional (or functional-genetic) explanation concerning phenomena independent of human consciousness is based on a fully objective determination. On the other hand, the functional (functional-genetic) determination of phenomena dependent on human consciousness inevitably involves the subjective factor (subjective determinant). We can put it differently by stating that the functional (functional-genetic) explanation in the humanities is not complete unless it involves – as its link – the humanistic interpretation.

It follows, I think that the functional (functional-genetic) determination of human activities is never fully objective, just because of this (necessary) subjective link taken into consideration by the humanistic interpretation. What follows is that in the biological sphere this type of determination is simpler than in the social sphere.

It is worth mentioning, by the way, that the conception of the functional-genetic determination of the development of science (of scientific practice understood as a kind of theoretical production), the determination in which the subjective factor plays no important role, was worked out by Althusser. According to that conception, the role of the observing subject (as of any subject acting in history) is fully (and univocally) determined by the functions performed by him/her in a given productive (socio-economic) structure.

A similar conception, clearly reducing – as J. Kmita shows – the role of subjectively-rational determination, was developed by O. Lange in his work *The Whole and Development in the Light of Cybernetics* (Lange 1962). Within the framework of that conception, the subjective determination of human activities is adjusted to the developmental direction of the socio-economic system. These activities may be intentional and conscious but at the same time the knowledge and norms subjectively determining them are prejudged by their functions, or more exactly, by the functions performed by the activities that are subjectively determined by these factors in the context of a given directionally organized system. That is why explaining human activities as rational by pointing to their subjective determination (the humanistic interpretation of these activities) becomes a research procedure of little cognitive importance, something of an addition to the basic type of a diachronically functional explanation (Kmita 1971, p. 165).

Such conceptions presupposing a specific "objectively subjective parallelism" were generally accepted, in Kmita's opinion, by the representatives of both Marxist "schools" existing in our country till recently: the "scientistic" and the "anthropological" one (Kmita 1971, p. 166).

According to historical epistemology, all the approaches leading to a reduction of the role of subjective determination in the development of science are wrong, mainly because without considering determination of such a kind it would be impossible to "epistemologically qualify particular, historically given states" of the consciousness "produced" by science, which would make the formulation of the "development of science as the development of scientific cognition" impossible (Kmita 1987a, p. 605).

> The fact that a certain type of social practice, scientific practice, is regulated in the socially subjective course by a specific collection of normative and directional opinions, means that in each period of the development of this realm of social practice specific norms and methodological directions are generally respected (in the group of researchers), epistemologically qualifying particular "products" of this practice, subjectively acknowledged (on a social scale) as cognitively legally valid research results to a degree in which they comply with these norms and directions (Kmita 1987a, p. 605).

Translated by Aleksandra Wach

RELATION OF CORRESPONDENCE
AND LOGICAL CONSEQUENCE*

I. Two Standpoints

The initial problem of almost all methodological controversies concerning the principle of correspondence is the issue of logical consequence of a former (corresponded) theory with the later (corresponding) one.[1] The way in which this problem is solved divides physicists and methodologists into two opposing groups. There is a group of "implicationists" who think that the former theory (law) in the correspondence chain of theories (laws) "is contained in" the latter theory (law) being a "particular case" of this latter theory (law). On the other hand, there is a group of their opponents – let us call them "anti-implicationists" – who argue that, strictly speaking, there is no question of "being contained" or being "a particular case" at the

---

* Translation of "Relacja korespondencji a wynikanie" published in: W. Krajewski, W. Mejbaum, J. Such (eds.) (1974), *Zasada korespondencji w fizyce a rozwój nauki* [The Principle of Correspondence in Physics and the Development of Science]. Warszawa: PWN, pp. 65-114.

[1] The problem of logical consequence also arises when the links of relation of correspondence are laws (generally: statements), and not only theories (generally: sets of statements, and in particular, of laws). However, it does not occur in case when concepts are in a relation of correspondence to each other. I think that in the latter case a relation of another kind is meant, which only due to certain analogies to the relation of correspondence, combining scientific laws and theories, is sometimes called the relation of the correspondence of concepts. What is mostly meant here are simply the concepts occurring in laws and theories between which the relation of correspondence occurs. For instance, P. Bridgman seems to refer the principle of correspondence, among other things, to the operation of measurement and the operationally defined concepts in case when they occur in laws or theories corresponding with each other. According to Bridgman, the Einsteinian length of bodies in motion and length in the elementary sense (i.e. length of resting bodies) are connected with each other; in both cases in the border when the velocity of the measuring system tends towards zero, operations aim at the operation of measuring length for the immovable object (Bridgman 1958, p. 12ff).

Since we are interested in the problem of the relation of correspondence to the relation of consequence, further in this work we shall totally leave out of consideration the problem of occurrence of the relation of correspondence between notions, and we shall focus our attention on the relation of correspondence whose links are laws of science and scientific theories.

correspondence transition. Members of the former group claim that the corresponded theory logically follows from the corresponding theory (and some additional premises which specify "the boundary conditions" under which both theories being in the relation of correspondence are non-emptily fulfilled). According to the latter group, by contrast, such consequence may occur at most either between the corresponding theory and a certain modification (which is an improvement from the point of view of progress of knowledge) of the corresponded theory or between a certain simplified version of the corresponding theory and the corresponded one.

Agreement between both standpoints actually concerns only one thing: that the relation of correspondence is a counter-symmetric relation, i.e. that the corresponded theory and the corresponding one which are in this relation to each other cannot swap places and that this is a transitive relation which results in the theories sometimes creating a correspondence sequence. The sequence consists of (three or more) theories which are in the relation of correspondence to one another and thus create the subsequent links of a given sequence.

Previously, the "implicational" conception of the principle of correspondence seemed to prevail both amongst physicists and methodologists. At present, the latter conception is distinctly dominant, both in terms of the number of its advocates and the number and the importance of arguments gathered. However, one should strongly emphasize in this connection that not everyone, who – while characterising the relation of correspondence – uses traditional descriptions suggesting that logical consequence occurs between theories connected by the relation of correspondence, deserves the name of an implicationist. What is meant are such expressions as: a latter theory constitutes "a correspondential generalization" or "rational generalization" (Bohr's expression) of the former one; or the earlier theory is a "boundary case" or "particular case" of the latter, "is contained" in the latter and the like. These descriptions are "widely" used by both adherents and opponents of the implicational conception of correspondence: in works by both groups they occur almost equally frequently. And sometimes "to separate" themselves from implicational conceptions scientists use more cautious descriptions of the kind: the former theory constitutes "approximately a special (boundary) case" of the latter one, or simply: the former theory "follows in approximation (but not strictly)" from the latter one. One can also encounter such expressions characterizing the relation of correspondence as "pragmatic consequence" or "(approximate) obtainment" of an former theory from the later one.[2] Application of this kind of ambiguous and not

---

[2] In one of the works on methodology of physics we can read, e.g.: correspondence of general relativity theory with Newtonian theory "may be obtained by assuming that world is

very lucid expressions indicates the main difficulty in the standpoint of "anti-implicationists": realizing that the relation of correspondence between theories or laws is not a relation of consequence, they are not able to answer exactly the "positive" question, i.e. what does this relation consist in, what it is in its essence.

II. Why the "Implicational" Conceptions of Correspondence Are Wrong

Before we pass on to characterizing the principle of correspondence and the relation of correspondence, let us try to further arguments which indicate (in some cases incontrovertibly, as it seems) that, strictly speaking, we cannot speak of "consequence" of the corresponded theory with the corresponding one, or about the former "being contained within" the latter.

When would this "consequence" or "being contained" occur? It would if each statement of the corresponded theory were either a statement of the corresponding theory, or a particular case of some statement of this theory, being different only in the degree of generality (range of applicability).[3] A precondition (although obviously not a sufficient one) of the occurrence of the relation of consequence between correspondentially related theories would be that the corresponded ("resultant") theory does not contain sentences that are false from the point of view of the corresponding theory. To put it differently, the truthfulness of the corresponding theory would have to involve the truthfulness of the corresponded ("resultant") theory.

Euclidean in its spatial infinity, and following Einstein, the *approximate* solution (assuming the weak gravitation field and slow motions of matter) of equations of the field under these boundary conditions. Then one obtains the Newtonian law of gravitation and the equations of Newtonian movement. Einstein showed that in the next step of approximation, certain conclusions result from this solution, which confirm the principle as to certain effects which may not occur in case of strict solutions of field equations" (Bażański and Demiański 1967, p. 150). As can be seen the authors understand "correspondence" as "obtaining" of the earlier theory from the latter one, emphasizing that this "obtaining" (deduction, on the basis of logical consequence) concerns not the corresponding theory but the theory being its essential simplification (modification). At the same time, the authors emphasize that the approximation of the new theory from which the earlier corresponded theory results may deprive the first one of certain significant features which are characteristic of it in the strict (primary) formulation.

[3] In reference to laws, this can be expressed as follows: the corresponded law would result from the corresponding law if it were a particular case of the latter, i.e. if these laws (in their implicational formulation) did not differ in consequents describing regularities but only in precedents, giving conditions of occurrence of these regularities, and if the difference in precedents consisted exclusively in the fact that the range of applicability of the corresponded law would be contained in the range of the corresponding law. To be more precise, the first one would be determined by a set being a subset proper of the set determining the range of the corresponding law.

Although the principle of correspondence was formulated (and the name itself introduced) only by Bohr in the early 1920s (cf. Lewenstam 1974; Niedźwiecki 1974), this principle had been applied intuitively much earlier, among others, by Newton, Fresnel, Planck or Einstein, and had steered the development of physics at least since the time of Galileo. Also, some arguments against its implicational interpretations were already known in Newton's times. They were connected with the problem of the "consequence" of Galileo's laws (e.g. his law of free falling) and Kepler's laws from the (relevant) laws of Newton. Physicists (and Newton himself) realised it very well since the very beginning that, speaking strictly, one cannot speak of any consequence in these cases. For example, Kepler's laws may be "deduced" from laws of motion and Newton's gravitation law only when it is assumed that the only interactions occurring between bodies of the planetary system are interactions of particular planets with the Sun. "Deducing" one or another of Kepler's laws one should thus abstract not only from the interactions of planets with one another but also from interactions of planets with all the other bodies in the Universe. Similarly, in order to "deduce" Galileo's law of the free fall of bodies near the Earth from Newton's laws one should abstract from the fact that the Earth gravitation power is the function of distance from the centre of the Earth (assuming that the Earth is an ideal sphere of a perfectly symmetrical, i.e. centrobaric distribution of mass), and also from a number of other facts.

This state of affairs has been comprehended by methodologists at least since the time of Mill, who knew, for example, that Kepler's first law resulted from Newton's laws only when one adopts the assumption (which contradictory with the law of gravitation, therefore, a counterfactual assumption) that the Sun is the only body influencing gravitationally on planets (Mill 1847). Mill also drew attention to the fact that from the latter law, being a "correspondential generalization" of the former one, there results – strictly speaking – not that former law but its certain modification (improvement) with which many contemporary authors agree, including Kemeny, Oppenheim, Feyerabend, Kuhn, Scheffner and others. A modification of this kind "surpasses" the law modified thanks to the fact that it takes into account certain dependencies which were abstracted from when formulating the latter law, which allows more description, more precise prediction and more thorough explanation of the investigated phenomena. It follows that in the case of "correspondential transition" the previously existing law by no means constitutes, strictly speaking, a case of a new law.

At most, one can recognize that the results (e.g. of explanations or prognoses) of the first one are approximate to (but not identical with) the results of the second within the definite limits (e.g. within the earlier experiments), therefore, under certain specific (boundary) conditions.

Apart from these conditions, divergencies between them may become optionally large, and grow to infinity. The corresponded law is therefore only an approximate (and not strict) special case of the corresponding law and by no means in the whole conjectural (i.e. primarily assumed) range of its applicability but only in a certain (usually drastically) limited area (the so-called boundary area).

Thus, if "obtaining" the corresponded law from the corresponding law is concerned, two possibilities exist: first, the possibility of the logical derivation from the corresponding law which is a modification (improvement) of the corresponded law (e.g. from Newton's laws "improved" laws of Kepler and Galileo it may be deduced and the laws deduced from Newton's laws will be analogical to Kepler's and Galileo's laws); second, the possibility of deducing the corresponded law from the law which is a simplification – through idealization, i.e. abstracting from certain factors influencing the regularity involved (as described by the law) – of the corresponding law or, which means exactly the same, the possibility of deducing of the corresponded law from the corresponding law assuming counterfactually (i.e. contrary to the knowledge acquired) that certain interactions which actually occur do not take place (e.g. from Newton's laws Kepler's laws may be derived assuming that the only power applied to planets is gravitation coming from the Sun; similarly, from a theory which is a simplification of the general relativity theory, Newtonian gravitation theory may be deduced). Therefore, the consequence that is meant here assumes at least one of the two: modification (improvement) of the corresponded law or modification (simplification this time) of the corresponding law.

On the other hand, resulting of the corresponded law (or theory) from the corresponding law (theory) does not occur on the grounds of the existing knowledge, i.e. without assuming (adopting) simplifying conditions.

Put more simply, the relation of correspondence that is not the relation of consequence is still something approximate to the relation of consequence: after all, it occurs between the laws (theories) one of which (law or theory) "results approximately" from another law or theory (to be more precise, resulting occurs between segments at least one of which is a modified version of a law or a theory being a link of the relation of correspondence). This means that the new law, while not containing the previous one with which it corresponds to as its particular case, nevertheless leads (or must lead), within certain limits, to results which differ to an infinitely small degree from the results of the other.

The principle of correspondence is, therefore, the principle of moderate cumulativism, the principle of dialectic development of scientific knowledge which "is accumulated" in such a manner that a new law (theory) is at the same time a continuation of the previously existing one (with one respect) and an overcoming thereof (with another respect).

If the hitherto existing law simply resulted from the newly adopted one, we would merely have a moment of continuation in the development: the increase of knowledge would not be accompanied by its modification leading to precization and making more thorough the existing knowledge; the development of knowledge would consist exclusively in its extension and extrapolation onto new areas and conditions. If, in turn, we could not speak of any, even approximate "resulting," the new knowledge would in no sense be a continuation of the previous one. "The development" of knowledge would simply consist in a total rejection of the old knowledge and its replacement by the new one, without any accumulation of knowledge. The proof that the progress in science does not simply consist in accumulation of the newer and newer results is the fact that the victory of the new theory is not a painless process. In this connection, some even think – and certainly there is some exaggeration in it – that great new ideas win in science only when their opponents (i.e. those who are the adherents of the old ideas) die.[4]

It can be seen from the above that extreme cumulativism – merely stating that new statements accepted in science do not lead to the modification or rejection of the hitherto existing ones but consist in adding new knowledge to the already existing knowledge being its extension and generalization onto new cases – is not a standpoint that adequately describes the process of developing scientific knowledge. As matter of fact, nothing in that knowledge is ever-lasting and everything is submitted to the process of improvement. Whereas it is difficult to perceive in the humanities and philosophy, it can be seen clearly in natural sciences that almost at every step one can observe cases of knowledge cumulation but also of knowledge deepening, which assumes the modification and limitation of the range of applicability of previously formulated statements.

It was observed long ago that what is accumulated in science and does not become obsolete are not so much explanations but descriptions. The

---

[4] This view, developed by Kuhn in his work *The Structure of the Scientific Revolutions* seems to have been started by Planck, who – based on his personal experience and also on an analysis of the problem of origin and influence of scientific ideas wrote: "The course of historical development described here, may at the same time be an example of a certain general truth, which may seem somewhat strange at first glance. A great new scientific idea does not usually clear a way for itself in the way that it opponents become gradually convinced and converted – the transformation from Saul to Paul is very rare indeed – but rather in this way that its opponents slowly die out and the growing generations get at once acquainted with and become familiar to the new idea. Also, the following rule holds in that area: future belongs to those who have the young on their side. That is why a proper setting of school teaching is one of the most important premises of the development of science. . . " (Planck 1970, p. 170). The last sentences of the above opinion sound very topical and deserve close attention, regardless of how we approach Planck.

two quantitative laws given which describe (at least partly) the same area of reality relatively reliably cannot lead, at least to some extent, to totally different results. Proximity between them must occur in the subrange of their applicability in which they are approximately compatible with experience (and consequently with reality). For instance, classical mechanics and relativist mechanics, which both describe bodies in slow motion (as compared with the speed of light) with great precision, correspond well with each other as referred to objects moving with a speed incomparably lower than that of light ($v \ll c$).

Laws which correspond with one another are usually compatible with each other only in one "boundary" point. Further on, there is a sphere of their increasing incompatibility with each other. Thus, this is a sphere in which at least one of these laws shows a growing disagreement with reality, the greater the more distant a given area of applications is from this "boundary point" of compatibility. New laws (and, in general, new scientific knowledge) appears in agreement with reality and with the old knowledge but only as far as the past experience (or somewhat further) in the range in which they have been tested (or a somewhat greater one) is concerned. Thus, they are adapted to both but only in a certain area, which delimited mainly by the scope of the earlier human practice. Whether a new law will be considered as rather compatible with the hitherto existing law concerning the same field, and also compatible with reality or rather incompatible with them will depend on which subrange of applicability of this law (mainly) is meant by us and in which subrange of their conjectural applicability we compare these laws with each other: in a subrange encompassed within the sphere of compatibility or in the sphere of incompatibility between them.

The above explains "the paradox" that the same scientist (Bohr) who formulated the principle of correspondence in physics is also the author of a completely different and, at first glance, totally contradictory methodological principle in physics, i.e. the principle of "strangeness" of a theory.[5] According to the principle, a new physical theory (especially a theory of elementary particles) must be "crazy" enough, i.e. must deviate far from the hitherto existing views, if it is to be considered as the real

---

[5] Cf. Tamm (1961, p. 234). Bohr expressed his view on the non-linear theory of the spinor field as formulated by Heisenberg in the late 1950s in the following way: "as for a new theory, Heisenberg's theory is not crazy enough...," i.e. it departs too little from the hitherto presentations to be able to constitute a probable solution of paradoxes and difficulties accumulated in the theory of elementary particles and in general in microphysics (p. 234). A discussion was started amongst physicists as to what the scope of applications of Bohr's principle of "strangeness" was to be. Some of them are of the opinion that it concerns exclusively new great theories of microphysics, others – that it may be used in the whole of physics or even in science in general. It seems that after this principle has been properly formulated it can be ascribed a universal scientific significance.

solution of these paradoxes and problems which have accumulated in a given field of physics (especially in microphysics). At present, searching for ideas that are "crazy" enough is a methodological requirement at least in certain (the most important to that) fields of physics, and even a specific auxiliary criterion of cognitive truthfulness and fruitfulness of a new theory. Although Bohr formulated his principle with reference to physics (or, as some think, having in mind only the theory of elementary particles and quantum theory of the field), it is undoubtedly of a general methodological significance. Postulating that the new theory deviates far enough from the previous views (theories), it constitutes as if a complementary "supplementation" of the principle of correspondence: if the principle of correspondence concerns the subrange of the applicability of theories that follow one another in which they are – within (the limits) of experiment – compatible with one another, then the principle of "strangeness" of a theory refers to the subrange in which the divergence between them is considerable. This kind of "complementary" approach may not be strange in the case of the author of the principle of complementarity in physics.

Another argument in favour of the incompatibility of a new theory with the previous one in the correspondence sequence is that, on the whole, at least certain terms occurring in both these theories have a different cognitive meaning. For example, in many laws of classical and relativistic mechanics there occurs the term "mass." However, relativistic mass (or if only resting mass) is not the same as mass in the classical sense. J. von Neumann also showed that the cognitive sense of many terms occurring in quantum mechanics is different than the cognitive sense of the same terms applied in such theories with which it corresponds, such as classical mechanics or Maxwell's electrodynamics. The change in meaning of terms caused by the development of knowledge makes it difficult to establish logical relations occurring between the laws and theories which follow each other (in particular, it makes it difficult to solve whether these are theories in disagreement with each other, or theories which are simply independent of each other). At the same time, however, it points to the irresolvability of one group of them to another group and the impossibility of deducing one from another. Since extensive discussion of this argument may be found in the works of Feyerabend, Kuhn and others[6], I am not going to deal with it.

---

[6] Cf. e.g. Feyerabend (1962, pp. 189-283) and (1963, pp. 3-39). See also Kuhn (1968). It is worth noting that these authors actually reject the principle of correspondence as a general methodological principle steering the development of scientific knowledge, ascribing to it only a "fragmentary" significance in the development of physics, i.e. recognizing that it played some role exclusively in reference to some physical theories. For instance, Feyerabend interprets it as a component of a programme of Bohr's quantum postulates, a programme which has already exhausted its heuristic possibilities long ago. On

But I shall present one more argument which seems to show incontrovertibly that in some cases at least disagreement between subsequent links of the correspondence sequence of laws or theories cannot be avoided and may be proved. These are cases of "correspondential transition" in which a given law (theory) corresponds directly with two rather than one previous law (theory). Cases of this kind exist, however, as can be presumed, they occur rather seldom.[7]

One of such cases was established by Planck in 1900 when he introduced a formula for the distribution of energy belonging to waves of different length λ of the radiating body (a perfectly black one) at absolute temperature T. Since an idealizational law was at issue, the above formula could not have been discovered through direct observation.[8] Planck derived it based on two other formulas (expressing laws) which had been known before, and which concerned the same problem: Wien's formula and that of Rayleigh. The former (Wien's law) was accurate – with good approximation – but only for small values of λ and T; at the transition to the larger values of these magnitudes (to put it more precisely, their metrical scales), the deviations from the values experimentally established became greater and greater. Rayleigh's formula (law), on the other hand, well agreed with experimental data for large values of λ and T but in the course of passing on to small values was increasingly unreliable. Therefore, the two laws are distinctly incompatible with each other in the whole range – both conjectural and real – of their applicability, i.e. there cannot be between them even the relation of correspondence, which, after all, determines "the approximate compatibility" between the laws (or theories) concerned in a certain range of their applicability. Planck, while searching for a formula which would encompass those formulae as approximate (at least for certain extreme values of variables which occur in them) boundary cases, discovered a new law describing faithfully the distribution of energy for all the accessible wave lengths and temperatures.

However, an attempt at a theoretical justification of this law (or, to put it differently, its explanation) showed that in the process of its derivation

---

the change in meaning of certain terms during transition from the classical theories to quantum mechanics see also von Neumann (1932).

[7] A question also arises whether a law (or a theory) may correspond with a number of previous laws (theories) that is greater than two. The question will be left unanswered. Theoretically, there is no impediment for this kind of cases to exist. But in practice they occur either very rarely or not at all. So far, I have not been able to find if only one such case. But it is worth pointing out that there are frequent cases when two or more latter laws (theories) correspond directly with some one previous law (one theory). For instance, numerous laws of both relativist mechanics and quantum mechanics correspond with the laws of classical mechanics.

[8] It concerned a "perfectly black" body (*"schwarze Strahlung"*) whose existence is (and was at the time when Planck formulated this law) excluded by existing physical knowledge.

one should adopt an assumption of the quantum character of radiation energy; to be more precise, the assumption that the exchange of energy between the radiating oscillators (and in any case its emission and absorption) may occur not in any small quantities. But it may also occur in strictly determined portions whose magnitude depends on wave length (these portions whole multiples of the $h \cdot v$ product, where h is Planck's constant, v – frequency of radiation).[9]

This means that the new law, while containing the description of the course of the relevant phenomena, which is partly in agreement with the descriptions contained in the previous laws, provides also totally different explanations of the causes of this course of phenomena. Its recognition constituted shifting from "classical" explanation to "quantum" explanation. This confirms the thesis that was formulated earlier, stating that laws (and theories), being subsequent links of the correspondential sequences, with partial proximity of descriptions which they provide, lead – at least in general – to completely divergent interpretations and explanations of the phenomena described.

However, we are interested in something else at this point, namely the problem of the agreement of the corresponding law with the corresponded law and the question of the possibility of logically deriving the latter from the former. It had been known from the start that Wien's and Rayleigh's laws were incompatible with each other and that they gave different results, including different prognoses. It follows that also Planck's law cannot be in agreement with both, at least in the whole range of their conjectural (initially assumed) applicability. At best, it might have been compatible with one of them. Assuming that the relation of correspondence between Planck's law and the laws of Wien and Rayleigh actually obtains, which does not raise any doubts, this proves that at least sometimes – as in this case – "the correspondence transition" occurs between laws that are incompatible with one another, which obviously excludes the possibility of a purely logical derivation of one from the other.

However, that is not all. There are sufficient reasons to recognize that incompatibility occurs here in both cases involved in "correspondential transition," thus both between Wien's law and Planck's law as well as between Rayleigh's law and Planck's law. After all, there are no grounds to think that one of those cases differs in this respect from another: the

---

[9] Later, the indispensability of assuming of the quantum character of radiation for derivation of Planck's formula was discussed many times. Among others, Hilbert confirmed the fact that this formula may be obtained only when non-continuity of energy transfer is assumed, while all attempts at deriving it based on assumptions of classical physics lead to other formulae which disagree with Planck's formula and, more importantly, with experience.

relationship (e.g. considered from the mathematical point of view) of Planck's law to both of them is the same.

If, in turn, we compare this relationship with the relationships which occur in other cases of correspondence then it seems to be typical, at least from the point of view of mathematics. Hence, we can conclude (with some degree of probability) that since there exists a relation of incompatibility in the cases considered between the corresponding law and the corresponded law then an analogical relation also occurs in many (or in all) the other cases of "correspondential transition."

Planck was in an unusual and propitious situation that he at once captured two relations of correspondence which his law of the distribution of radiation energy enters. As has already been mentioned, it is a rare case but not a unique one. When does it occur? It occurs when the regularity is at the same time accessible to theoretical examination and experimental testing not in one "boundary point" but in a wide area of its occurrence or at least in two various "boundary points" that are distant from each other (but not necessarily located "on the opposite ends" of its occurrence, e.g. for the largest and the smallest values of certain magnitudes which characterize this regularity). In the case at hand, the distribution of the radiation of bodies being in a wide range of temperature scale and sending electromagnetic waves of a very wide range of frequency was investigated. And although initially there was no success – on the grounds of assumptions of classical electromagnetics it had been impossible – in establishing of one law for the whole range of temperatures and waves studied it still appeared possible to derive two formulae adequately in certain points (areas) (each in a different boundary), albeit in disagreement with each other and with experience when they were extrapolated beyond those areas. This is an illustration of the fact that it is only a certain (one or more) threshold area of the conjectural applicability of a given law that is not always accessible to experience. It occurs so usually at the beginning of a study of a given regularity; with the more advanced studies, the area of experimental exploration is extended but usually never, at no stage, does it encompass the whole range of occurrence of the dependence tested. That is why an extension of the area that has been studied experimentally is the area being studied – at least at a given stage – exclusively theoretically; this area may be called an area of extrapolation.

The case of Planck's law is also an illustration of a thesis that the relation of correspondence usually occurs when a given initial law of a quantitative character, which is mathematically formulated is unreliable to a greater and greater extent for certain extreme (boundary) values, and to a lesser and lesser degree for other extreme (boundary) values (usually "opposite" ones, i.e. those from the other end). Then the new corresponding law, which takes into account a certain factor (or factors) which are essential for the dependence studied in the area in which the hitherto existing law (the corresponded one) appeared to be unreliable, becomes a better approximation of reality in this area.

Another instance of the same kind as Planck's case, this time, however, concerning a theory, is the one established by Dirac. His relativist quantum mechanics of 1928 constitutes a "correspondential generalization" of two theories that incompatible and non-corresponding with each other, i.e. relativistic mechanics and (ordinary) non-relativist quantum mechanics, which, in turn, are in the relation of correspondence to Newtonian classical mechanics. For small velocities, or more precisely when $v \to 0$, Dirac's equation on electron is transferred to Schrödinger's equation of quantum mechanics. By contrast, for objects that are large enough, precisely when $h \to 0$, Dirac's equation comes into a relativist equivalent of Hamilton-Jacobi equation of classical mechanics (when both the above conditions are fulfilled, i.e. when $v \to 0$ and $h \to 0$, Dirac's equation is transferred into "classical" equation of Hamilton-Jacobi). As in the case of Planck's law we can conclude that since both corresponded theories are incompatible with each other then there is also an incompatibility of each of them with the corresponding theory, i.e. Dirac's theory. It should also be stated that it is a typical case for theories which correspond with each other, so that such theories are always in the relation of disagreement with each other.

III. The Principle of Correspondence and Testing Laws and Theories in Different Dimensions and Aspects of Their Generality

Why it happens so that a new law or theory is (approximately) compatible with both the reality and with the hitherto existing knowledge usually merely in a certain boundary area of their applicability? It follows, among other things, that usually "we generalize quantitatively" (we formulate quantitative laws) on the basis of certain "boundary" cases, i.e. the ones which are placed in the narrow interval of the range of changeability of variables. They occur in the formulation of the quantitative law which is being constructed, and generally in the interval of the greatest or smallest ("boundary") values of some of these variables. At any rate, generalization leading to laws usually starts from (or takes place on the basis) of boundary cases and at a given stage of the development of science often does not exceed beyond it, in this sense constituting a one-sided induction; going beyond them often means a new phase in the development of a given discipline. After all, we usually apply and test laws also only within a very narrow (often almost threshold) area of their conjectural applicability. This area usually fits within only one "dimension" of generality of a law.[10] That

---

[10] The distinction between various dimensions and aspects of the generality of laws and theories is of great importance for the problem of testing theoretical knowledge. On various

is why a law tested "in one respect" ("from one point of view"), i.e. in one dimension (aspect) of its generality may appear not to be tested from another point of view (in another respect), and consequently not tested in its whole range. Hence the postulate that all dimensions and aspects of generality which are ascribed to a law be taken into account as far as possible in the process of testing the law: both spatio-temporal universality and various dimensions of conditional generality dependent on the magnitude of an area and the frequency of the occurrence of conditions described in the antecedent, and also the degree of multilevel and multiaspectual generality of a law. If we test a quantitative law, then considering in the testing process all dimensions and aspects of its generality is geared to establishing its truthfulness for the whole range of changeability of variables, i.e. for all the possible values of variables occurring in that formulation.

However, the existence of a correspondential sequence, i.e. such ordered series of laws of which every former one is to some extent an approximate case of the next, and also the occurrence of cases of "double" correspondence, established by Planck or Dirac, indicate that although we usually begin testing a law starting from a certain "boundary" zone of its total range of applicability, its range is not always so wide, particularly as far as one-level laws (e.g. macroscopic) are concerned. In spite of the extension of the applications and the scope of confirmation of a law, we still remain in the boundary zone and must restrict ourselves to the study of non-representative cases for the whole range of the law. It often happens that we unexpectedly encounter boundaries of applicability of a given law by no means after having studied a considerable region from the range of the conjectural applicability of a given law, which had previously been ascribed to it. Often a quantitative law appears to be strictly fulfilled only in this boundary area the examination of which suggested an idea about a given law and for which it had previously been tested with maximum exactness, as great as it was possible at a given stage.

The occurrence of various universal constants, such as h, e, c, $-273°C$, and the like, usually leads to the restriction of the range of some laws, as compared with the range which was ascribed to them in the period in which those constants were not known or in which it was not possible to "correlate" them with the respective laws.

The occurrence of those constants results in the non-existence of whole domains of phenomena that were initially believed to exist (temperatures lower than the absolute zero, velocities greater than the speed of light, portions of action smaller than Planck's constant, electric charges that are not fractions or

---

dimensions of generality of law see Such (1972), Chapter IV: *Wymiary universalności oraz szczeble ogólności praw nauki* [Dimensions of Universality and Levels of Generality of the Laws of Science], pp. 238-361.

multiples of the elementary charge, and the like), the fields to which one or another law was to apply. Moreover, it follows that some laws appear not to be fulfilled even for some real values, when only these values are approximate these theoretical constants. To give an example, laws of classical mechanics, thanks to the existence of the c constant (constant velocity of light) not only do not concern – in spite of the primary assumptions – objects moving with speeds higher than that of light (the hypothetical tachyons) but also objects whose speed is comparable with that of light. This leads to the restriction of the range of the conjectural applicability of many laws, i.e. to the necessity of narrowing down the antecedent and to the formulation of more exact and more general laws which often encompass new levels of reality and constitute new links in correspondential law chains. The fact that those links continuously grow indicates that even a drastically limited range of the initial law, which is the first link of the correspondential sequence, is submitted to further limitation. As knowledge develops, Newton's laws of mechanics appeared to be more and more limited by special and general relativity theory, quantum mechanics, Dirac's theory, and the quantum theory of field. This range has been narrowed down and will certainly be further narrowed down by new general physical theories. In turn, also the conjectural ranges of laws formulated by the ever newer theories – thus of laws which constitute further links of the correspondential chain – undergo limitations from time to time. So, as in the process of testing a law we pass on from its boundary cases to those which represent the ever and newer phenomena which fall under its conjectural range. The latter one is usually narrowed down considerably: a quantitative law usually appears to be strictly fulfilled only in a certain threshold area which, after all, does not have to be identical with the initial area of applications of a law, i.e. a field for the description of which a law was formulated and in which it has been tested (with a certain degree of precision).

The conviction of the old mechanicists about the infallibility and the unrestricted applicability of laws of mechanics encompassing all phenomena and all aspects of phenomena was initially based on the belief that every particular case falling under a given law "represents" the whole scope of its conjectural applicability. And by this it is an evidence of an unlimited universality of this law. So it suffices to test the law in a given area and on a given level of reality to be able to claim its unlimited applicability. It is just what had been thought before Newton. For instance, Descartes thought that the extrapolation of laws established under terrestrial conditions did not require "any additional" testing of them in other areas and on other levels of reality that is, let us say, in the microcosm or under cosmic conditions. But such ideas were also upheld after Newton, although Newton himself thought that extrapolation of laws onto new levels of reality should be based on their previous testing on these new levels. At present, it is obvious that any extrapolation of law onto new fields requires confirmation and that a given law, in its various aspects, is generally true, strict or reliable to an uneven degree since not all these aspects have been tested with the same degree of precision. It follows that the reliability of prognoses or explanations which we obtain on the basis of a given law depends not only on its reliability as such, i.e. reliability "in general," but on the reliability of those of its dimensions (aspects) of generality which can be applied in the case at hand, i.e. under which the phenomenon being predicted or explained falls.

All this explains sufficiently why the new law that compatible with reality and with the hitherto existing law in certain narrow subranges of its applicability appears to be in disagreement with them in other subranges. In its descriptive and also prognostic function it is partly ever-lasting, while explanations which it provides are usually replaced by deeper and more adequate ones. In the sphere of explanations, and also in the descriptive and prognostic spheres, which extend far beyond the framework of earlier experiments, the new law usually reveals considerable deviations from the hitherto existing knowledge, it reveals its "strangeness" in the light of what we have become used to. To put it differently, the practical results to which the laws corresponding with each other lead are – to some extent – approximate, almost the same, whereas the theoretical results, especially those beyond that extent differ significantly from each other.

However, when it is emphasized that in the correspondential chain of laws (which may consist of many laws which are in the relation of correspondence with one another as this relation is transitive), the law being an earlier link of this chain never constitutes, strictly speaking, a particular case (or consequence) of the latter law, then what is meant is not only that between these laws there occurs a certain (generally wide) "scope of incompatibility" but also that in the "scope of compatibility" too the divergence between them actually occurs. However, due to its minuteness it may be practically unimportant or even imperceptible for the measuring technique at our disposal. This is because there is a continuity rather than a sharp borderline between those scopes: one scope gradually turns into another and the decision where the "scope of incompatibility" ends and the "scope of compatibility" begins is to some extent arbitrary and is partly determined by practical reasons such as a degree of exactness of results obtained on the basis of a given law or degree of precision of the experimental technique, required for carrying out of a given goal which we have at our disposal in a given field. At any rate, the practical (observational) similarity of the results of applications (descriptions and prognoses) of laws corresponding with each other in a certain field does not allow to "level out" their theoretical incompatibility in the whole range of their applicability, since they provide competitive explanations of the studied phenomena.

IV. The Principle of Correspondence and Testing Laws and Theories on Various Levels of the Theoretization of Knowledge

The relation of correspondence plays an important role in the process of testing both the hitherto accepted and new theoretical knowledge. The principle of correspondence probably would not be formulated at all by physicists if it had not been for the difficulties in the process of testing new laws and theories through direct confrontation with the experimental material.[11] It was Bohr, the founder of that principle who used it first and, as it were, perceived its great heuristic significance only when he could not count on the direct testing of his new quantum theory of atom and had – as a certain kind of mediator between this theory and experiment – to refer to the classical theory of radiation. It allowed him to establish relationships of the so-called Bohr's quantum postulates with experience as it combined certain quantum magnitudes with classical ones. The classical theory of radiation served as a link mediating in the process of interpretation of Bohr's new theory. Without that "mediator," observational interpretation of the latter was practically impossible, at least at that time.[12]

The same motive occurs many times in the works of Einstein and Heisenberg. It should be emphasized that the latter used the principle masterfully both at the stage of constructing new theories and of their initial testing by means of theoretical confrontation with the hitherto existing theories and, by their mediation, with experience. In the last case, the "empiricization" of the new theories was also meant, giving them observational meaning , which often is simply impossible, especially at the beginning, directly after the formulation of a new theory, without the participation of the previously constructed theory. For instance, when creating the general relativity theory, Einstein aimed first of all at "testing" its equations with Newton's equations (or to their equivalent – Poisson's equations) as to the first approximation or was guided by the principle of correspondence.

---

[11] Although it is probable that in this case it would have been formulated by philosophers of science who deal with the question of development of knowledge since it constitutes one of the principles "of dialectical development of knowledge," principles steering transformations in knowledge which occur most visibly on the grounds of physics.

[12] Cf. Bohr (1913, pp. 1-25), (1923) and (1920, pp. 423-469). In this connection, it is worth emphasizing that the etymological kinship of such expressions as "the principle of correspondence," or "the relation of correspondence" on the one hand, and "the rules of correspondence" (co-ordination, subordination) on the other is not accidental. The principle of correspondence may be, and actually was, interpreted by Bohr as a meta-statement, indicating in what way theoretical terms of a new quantum theory (e.g. Bohr's atomic theory) may be connected with observational terms of the hitherto existing theory, e.g. the classical theory of radiation. See Bohr (1924) and (1934). See Niedźwiedzki (1974), Part II, Chapter 1.

The indication that in certain simple (approximate) applications general relativity theory leads to the same results as Newtonian theory, constituted a confirmation of the former. It was only later on – when in a certain range of its applicability it appeared compatible as a good approximation with Newtonian theory – that scientists began to try to confirm it experimentally on a new empirical material, which could not fit in theoretical schemes developed by Newton. Also, without establishing the fact that the general relativity theory corresponds with Newtonian theory of gravitation, Einstein would not have been able (at least initially, i.e. immediately after having formulated his theory) to combine it with experience, giving some of its terms an observational sense.

The problem of the formulation of matrix mechanics by Heisenberg is a similar one. In the process of constructing the theory, Heisenberg consciously applied Bohr's principle of correspondence, approaching classical mechanics as a model and at the same time as an (approximate) case of quantum mechanics, and interpreting the latter as formally and physically analogous to classical mechanics.

Physical theory usually goes through "the baptism of fire" in theoretical confrontation with the hitherto existing theories. Only after having successfully gone through the stage of confrontation with the knowledge accumulated in a given field is a new theory (a candidate for a theory) treated seriously and it is recognized as worth experimental testing by experiments conducted especially to achieve this aim. However, if the initial theoretical confrontation is usually conducted, just on the basis of the principle of correspondence, by the creator of a new theory him/herself, then confrontation of the theory with experiment is already, at the present stage of development of physics, the job of other physicists, i.e. experimenters.

The area of experimental confrontation usually depends, at least at first, on the scope of theoretical confrontation, particularly on the scope in which the new theory and the previous one lead to experimental results that are incompatible but at the same time accessible to the existing measuring and experimental technique, and which can be compared with one another.

That is why the initial testing of the general relativity theory through observation was restricted to certain effects induced by local fields (which are the fields of separate stars or the field of the solar system taken as a whole in contradistinction to the fields of galaxies or clusters of galaxies) as only these effects which "differentiate" both theories were available for observation at that time.

However, if what is important in the theoretical confrontation, conducted by means of the rule of correspondence, is mainly showing that within the limits in (and the precision with) which the hitherto existing theory was confirmed in experiment, the results of the new theory are in line with the results of the former, then what is meant in experimental confrontation is a demonstration of the prevalence of the new theory wherever the old theory does not give quite satisfying results and wherever the new theory is in disagreement with it. That is just why scientists who tested experimentally the general relativity theory focused their attention on those astronomical phenomena[13] as to which relativity theory and the Newtonian gravity theory differed essentially in their prediction.

All this explains the deduction of the Newtonian theory from the general relativity theory (when certain simplifying conditions are adopted), which was confirmed by those phenomena. It also explains Einstein's establishing that there is a relation of correspondence between them. It simultaneously confirmed and refuted the Newtonian theory: it confirmed it as approximation, finding many deviations from it, just as deducing Kepler's laws by Newton from the tested (but simplified) law of gravitation and the laws of mechanics, and establishing the relation of correspondence between them, both confirmed and rejected Kepler's laws in the same sense. In my view, this is the case wherever the principle of correspondence is applied: the corresponding theory (or law) constitutes a confirmation of the corresponded theory (law) but not within the boundaries of its whole conjectural (i.e. initially assumed) range of applicability but only in certain narrow range in which results of both theories are to a great extent in agreement with each other, albeit never completely.

It is also worth noting that in the process of testing new knowledge, in some cases at least, what may be involved is the problem of correspondence of new knowledge not merely with the previous knowledge gained in a given field but also with knowledge accumulated in similar fields and even the common knowledge.[14] For instance, the metrics of space and time on the level of microworld cannot be completely optional but must in some way be approximate to the metrics on the level

---

[13] What was meant here were the astronomical observations, because under the "earthly conditions" the general relativity theory could not have been tested until the 1960s when the so-called Mössbauer effect was used for showing dilatation (slowing down) of time in the gravitational field of the Earth, as its only observational consequences setting it apart from the Newtonian theory while being accessible to measurement technology were some astronomical phenomena.

[14] The question of the correspondence of scientific knowledge with common knowledge has been taken up by M. Popowicz with reference to mathematics and logic. Among other things, he writes: "It is not so that the founders of formalized systems may freely choose optional representations of truthfulness and provability. For every theory of proof, as for the formal theory of truth or for a formal theory of sense and meaning, a specific restriction is imposed: in a certain boundary case such a theory should provide results that are completely compatible with the "naive" point of view of common sense on the provability, truthfulness, non-contradictoriness and the like" (Popowicz 1969, pp. 104-105).

of the macroworld. That is why the newly formulated theories of microphysics (quantum mechanics including) must be in the relation of correspondence to the relevant theories of macrophysics (including classical and relativistic mechanics, classical electrodynamics and classical thermodynamics).

However, the principle of correspondence does not only serve to test new theoretical knowledge. It imposes certain conditions on the methods of testing (especially on falsification) of the hitherto existing scientific laws and theories. A well confirmed law or theory even to a narrow (boundary) range, which in no way is in line with its conjectured range of applicability, cannot then be totally rejected by the new experimental data or theoretical facts. For their proper evaluation, especially for the assessment of the scope of their real applicability one should have access to a new, more exact and more general knowledge, which is only provided by laws and theories which correspond with them. The latter ones are characterized by at least the same degree of theoretization and by a higher degree of informational content (determined by the level of generality and exactness of knowledge) than the laws and theories being falsified. It can be seen from the above that testing laws and theories does not simply consist in the confrontation of their observational consequences (i.e. predictions, prognostic sentences) with experimental data, without confronting their consequences with the whole remaining science, including theoretical confrontation. The confrontation occurs not only on all the levels of generality and theoretization lower than the degree represented by a law (theory) which is being tested, but also on the same level which that law or theory represents, and sometimes even on a yet higher level. The latter case occurs when a law (theory) which corresponds with the law (theory) being tested (rejected) represents a higher level of theoretization, or when such a higher level of theoretization is represented by a law (theory) from which a corresponding law (theory) is derived, and which is always more abstract than the latter, i.e. it contains more idealizing assumptions in its precedent.[15]

However, why is it that the refutation of a given law (theory) which has been tested to some extent (even if only one-sidedly, i.e. on a narrow empirical basis which is not representative of the whole range of its applicability) requires a corresponding law (theory) of higher informational (logical and empirical) content? First of all, it is necessary in order to show that experimental anomalies which are *prima facie* in disagreement with a given law (theory) are actually counter-cases that

---

[15] See the next paragraph for more abstract laws of that type, describing ideal types of the higher order, the laws from which (together with appropriate principles of co-ordination) are laws derived deductively (of a lower level of abstraction) corresponding with the hitherto existing laws.

falsify that law (theory), and then in order to assess to what degree they falsify the law that has been rejected, i.e. to delimit the scope of its falsification. As it is, empirical facts (including experimental anomalies) may be differently interpreted and it can be shown that the correct interpretation is the one which makes them incompatible with the knowledge being tested, i.e. makes them merely counterexamples.

That is why, for example, an experimentally established fact of the dependence of the mass of an electron on its movement did not by the same token ("automatically," as it were) reject the Newtonian classical mechanics (based on the assumption that the mass of bodies is constant, independent on movement). It was Lorentz who supplied an interpretation of that fact (and of a number of other new facts) under which it was possible to "reconcile" it with classical mechanics. Only the new interpretation of facts of that type given by Einstein in his special relativity theory revealed their genuine incompatibility with classical mechanics and determined the range of that incompatibility. It became possible because the new theory established certain new universal dependencies which had not been taken into consideration by the hitherto existing theory (among other things, the functional dependence of the mass of all objects on their relative velocity and the general proportional relationship between mass and energy).

However, not only for theoretical reasons, in physics we renounce a previous law (theory) only when we establish an appropriate corresponding law (theory) of the higher informational content. Also, practice requires that a given law (theory), tested to some extent (thus relatively true) be used in the capacity of a theoretical instrument of practical activity at least until we obtain some better theoretical instrument of efficient activity in a given field.

Does that mean that in physics (or, broadly speaking, in physical sciences) we do not know "purely" experimental cases of rejecting a law or a theory (a purely experimental one, that is the rejection on the basis of experimental facts alone, without resorting to a theory)? It seems that such cases do exist and that, at any rate, there are possible cases of rejecting a statement considered to be a law without establishing a law that would correspond to it.

However, I believe that the above is restricted exclusively to situations whereby a given law has been completely falsified, and therefore it turns out to be a statement which had been mistakenly considered to be a law of science (since it does not reflect, even in approximation, any regularity of nature). It seems that astronomers have encountered such a case in connection with the so-called Titius-Bode (or Bode's) law. It was published in 1772 by J. Bode, who – while studying distances from the Sun of the then known planets – noticed that if one adds 4 to each number in a simple set of 0, 3, 6, 12, 24, 48, 96..., each of which is twice as large as the preceding one, then in the set thus obtained (4, 7, 10, 16, 28, 52, 100...) each subsequent element of the set (except 28, which can however be attributed to the hypothetical planet under which the planetoids circulating between the orbit of Mars and that of Jupiter can be subsumed as a single entity) corresponds quite exactly to the relative distance of one of the planets from the Sun. That "law" was initially confirmed by a new fact since the distance of the next planet (Uranus) to be discovered after the law had been formulated appeared to be compatible with the next number (196) of Bode's sequence. However, for Neptune the next number of the sequence (388) appeared to be much too large[16] (whereby it is the number 300 that approximately corresponds to that planet). Since none of the numbers of Bode's sequence corresponds to Neptune, Bode's law has been rejected. This seems to be an example of completely rejecting a law without establishing any better law for the aspect under study (the location of planets in this case).

Was that law rejected on the basis of "experimental facts alone," without reference to a more general and more precise law which might have replaced it? I think that there should be a positive answer to this question. So far, we do not know a law which might be recognized as "a correspodential generalization" of Bode's "law." Such cases of "purely" experimental refutation of a law occur, as it seems, only when a given "law" appears to be an apparent law (describing a certain accidental regularity and recurrence and not the regularity of nature) and as a law is submitted to total falsification (i.e. rejection in the whole range of its presumed applicability). One can conclude from the above that no real law can be falsified in a purely experimental manner.

After all, even in the process of the falsifying Bode's law, theory played a definite, albeit indirect, role. What is meant here is the Newtonian theory of gravitation, without which neither discovering new planets nor establishing their distance from one another would be possible at that time. However, this was the case of an indirect role in the sense that theory served not to occupy "an empty" place of a law that has been refuted but to establish facts which helped to refute it. After all, some scientists have doubts to this day whether Bode's law has really been falsified in its whole range, and whether it has been definitely refuted. The thing is that it displayed certain (however insignificant) heuristic productiveness and in this connection some scientists hope that in future it can prove to be an approximation of a more precise (although mathematically more complicated) law. Let us now describe in greater detail what the relation of correspondence between laws consists in.

---

[16] But it is conformable approximately with the distance of Pluto.

V. What Does the Relation of Correspondence Between the Laws of Physics Consist in?

In her unpublished Ph.D. dissertation, *Zasada korespondencji w fizyce* [The Principle of Correspondence in Physics], I. Nowakowa showed that "implicational" conceptions of correspondence such as all the other hitherto developed versions of the principle of correspondence are inadequate.[17] Their inadequacy results mainly from the misunderstanding of the idealizational character of scientific laws and theories. No wonder then that they dominated in the period when the understanding of the idealizational character of laws and theories in physics was insufficient. Implicational conceptions came into prominence particularly at the time when positivist trends dominated in the methodology of physics. Those trends proclaimed extreme cumulativism, according to which the subsequently formulated statements, laws and theories are in agreement (not in contradiction) with the previous ones.

It is my opinion that I. Nowakowa's dissertation contains a description which, in its general outline, is a proper description of the relation of correspondence, i.e. a relation obtaining between laws or theories the development of which is guided by the principle of correspondence. The description is based just on such a "basic intuition": " the later theory corresponds with the former theory when it contains more idealizing assumptions than the previous one and when it is submitted to concretization because of those assumptions" (Nowakowa 1972, pp. 47-48).

I. Nowakowa elaborates on that intuition when applied to laws as follows:

> The latter law *T'* corresponds to the former law *T* if it is more abstract than *T* (i.e. contains more idealizing assumptions) and when statement *T'* may be concretized with respect to the newly added idealizing assumptions, obtaining statement *T"* which is equally abstract (i.e. has the same number of idealizing assumptions) as *T* but explains a larger area of observational sentences than T does, although with no less degree of accuracy (p. 49).

That intuition has been explicated in the thesis. According to that explication, "the general definition of the concept of correspondence of laws is as follows: the idealizational law *T'* corresponds dialectically with the idealizational law *T* then and only then when:

1) law *T* is formally equivalent to some logical consequence *T"* of law *T'* (in a particular case – with law *T'* itself);[18]

---

[17] Cf. Nowakowa (1972). Nowakowa develops her standpoint on the question of the principle of correspondence in her works (1975a, 1975b and 1994).

[18] In the work discussed here we find the following definition of the concept of formal equivalence of statements. Two physical statements (especially idealizational laws):

1) $A_1(x) \wedge \ldots \wedge A_n(x) \rightarrow B(x) = f[C_1(x), \ldots, C_k(x)]$

2) $D_1(x) \wedge \ldots \wedge D_m(x) \rightarrow E(x) = f[F_1(x), \ldots, F_k(x)]$

2) realistic assumptions of law T' are equal to some realistic assumptions of law T;

3) a sequence of idealizing assumptions of law T is a subsequence proper of the sequence of idealizing assumptions of law T';

4) concretization T'' of law T' of the same number of idealizing assumptions may be derived as those of law T (pp. 50-51).

According to her, the relation of dialectical correspondence in her version holds between the law of the fall of bodies by Galileo and the analogous law of falling in Newtonian classical mechanics. The first one has the following form:

(1) $[ob(x) \wedge s(x) \wedge h(x) \ll R_e \wedge V_0(x) = 0 \wedge g = const \wedge F_{tr} = 0] \rightarrow$
   $\rightarrow S(x) = gt^2(x)/2$

where h is a height from which the body falls, t is the time of fall, R_e is a radius of the Earth, V_0 is the starting velocity, F_{tr} is the force of resistance of the environment, S is the way the body travelled, g is gravitational acceleration, ob stands for the predicate "... is a body," and s stands for the predicate "... is falling."

The above law is based on the four following assumptions, of which only one (the fourth one) is an idealizing one according to Galileo:

1) the body x falls in the vicinity of the Earth;
2) the starting velocity of the body x is equal to zero;
3) the gravitational force of the Earth (g) is constant;
4) the forces of the environment's resistance acting upon the body x are equal to zero.

The second law is the following one:

(2) $\{ob(x) \wedge s(x) \wedge V_0(x) = 0 \wedge F_{tr}(x) = 0 \wedge F_{gr}(x) = 0 \wedge F_{el}(x) = 0 \wedge$
   $\wedge V_e = const \wedge P_e = const \wedge W_e = 4\pi R^3/3 \wedge we = 0 \wedge W(x) = 0 \wedge$
   $\wedge g = const \wedge gx = 0\} \rightarrow S(x) = gt^2(x)/2.$

It is based on the following assumptions of which all but the first three are idealizing in nature within classical mechanics:

1) x is a physical object (body): $ob(x)$;
2) x falls in the direction of the Earth: $s(x)$;

---

are *formally equivalent* with respect to the mathematical formula $X = f(Y_1, \ldots, Y_k)$ if and only if:

1. variable "x" in the expression (1) runs through a set of physical objects of a U-type, and in expression (2) a set of physical objects of a W-type,

2. $A_1, \ldots, A_n, B, C_1, \ldots, C_k$ are physical magnitudes defined on the set U; $D_1, \ldots, D_n$, $E, F_1, \ldots, F_k$ are physical magnitudes defined on the set W,

3. variables $X, Y_1, \ldots, Y_k$ range over the sets of numbers which constitute respectively the values for physical magnitudes $B, C_1, \ldots, C_k$ and for physical magnitudes $E, F_1, \ldots, F_k$ (Nowakowa 1972, pp. 49-50).

3) the starting velocity of the body is equal to zero: $V_0(x) = 0$;
4) the resistance of the environment active on the body x is equal to zero: $F_{tr} = 0$;
5) the non-earthly gravitational forces acting upon the body x are equal to zero: $F_{gr} = 0$;
6) the forces of any other nature acting upon the body are equal to zero: $F_{el} = 0$;
7) the Earth is inert (its vector of velocity is constant): $V_e = const.$;
8) the density of the Earth is constant: $P_e = const.$;
9) the Earth is an ideal sphere: $W_e = 4\pi R^3/3$;
10) the Earth does not revolve, its angular velocity is zero: $w_e = 0$;
11) the volume of the body is equal to zero: $W(x) = 0$;
12) the acceleration imposed on the Earth by the falling body is equal to zero: $g(x) = 0$;
13) the gravitation of Earth is constant: $g = const.$;

According to I. Nowakowa, correspondence holds between these laws because:

1) the laws (2) and (1) are equivalent, formally speaking, because of the common mathematical formula assumed by their consequents: $A = BC^2/2$, where B is a constant function, and A and C are the functions with changeable values;
2) the realistic assumptions of the laws (2) and (1) are equipollent, since all the realistic assumptions of the law (2) have their equivalents between the realistic assumptions of the law (1);
3) the sequence of idealizing assumptions of the law (1) is the subsequence proper of a sequence of the idealizing assumptions for the law (2) since the only idealizing assumption of the law (1), i.e. the assumption $F_{tr} = 0$ is one of the 10 idealizing assumptions of the law (2);
4) the law (2) can be concretized because of the newly added idealizing assumption 5.-13. which gives us a law that is as abstract as law (1).

I think that the concept of dialectical correspondence as presented above is correct in its outline and makes for the best reconstruction of the idea of correspondence, which plays such an important part in contemporary physics. At the same time it seems that the concept requires a few slight modifications. It does not, for instance, take into account some important intuitions of the contemporary physicists. The basic intuition, according to the author, is that the later theory corresponds to the former one, if it is the former's idealization, and what is more, the later theory is being concretized because of the new idealizing assumptions (Nowakowa 1972, p. 46). The above-mentioned intuition has prompted the author to grant the status of corresponding law only to those laws which are the same in consequents and differ in precedents alone: as the first one has less idealizing assumptions than the second one (strictly speaking, a set of idealizing assumptions of the first law, which can be empty, is always a proper subset of the set of idealizing assumptions of the corresponding law, which is never an empty one). Therefore, for example, the Newtonian law of the falling bodies, corresponding to the law of falling as given by Galileo differs only in the (greater) number of idealizing assumptions. In Mejbaum's terms (cf. Mejbaum 1964) these are not even different laws but different formulae of the same law (as the laws differ only in consequents, conditional clauses which differ only in precedents can at most pass for different formulae of the same law).

The basic question is thus as follows: what are the elements that enter the relation of correspondence and in what respect are they different?

In the physical and methodological literature on the subject, the following objects are usually mentioned as the elements of the relation of correspondence: a) pairs of concepts, b) pairs of laws, c) pairs of physical dependencies, and finally, d) pairs of theories. It seems that one of the basic intuitions of the physicists as far as the relation of correspondence is concerned is that the difference occurs not only between the precedents but also between the consequents of the laws which enter the relation in question. The difference has also to occur between the systematised sets of such laws, i.e. between theories. That is why the author's favourite intuition does not seem to be the physicist's one, though some methodologists are tempted to embrace it.

If a corresponding T' law is said to result from the abstractization of the first law T (idealization i.e. acceptance of the new idealizing assumptions) provided that a concretization is achieved from T' by abolishing the new idealizing assumptions and introducing the respective corrections to the consequent of the law in order to account for the factors which were abstracted from when idealizing assumptions were made. If so, then a relation of correspondence holds between T and T'' which differ not only in a precedent but also in a consequent. Coming back to the example in question, the relation of correspondence holds not between Galileo's law of falling and the law which differs from it in a number of idealizing assumptions (precedent alone) but between the former law and the law which is one of the concretizations of the latter, thus not between the law of a consequent $S = gt^2/2$ but, for instance, a law with a Newtonian correction taking into account the dependence of the earthly acceleration g

$$S = \frac{G\,M_e\,t^2}{2\,R^2}$$

on the distance from the centre of the Earth:
according to the formula $g = GM_e/R^2$, where G is a universal constant of gravitation, M_e is a mass of the Earth, and R is a distance of the falling body from the centre of the Earth. Similarly, as another example, the

relation of correspondence holds not between Newton's second principle
of dynamics and its abstractization which differs only in a precedent but
between the principle with a $F = ma$ consequent and a respective relativist

$$F = \frac{ma}{\sqrt{1 - \dfrac{v^2}{c^2}}}$$

principle with the following consequent:

What is the difference between the relation of correspondence thus
understood and a relation of dialectical correspondence as understood by I.
Nowakowa? The difference is that the first one is a logical product (a
relative one) of two relations: the relation of dialectical correspondence
and the relation of concretization (if the relation of dialectical
correspondence has to be termed R, and the relation of concretization K,
then the relation of correspondence S understood so will be a relative
product $R \times K$, or $S = R \times K$).

It is interesting to compare I. Nowakowa's and W. Mejbaum's
concepts in this respect. If Mejbaum (1964) does not define the very
relation of correspondence properly (among other reasons, the failure to
give appropriate room for the idealizational character of the laws of
physics can be blamed for it) but does point out the elements of the
relation, I. Nowakowa does the opposite. In principle, she characterizes
properly the relation of correspondence and fails to name its elements.
Thus, for instance, according to Mejbaum the relation of correspondence
holds between the following statements (strictly speaking, dependencies),
of which the first one is Ohm's law (strictly speaking, its consequent):

(1) $E = RI$

(2) $E = RI + LdI/dt$

(3) $E = RI + dI/dt + Q/C$

where E is an electromotive force, R is resistance, I is the intensity of the
current, C is the condenser's capacity, Q is the condenser's charge
(Nowakowa 1972, p. 46).

I. Nowakowa shows that a relation that in fact holds between these
laws (if they are supplied with their precedents in a logical way) is not a
relation of correspondence in Mejbaum's sense but rather a relation of
idealization and concretization (strictly speaking, the relative product of
those relations). The relation holding between the above-mentioned
relations, which Mejbaum improperly describes but properly notices as a
correspondence relation, does in fact merit the name. However, it should
be adequately described, which I. Nowakowa does, although give a
different relation the name of a dialectical relation of correspondence. It is
a relation composed of (i.e. being a logical product) the relation of

idealization (terms relation of dialectical correspondence in I. Nowakowa's) and the relation of strict concretization.

In order to "pass" from the given law (theory) to the law (theory) which corresponds to it, the following steps must be undertaken: (1) the basic law must be made more abstract, i.e. one or more idealizing assumptions must be accepted, abstracting from the factor influencing the relation described by law's consequent, and (2) concretization must be performed in order to abolish that (those) assumption(s) and introducing the correction to the consequent, whereby the influence of the given factor on the regularity in question is taken into account.

In order to perform such a transition, wherewith a more perfect law or theory is established, two scientific discoveries must be made: (1) a qualitative one, when one of the side factors is discovered, and then abstracted from if a respective idealizing assumption is accepted (the discovery itself then consists of a proper idealization) and (2) a quantitative one, when the influence of that factor upon the regularity in question is accounted for (the discovery consists in conducting a proper concretization). Without a proper idealization (an elimination of a really functioning factor in a purely intellectual way) and a proper concretization (accurate estimation of an influence which the factor has upon a regularity in question) the law which has just been obtained will not be a more perfect description of the regularity investigated.

If none of the two mistakes which are possible here occurs, what we obtain is a more perfect law or a more precise theory which is a "corresponding" generalization of a law or theory accepted so far, and which itself is a link in the correspondential chain of subsequent laws or theories in a given field.

A question arises which of the two approaches is better. In my view, it is a modification (a slight one from the formal point of view) of I. Nowakowa's approach. The modification which plays an essential role as far as the central aspects of the issue are concerned.

This is a result of the different emphasis on the intuitions considered, the way physicists speak, and different views on the role of the stages of idealization and the discoveries which accompany them.

For I. Nowakowa, a principal stage, at least when a logical reconstruction of the physicist's work is undertaken, consists of a formulation of a new law T' by making the basic law T more abstract. For me, the essential part consists not of formulating a more abstract law T' but of a discovery of a "concretization formula" (the co-ordination principle in L. Nowak's terms) and of forming a concretized law T'' on its basis. In other words, an essential discovery in physics is not a discovery of a new factor which influences the dependence (regularity) in question (a qualitative one) but the estimation of this influence. This allows forming a new physical theory which is always quantitative in nature. From this point of view, it was less important that g (earthly acceleration) has been discovered to be a variable magnitude than a discovery of a function which described the changes of this magnitude. The first discovery could be of an experimental kind, whereas the second one was possible only due to the new theory (the Newtonian theory of gravitation). Likewise, the discovery that the mass of an electron depends on its velocity that the special theory of relativity has made, while helpful as far as the forming of a new theory was considered, did not achieve a breakthrough which would consist in forming a new theory. Instead, the breakthrough consisted in formulating a few functional regularities, especially in a relativistic formula that the mass of a moving body depends on its velocity. Therefore, concretization rather than idealization is a basic stage of constructing the new theory. This is the case at least from the point of view of physics, i.e. of the physicist who makes the discoveries.

The above explanation of what the physicist does is close to his/her own intuitions but can be opposed by a methodologist who is logically reconstructing the physicist's procedures. Such a reconstruction shows that the concretized law T'' is a logical inference from the more abstract law T' (and the respective principle of co-ordination) which cannot be deduced from the basic law T. It is here that the methodologist (who logically reconstructs the physicist's procedures) can be different from the physicist (who is busy making physical discoveries) as far as the estimation of the value of those two stages (discoveries) is concerned.

This influences the terminological decisions, especially as far as naming the relations considered is concerned (which is to be called a relation of correspondence). The argument also involves another one, which touches upon a question of the aims of the methodologist and the theoretician of science, his/her "privileges," and how autonomous his/her logical reconstruction of scientific procedures can be in relation to what a scientist really thinks and does. This is at the same time an argument about the status of methodology as a science.

While rejecting a concept of the so-called phenomenological methodology (methodological phenomenalism) which is a trivial and down-to-earth description of what a scientist does, I opt for a reconstructing methodology (cf. Kmita 1972), while thinking that, assuming a relation of correspondence holds between the laws of the same consequent, we are too far away from the physicists' intuitions and their own comments on what they are doing.

On the other hand, I would like to stress that a transition from the concept of the principle of correspondence as presented by I. Nowakowa to the principle of correspondence which I prefer can be achieved without any great effort and it actually means a "change of language" and a slight modification of the conceptual framework presented above.

VI. *Correspondence and Logical Consequence*

To conclude, let us ponder on what the problem logical consequence of the corresponded law from the corresponding law looks like in the above explication of the relation of correspondence. Let us return to the example of Ohm's law and the laws which correspond with it. Ohm's law reads as follows:

$$(1) \quad P(x) \rightarrow E(x) = R(x)I(x).$$

It is a factual law since only one assumption occurs in its precedent and this is an assumption of a realistic (non-idealizing) nature P(x), whereby P is a predicate "the electric conductor in which (electric) current flows."

The law which corresponds directly with law (1) will be obtained thanks to two discoveries:

1. the "qualitative" discovery, consisting in showing that law (1) is false since the dependence between tension *E*, current intensity *I* and resistance *R* described by this law and applied to electric current in the conductor is influenced by the (finite) capacity of the condenser switched into the circuit. Leaving that influence out of consideration, i.e. assuming the idealizing assumption that the capacity of the condenser switched into the circuit is infinite, $C(x) = \infty$, we obtain an idealizational law (of the first degree) which is as follows:

$$(2) \quad P(x) \wedge C(x) = \infty \rightarrow E(x) = I(x)R(x); \text{ and}$$

2. the "quantitative" discovery, consisting in evaluating the magnitude of influence exerted on the dependence described in Ohm's law by the restricted capacity of the condenser as a result of which we obtain an appropriate principle of co-ordination (or, to put it differently, of strict concretization).

From the idealizational law (2) and the principle of co-ordination, we obtain just a factual law:

$$(3) \quad P(x) \wedge C(x) < \infty \rightarrow E(x) = I(x)R(x) + Q(x)/C(x)$$

being in the (direct) relation of correspondence to the factual law (1), or Ohm's law.[19]

Laws (1) and (3), just as all laws which correspond with each other, are at the same level of abstraction, i.e. they contain the same number of idealizing assumptions. In this case, these are factual laws (of the zero degree of idealization) which only possess realistic assumptions.

---

[19] The procedure of idealization and concretization as applied for various kinds of laws and theories is described in detail by L. Nowak in his works. See e.g. Nowak (1971).

The relation of correspondence occurring between them is not a relation of logical consequence, nor is it (which is mostly meant here) its reverse: after all, law (1) does not follow from law (3), neither is it the other way round. Logical consequence which occurs here is the consequence of law (3) from law (2) and an appropriate principle of concretization. On the other hand, the consequence of law (2) from law (1) does not occur. This means that law (1) can be neither directly nor indirectly derived (i.e. neither directly nor indirectly results) from law (3).

Moreover, one can speak of "approximate" or "pragmatic" consequence of the corresponded law (1) from the corresponding law (3), in the sense that if we leave out the last part, $Q(x)/C(x)$, of law (3), i.e. we adopt the "simplifying condition" that $C(x) = \infty$ (or, which boils down to the same thing, that $Q(x) = 0$) then the consequent of law (3) is transferred (or reduced) into a consequent of law (1).

Things are similar in the case of resulting between further links of the correspondential sequence of laws whose first link is Ohm's law.

When also law (3), and consequently law (2) too, appeared to be false with respect to the rise of electromotive force $e(x)$ in the circuit, a new corresponding law was formulated, formally speaking, in the same indirect manner. Therefore, it again appeared indispensable to take two steps: first on the basis of the following idealizing assumption: in circuit x in which (electric) current flows, electromotive force does not arise: $e(x) = 0$, an idealizational law was formulated:

$$(4) \quad P(x) \wedge C(x) = \infty \wedge e(x) = 0 \rightarrow E(x) = I(x)R(x);$$

second, assuming proper principles of strict concretization from law (4) a factual law was obtained through its strict concretization conducted with respect to both idealizing assumptions

$$(5)\ P(x) \wedge C(x) < \infty \wedge e(x) \neq 0 \rightarrow E(x) = I(x)R(x) + e(x) + Q(x)/C(x).$$

This law corresponds directly with law (3).

When also law (5) appeared to be false, proving indirectly the falseness of law (4), the next corresponding law was formulated in the same way from the point of view of logic.

The first step was again the adoption of an idealizing assumption, this time one stating that the coefficient of auto-induction of circuit x is zero: $L(x) = 0$, thus the assumptions "lifting" with the influence of inductiveness on the dependence studied. As a result, an idealizational law of the following form was obtained:

$$(6) \quad P(x) \wedge C(x) = \infty \wedge e(x) = 0 \wedge L(x) = 0 \rightarrow E(x) = I(x)P(x).$$

With respect to all three idealizing assumption the strict concretization led, in the second step, to a corresponding law of the following form:

(7) $P(x) \wedge C(x) < \infty \wedge e(x) \neq 0 \wedge L(x) > 0 \rightarrow E(x) = I(x)R(x) +$
$+ I(x)dI(x)/dt + e(x) + Q(x)/C(x).$[20]

Factual laws (1), (3), (5) and (7) form subsequent links of the correspondential sequence, describing "Ohm's dependence" more and more exactly. No earlier link of the sequence results from the next one, although we can speak here (not strictly) about "pragmatic consequence," "approximate consequence" or "obtaining" a law being the former link of a sequence from a law being (an optional) latter link of the same sequence. "The logical mediator" between the corresponded and the corresponding law is always an idealizational law of a higher (at least by one level) degree of theoretization from which the corresponding law results.

Obviously, the laws remaining in the relation of correspondence understood in such a way must not be factual laws. Also idealizational laws form corresppondential sequences of laws, and analogous logical relations occur between those laws. However, if the explication of the relation of correspondence supplied here is correct, the correspondential sequences of laws contain homogeneous laws because of the degree of idealization (abstraction): any given correspondential sequence of laws includes exclusively laws of the same degree of idealization, which are in the relation of correspondence to one another. Thus, we have correspondential sequences of laws of the zero degree of abstraction (factual laws), of the first, second and of the ever higher levels of abstraction (idealization).

I think that the same concerns also sequences of correspondential theories. However, in this case, new complicated problems arise which must be solved.

---

[20] All the "transformations" and "modifications" of Ohm's law are precisely described in I. Nowakowa's work discussed above, pp. 65-70.

EXPERIMENT AND SCIENCE*

1. The apparently obvious claim that modern (as distinct from ancient and medieval) sciences have become experimental sciences thanks to the experimental method is in fact misleading, since only some of the empirical sciences have been able to apply experimental methods in their areas of investigation. Considerable areas of scientific study, including empirical ones, have been doing well to this day without experimental and laboratory techniques, which demonstrates that experiment caused a split of the modern science into two camps, whose stand on experiment is different. Obviously, this does not mean that experiment has exerted influence only on the "part" of science in which it has been applied. It changed the picture of the whole science, defining anew, as it were, its status as well as its goals and social functions, mainly through bringing science (especially some of its fields) closer and making it more similar to industry and other branches of material production, in particular to those that also apply (technical, industrial) experimentation. It can be argued that it is just thanks to experimental studies that modern science, and in particular contemporary science, has achieved the high rank and social importance that it enjoys at present.

2. One should also take into consideration the fact that, thanks to experimentation, the development of certain sciences has been accelerated, which caused the experimental natural sciences to come to the fore of scientific experimental studies and, in consequence, become a model of scientific procedures. Apart from that, experimentation surely contributed significantly to the increase in the role of mathematics, measurement and quantitative studies in science, which was reflected in the fact that experimental natural sciences became as a matter of fact identical with mathematical natural sciences. It is thus to experiment (and to the measurement) that we owe further (because it occurred as early as antiquity and the Middle Ages) approximation of mathematics and empirical sciences. In tune with the increase in the role of mathematics in empirical sciences, caused partly by experiment, the role of modelling and idealization in modern and contemporary science increased, which means that experimentation significantly contributed to the modification and

* Translation of "Eksperyment i nauka," published in D. Sobczyńska, E. Zielonacka-Lis and J. Szymański (eds.) (1995), *Teoria – Technika – Eksperyment* [Theory – Technique – Experiment]. Poznań: Wydawnictwo Naukowe IF UAM, pp. 41-46.

development of the methods of scientific cognition not only in their experimental dimension but also in the theoretical one. This is even more justified that the role of experiment is to a certain extent analogous to the role of idealization: both allow the consideration of the phenomena under study in their "pure" form, deformed to the least possible extent by external, random influences.[1]

3. As far as the aims of scientific cognition are concerned, it is obvious that experimentation fundamentally influenced social functions (also called "external goals") of mathematical, natural and technical sciences, causing those sciences to begin carrying out – to use Habermas' term – "the technical interest guiding knowledge." It is a matter of discussion to what degree experiment also contributed to the change of the so-called internal goals of science, cognitive aims delimiting the character of truth at which science aims. I am inclined to accept the conviction that (as observed already by F. Bacon) the inclination of modern and contemporary science to transform reality, also influenced significantly the nature of scientific knowledge, and not just the increase in its social role. Experimentation also strengthened the kind of rationality which is commonly ascribed to science (especially to the complex of mathematical, natural and technical sciences), and which consists in determining efficient means of realization of goals, in contradistinction to the "axiological rationality," which aims at the setting of the goals based on merits.

4. Emphasizing the importance of experiment in the formation of both the picture and the place and role of science in modern times we cannot help posing a question of whether all those changes deserve positive evaluation. The question is a part of a large issue, as it concerns the evaluation of the role of technology in science and the significance of industry (and even of material production in general) for transformations which occur in science. On the one hand, an advocate of the conception of "pure cognition" (cognition for its own sake) may think that the combination of science and technology and production, which is the basis of interpretation of science

---

[1] It was already Bacon who perceived that harnessing the experiment for the purposes of scientific investigation leads to the modification of methods co-ordinating theories with facts. However, according to Bacon, the main modification consists in replacing induction by simple enumeration (enumerative induction) – eliminative induction. The first one ("the induction of the ancients" as it was later called by Mill) was based on observation and common experience and, according to Bacon, did not protect against mental speculation. The second one, based on measurement and scientific experiment, on "artificial creation" of facts constituted – according to Bacon – a considerably "more powerful method" of studies. Its power was to consist in the fact that it is a method of eliminating false hypotheses, and not only of confirming hypotheses, and in that it is a method that is to a large extent quantitative, using also mathematics, while enumerative induction was only a qualitative method of confirming hypotheses.

as a component of the "productive powers" of modern society, undermined the "internal logic" of the development of science and deformed its goals so much that science lost its identity. On the other hand, however, science never – even in antiquity, which castigated practical occupations of free people – has been completely devoid of its practical functions, in spite of the fact that science as such arose in Greek antiquity, at the time when cognition obtained certain autonomy with respect to practical needs, thanks to which it rose to the rank of theoretical cognition. If one agreed with the opinion that science, in its whole history, has performed in human culture "the function of combining in a rational whole the practical knowledge with 'cosmology', episteme and tekne" (Amsterdamski 1973, p. 68)[2], then we cannot speak of the loss by the modern and contemporary science of its identity. This is even more obvious if we compare the "scientific attitude" with the so-called traditional attitude, characteristic of the so-called pre–scientific cultures (e.g. African ones). As R. Horton writes, people who think in a traditional way assiduously try to erase the passing of time, while scientists madly try to accelerate it, and their strong attachment to experimental methods makes them try to create ever new situations which nature, when left to itself, would create more slowly, or would not create at all (Horton 1992, p. 447).[3] Therefore, experiment strengthened the attitude of searching and openness to new problems, new situations and new solutions, which had already been proper to science.

5. By determining the above mentioned division of empirical sciences into "experimental" and "observational" ones, experiment thus brought the experimental sciences closer to technology and industry. As its result, no reasonable model of development of the bloc of mathematical, natural and technical sciences in modern and contemporary times can disregard the combination of science and technology. On the other hand, the bloc of all the other sciences – social sciences and the humanities – seems to be so closely connected with socio-political and cultural life that the model of development of this bloc cannot be considered to be realistic if one does not take into consideration the combination of humanities with the sphere of culture and ideology. In this way the unity of science itself seems to be

---

[2] If prediction and explanation are external aims of science then, following Popper's suggestions, I include among the internal goals those values of truth which make it a non-banal truth, i.e. such that is theoretically interesting and practically useful. These are: generality, exactness, great information content (empirical content), logical simplicity and epistemological certainty (the degree of confirmation) of knowledge. Popper puts information content (empirical content) at the fore of these values since it determines the degree of testability (falsifiability) of knowledge.

[3] Horton is right when he indicates that the scientific attitude even now in the West is rarely an intellectual attitude of the general public. He argues that in spite of the apparent modernity of world views, a contemporary Western layperson is rarely more "open" or scientific in his/her views on the world than an African peasant (Horton, p. 450).

endangered, the responsibility for which, according to the advocates of the unity of science, should be borne by experiment. Personally, I am inclined to think that such general names, denoting extremely extensive fields of human activity (and of its products) such as "science," "art" or "religion" encompass each one of them (and its results), which are so very varied and more and more different that, in the course of time, they become more and more inadequate and misleading. Therefore, I am not worried by the very fact of "the internal rupture of science," nor am I worried that various sciences serve "various interests guiding cognition" (Habermas) but the fact that the general name ("science") obscures that fact and that variety of interests, which are hidden behind cognition.

6. In connection with the influence of experiment on scientific cognition, and in particular on the division of sciences into experimental and other ones, there is also the question of the status of sciences such as astrophysics, astronomy or cosmology, which – although not of experimental character – are in some manner "superimposed" on the experimental science, which in this case is physics.[4] They are not experimental sciences in the strict sense of the word since they do not apply (although, perhaps, they have already started to apply) experimental methods of studies in their subject areas. However, their close dependence on physics (as their mother science) – meaning their use of various results obtained in physics, including those obtained thanks to applying the experimental method of studies – and also their immediate methodological dependence, consisting in their application of methods borrowed from physics, ultimately result in the knowledge which they formulate being indirectly dependent on experiment. It looks as though there exist such sciences that are not experimental themselves (in the sense of the methods which they apply), but which still they supply knowledge that is experimental in nature (in the sense that it would not have been possible to obtain if other sciences had not used the experimental method of studies).

7. The above discussion outlined versatile influence of experiment on the image and function of science. The influence, which put experiment at the very centre of the developing science, sufficiently explains the emergence of "new experimentalism," the trend which has quickly become renowned. Certainly, it remains an open question whether – apart from the charge of insufficiently appreciating the role of experiment in methodological studies on science – a stronger accusation is right. The latter is formulated by "new experimentalism" and addressed to the dominant view in the

---

[4] The possible inclusion of those sciences in disciplines of physics (and in physics in a broad sense) – which seems to be a right move – does not resolve the problem since it can be easily reformulated into the question of whether the whole physics (all its parts) is an experimental science.

philosophy of science ("theoreticism"). It claims that it is "the theory" rather than the "the experiment" that is the concept which is of key importance for the understanding of science and of its development. One of the arguments supporting the view that the superior role is played by "the theory" after all is that scientific theories are constructed either in all scientific disciplines (which deserve this name) or, in any case, in a greater number of them (namely, in all sciences if only partly nomothetical, i.e. those which are not purely idiographic in nature, if such exist at all) than the number of those disciplines which are subsumed within the bloc of experimental sciences. However, if the problem were restricted to latter bloc, the argument may appear to be of no significance.

HEGEL'S HISTORICISM AND CONTEMPORARY CONCEPTIONS OF THE DEVELOPMENT OF SCIENCE*

I. Introduction

As is well known, Hegel did not attach much importance to science in the modern sense of the word, i.e. to the special sciences. This was due to the fact that he assigned these sciences to the level of common-sense cognition ("reflective reasoning") that uses rigid, opposing categories of metaphysical thinking. The higher level of intellectual cognition ("conceptual thinking") that uses flexible dialectic concepts is represented by philosophy (although not any and every philosophy, and to be more exact, not any and every knowledge that occurred in the history of philosophy under this name).

In this connection, also the development of science (of special sciences) has not been a subject of Hegel's special interest, who, when working out his famous theory of development, focused most of his attention on the development of society (the spirit) in its various dimensions, and in particular on the development of religion and philosophy. He thought that the history of just those two fields of knowledge (religion and philosophy) concur with the general history of the world and constitute both their core (the deepest content) as well as their final manifestation.

This explains why Hegel's theory of the development of the society (historiosophy) as well as his conception of the development of philosophy (history of philosophy) and religion (history of religion) rather than the conception of the development of science are of greatest interest for the modern considerations on science and its development, even if Hegel himself was convinced that the course of history of philosophy, religion and special sciences is basically different. While the progress in philosophy consists in a subsequent total renewal of the whole of philosophy, and the history of religion means duration in time and changing circumstances of a definite, unchanging set of beliefs then the development of special sciences is realized due to incessant addition of

---

* First published in A. Arndt, K. Bal, H. Ottmann (eds.) (1999), *Hegel – Jahrbuch 1998: Hegel und die Geschichte der Philosophie*, Zweiter Teil. Berlin: Akademie Verlag, pp. 142-146. This text is reprinted here with the kind permission of Akademie Verlag.

new achievements to the framework of the already existing knowledge (Hegel 1965, vol. XVII, pp. 37-38).

After all, according to Hegel, the history of philosophy is closer to the history of religion (and of art) than to the history of special sciences. This is because philosophical knowledge is by its very essence completely beyond the limits of time so that philosophical truth is not dependent on any historical conditions, in view of which both religion and philosophy (as well as art) possess only the history concerning unimportant matters and conditions ("eine äusserliche Geschichte") (Hegel 1965, vol. XVII, pp. 32, 36, 86). According to Hegel it does not mean that there is no room for development in philosophy. Hegel emphasized repeatedly that in philosophy there occurs the development of thought which realizes subsequent "degrees of a logical idea" in the form of logical systems which follow one another. However, this development is devoid of any "historical outwardness" (geschichtliche Äusserlichkeit) (Hegel 1965, vol. VIII, pp. 60, 204).

In this essay, I shall confront Hegel's historicism, and his conception of the development of philosophy in particular, with some modern theories of the development of science.

II. Hegel's Historicism

At present both in science and philosophy three categories are quite clearly distinguished, which represent dynamics of reality: motion, development and progress. Motion is generally understood as any change, and development as directed change which is ordered in some way (monotonic with respect to some parameter); finally, progress is understood as a progressive change which consists in perfecting organization, thus in the transition onto a higher level. In this interpretation, development (directed change) is a specific case of motion (change in general) and progress (a perfecting change) is a specific case of development.

On the other hand, Hegel's theory of development, based on the idea of the "unity of opposites" and searching unity or even identity in that which is different, aims at a far-reaching identification of the three concepts mentioned above. The dynamics of the universe (of the Absolute Idea) in Hegels' interpretation, modelled logically as the development of a concept, assumes a progressive movement encompassing the whole reality, thus movement is here identified – at least in its general outline – not only with development but also with progress. Admittedly, not any and every change that occurs in the universe is a process of perfecting; however, all that according to Hegel deserves the name of true reality (apart from the so-called inorganic sphere of nature) is submitted to perfecting as a component of the Absolute Idea.

From this point of view, Hegel's theory of development is actually a theory of progress (perfecting change). It is because it gives the category of progress a universal, all-embracing dimension.

Hegel's rationalistic and perfectionist conception of development does not recognize in particular the occurrence in the history of any discipline of that which is called the revival. It is because a revival is a consequence of retreating in the development, a return to the previous stage, a rehashing of that which had already been dead. Hegel says that a man cannot become a child again and any attempts at "rehashing" of what is bygone are devoid of any originality (Hegel 1965, vol. XVII, pp. 76-77).

In reference to philosophy, this means that every philosophy is, by necessity, a fruit of its time and any revival of old philosophical systems occurring under the name of "neo" is pointless. In this connection in the history of philosophy, as in the history of any other sphere of being, Hegel distinguishes true reality which he opposes to apparent phenomena. True reality is represented by one and only one philosophy, and multitude and variety of philosophies only represent a semblance, or in the best case, particular phases of development of that only true philosophy, i.e. eternal philosophy. From this point of view, historical philosophies are as if numerous ripening fruits on one tree of eternal philosophy (*philosophie perennis*), which in the course of history is constantly strengthened, expanded and perfected. Within that one authentic philosophy only true statements and conceptions are formulated, and a mistake is only possible in the sphere of semblance.

In this way, the differentiation between reality and semblance enables Hegel to formulate an extremely cumulativist conception of the development of philosophy. It also allows opposing the historiography of philosophy proper which constitutes a thorough

study of one, only and true philosophy to such a philosophical historiography which unnecessarily fritters away its energy on secondary, external matters which are totally unimportant for understanding this one philosophy that has been maturing through the ages, simply on matters which belong to the external sphere of the history of thought (Swieżawski 1966, pp. 299-300).

This external sphere, unworthy of philosophy, which constitutes an objective science on truth, is a "burial ground of views," something in the shape of a "gallery of follies," signifying a total dispersion of thought (see Hegel 1965, vol. XVII, pp. 39-47).

Obviously, this kind of evolutionist approach to the history of any sphere of reality assumes that the particular stages and manifestations of what evolves are not important but what is important is that which is preserved through all the stages of the evolution, which constitutes as if suprahistorical content and its mutual relations.

According to this, a well-known aporia of philosophy's unity and multitude (and its aporia of temporariness and atemporariness connected with the former aporia) as a subject of historical studies is resolved by

Hegel – in what looks like a paradox – in an *a priori*, and as if ahistorical manner – devoid of the feeling of historicity. Actually, he speaks in favour of a certain kind of eternal philosophy (this term, however, has not been used by him), i.e. a universal philosophy (since it constitutes a collection of assumptions of all the transient philosophies which only create its moments), the only (although representing multitude), constant and fully true one.

Let us, therefore, pose a question: what, according to Hegel, does development (progress) consist in? In reference to the whole reality (the Absolute Idea), it consists in a constant process of self-consciousness of the spirit or, according to a somewhat different interpretation, in developing and perfecting the concept of freedom. Thus, progress in the development of the world is a progress in the self-consciousness of the spirit and the understanding of freedom. At the some time the ever great freedom is being realized through the necessary developmental process.

The development of any specific field consists in a logical process of the emergence of the consequences from assumptions inherent in the notion expressing the essence of the field. Also, assumptions are always most general and abstract, and the assumptions that follow constitute a more and more advanced concretization of those assumptions. Thus, one could say that the development consists in concretizing the notion of the spirit whose final and at the same time most concrete realization is the absolute spirit (the Absolute).

Every subsequent stage is only a partial realization of truth contained in the concept of the developing field. That is why every such realization is beforehand bound to be rejected and replaced by another, more perfect expression of truth. However, at the same time an optional, partial manifestation of truth is preserved in its every subsequent manifestation; therefore, it retains a permanent value – a value of a representative of absolute truth.

That which constituted a separate stage of development at a given time becomes at every next stage only its not independent moment. For instance, in general history particular spirits are only moments in the development of the general idea of the spirit in its real content. As a result, every subsequent step constitutes the unity of previous steps, especially those that precede it directly. In this manner, philosophy is a unity of art and religion as its moments. This means that the next step constitutes freeing the previous steps from the one-sidedness, that it constitutes their concept, reality, truth. It constitutes their unity; however, it is not an abstract unity but the one that accumulates in itself multiform kinds of its definiteness.

That the development process is enriched with content is a result of the fact that becoming occurs in the process of "removing" (*die Aufhebung*) of the hitherto existing content. "Removing" does not mean a (full) annihilation but consists in "preserving in a changed form." Two meanings are contained in the word *aufheben*: to put an end to something (*aufhören lassen*) and to preserve something (*aufbewahren, erhalten*). Therefore, according to Hegel that which has been "removed" is something which has at the same time been preserved, which only lost its immediacy but which has not been annihilated at the same time. Thus, for instance, in the spirit which is on a higher level than another one – as Hegel says in his Preface to *The Phenomenology of the Spirit* – a lower concrete being falls to the level of an inconspicuous moment. That which was previously the thing itself now is merely an indistinct trace. An individual whose substance is the spirit standing above, goes through his whole past in such a way that the one that sets about practising some higher science goes through all the initial knowledge which has been assimilated long ago in order to become aware of its content; he goes back to it in his memory but his interest is not focused on it. (According to Hegel, a certain important exception are only changes in the inorganic sphere, including mechanics and physics) since the negativeness in this area cannot be interpreted as the movement of the process but as the quietened unity which makes it that the thing is opposed to the process and is preserved in the state of indifference to the process. Thus, changes of this kind do not consist in – as it is in the process – the hitherto independent things being transformed into particular non-self-dependent moments (properties, aspects, sides) but in this that they preserve their status of objects that are mechanically included into new groups of things. That is why, for instance, the sphere of mechanics (including time and space, movement and gravitation) is defined by Hegel as "mutual existence in its outside" (*das Aussereinander*).

Also, negation characteristic of consciousness means that if consciousness removes something it does it so that what is removed is preserved and maintained, and thanks to this it still lasts despite being removed. However, it is preserved in a new "removed" form, i.e. it is transformed from an independent object into a certain specific moment, into one point of the new whole. Thus, this whole constitutes a certain unity which consists of numerous mutually removing moments.

Hence, it follows that the negativeness contained in the development is always a definite negativeness, and so it is also a positive content.

I shall now try to confront Hegel's theory of development with the studies of Lakatos, Feyerabend and Kuhn, in each case focusing my attention either on some interesting similarity or on some basic difference (or differences) between the theories compared.

*III. Hegel's Conception of the Development of Philosophy and Imre
Lakatos's Conception of the Development of Science*

From a certain important point of view, there is a striking similarity
between Hegel's conception of development of philosophy (and
development of consciousness in general) and Lakatos's conception of
scientific research programs that models the development of science.

This similarity consists in the distinct differentiation of two kinds of
history of a given cognitive discipline: the so-called internal history and
external history. In Hegel's conception we can find an opposition between
the "historiography of philosophy proper" (*"Geschichte des Gegenstandes
selbst"*), which is a thorough study of "one and only true philosophy" and
a "philosophical historiography" (*"Geschichte der äusseren Schicksale"*)
which fritters away its energy on questions that are of secondary
importance for philosophy, i.e. those which are external and completely
unessential for understanding of philosophy proper. The former may be
called a "philosophical" historiography of philosophy, and the latter a
"historical" historiography of philosophy.

Hegel was aware of the fact that the actual time sequence of systems is
not fully identified with the stages in the development of philosophical
concepts, and that is why one of the tasks of the historiography of
philosophy is to study to what an extent there is a divergence between
those two orders of sequences (Hegel 1965, vol. XVII, p. 59; Hegel 1929,
vol. VIII, p. 206). From this point of view, the historical study of
philosophical writings differs from the logical (philosophical) analysis
only in that the latter occurs beyond time and beyond space, and the
historical study concerns the same as far as its content but in concrete,
time-space conditions (see Hegel 1965, vol. XVII, p. 59; Swieżawski
1966, pp. 400-401). To put it more vividly, what is meant in the
philosophical historiography of philosophy is obviously also the
development of philosophy but the development which recreates on the
screen of time and space the stages of the logical sequence of steps in
reasoning which constitute the evolution of impersonal and pure
philosophical content. On the other hand, what is meant in historical
historiography is such a development that actually took place in history
and which occurred through errors, disrupted motifs that were taken up
unexpectedly, sometimes after whole centuries had passed, problems,
instances of negligence, and revivals (see Swieżawski 1966, pp. 410-411).

Adopting such an approach, Hegel actually identifies philosophy with
its historiography or the history of philosophy (practised philosophically),
and claims unequivocally that "the study of the history of philosophy is
the study of philosophy itself" (Hegel 1965, vol. XVII, pp. 59-60). To him,
the subject of proper, thorough and not merely external, erudite and
source-based historiography of philosophy is an eternal philosophy which

is at the same time always contemporary since it appears in all the (important) historical philosophies, each of which expresses only one aspect or moment in the development of this *philosophia perennis* (Swieżawski 1966, pp. 631-632). On the other hand, the most recent (important) philosophy which each time is the most complete expression of the whole hitherto development is of necessity the best developed and the most perfect one (Hegel 1929, vol. VIII, p. 59; Hegel 1965, vol. XVII, p. 71).

In Lakatos's work, we have a not much less distinct, analogous division into "internal history of science" and "external history of science" (Lakatos 1978). The former takes into consideration exclusively the internal factors which determine the development of science, i.e. factors of cognitive, rational character. Their importance is emphasized by the so-called internalism. All the cognitive reasons belong to them and any scientist is guided by them in his research practice. Here belong both empirical reasons (such as facts which constitute the results of application of experimental procedures, especially the so-called anomalies), and theoretical-methodological and logical (such as a consideration of truthfulness, non-contradictoriness, heuristic productivity, generality, exactness, high informational content of knowledge and the like). In this manner, abstracting from the external (non-cognitive) determinants, the internal history constitutes a "logical reconstruction" of the development of science, free from factors which influence science destructively and which obscure the picture of its evolution.

On the other hand, the external history of science considers the influence on the development of science of various external (non-cognitive) factors of a historical, social and psychological character which contribute irrational elements to the development of science. Their importance is emphasized by externalism.

It should be stressed that in history (historiography) of science the division into "internal history," usually understood as intellectual history, and "external history," defined as social history, could also be found before Lakatos. It is used, for instance, by T.S. Kuhn, who speaks of two approaches, or two different manners of practising the history of science. In an approach which prevails now and which is often called an internal one – Kuhn says – attention is focused mainly on the content of science as knowledge and on the intellectual determinants which form it. Its younger rival, an "external" approach is characterized by the interest in the activity of scientists as a social group in a wider cultural context, which requires taking into consideration non-intellectual factors influencing the development of science, especially institutional and socio-economic ones. Kuhn also thinks that the combination of both approaches seems today to be the greatest challenge that faces the history of science (see Kuhn 1977).

I. Lakatos stresses that a distinct division into "internal" and "external" history occurs in every conception of the philosophy of science but that it takes a different course in each of them. Also, his methodology of research programmes, as any other theory of scientific rationality, must be supplemented with the empirical-external history. As it is, the theory of rationality will not solve the problems of a kind: Why did Mendel's genetics disappear in the Soviet Union in the 1950s? Lakatos calls his differentiation between the two histories a "new" and "unorthodox" one. At the same time he notes that the way for this differentiation was paved by K.R. Popper, who described, better than anyone before him, the rupture which occurred between objective knowledge (in his "third world") and its distorted reflections in the minds of individuals (Lakatos 1978).

Lakatos sees the superiority of his differentiation in the fact that the methodology of research programmes marks a boundary between internal and external history, which is distinctly different from that marked by other theories of rationality. It transforms many problems that were external in other historiographies into internal problems (while the boundary line is seldom shifted into the opposite direction). And the rational reconstruction is the better the larger part of the actual science may be reconstructed as rational one (see Lakatos 1978). On the other hand, a touchstone of a relatively poor internal history (in the categories of which the larger part of the actual history is either unexplainable or anomalous) is that it leaves too much to the external history to explain.

Lakatos states that this does not mean that such a reconstruction of science is possible which will free us completely from historiographic anomalies. The actual history is always richer than the internal history. This springs from the fact that human nature is not woven from only rational motives and that is why no set of human judgements is fully rational either. Therefore, neither will any rational reconstruction agree with the actual history, and in this sense it is "falsifiable" (Lakatos 1978).

The similarity of Hegel's and Lakatos's conceptions discussed here undoubtedly arises from the circumstances that both represent a distinctly rational trend in philosophy. What both have in common is internalism: a conviction that in the development of science or philosophy a decisive role is played by internal determinants; one could also say that a decisive role is played by an internal logic of a given discipline. In this connection what is also common to both conceptions is their contemptuous (almost nihilistic) attitude to external history, which is to be of no importance for understanding of science or philosophy.

On the other hand, the basic difference is that, according to Hegel, the development of philosophy is in principle free of any error, while according to the point of view of Lakatos's critical rationalism methodological principles by no means constitute a guarantor of truth. That is why Hegel represents a rather extreme cumulativism which assumes the process of a principally error-free growth of truth in philosophy (realized by its most outstanding representatives). While in Lakatos's conception there is also room in science for basic errors, corresponding transitions (in Bohr's sense) and revolutions (consisting in abandoning one research programme for the sake of another), which fundamentally disturb what has been achieved so far.

The above difference is also manifested in the fact that while Hegel solves the aporia of unity and multitude of philosophy – as we could see – by speaking in favour of one true philosophy, free of any "contingency" of philosophizing, Lakatos's conception of scientific research programmes assumes that the development occurs within the competition among various research programmes with one another, and thus from time to time it requires the rejection of the earlier programme and speaking in favour of a new one. In this way, we have "theoretical monism" in Hegel, and "theoretical pluralism" in Lakatos.

IV. Hegel and Feyerabend. The Problem of Theoretical Pluralism

The conception of theoretical pluralism (proliferation) is defended even more vehemently by Paul Feyerabend, so the difference between Hegel and Feyerabend is even greater than that between the views of Hegel and Lakatos.

Hegel treats the multitude of actual philosophies as an illusionary, transitory and, at any rate, unimportant factor in the history of philosophical thought. This means that philosophical systems developed by concrete philosophers should be considered as manifestations of one philosophy which contains the principles of all the particular philosophies, becoming as if their common denominator. What is more, Hegel thinks that those concrete systems, actually contribute to the strengthening of the unity of philosophy due to the very fact that they emphasize the variety of manifestations of one true philosophy (see Hegel 1965, vol. XVII, pp. 47-67; see also Świeżawski 1966, pp. 375-376). In this manner, the multitude of philosophical systems properly expressed – i.e. understood as an evidence of constant development of philosophy as such – points to the unity of philosophy (cf. Świeżawski 1966, p. 376).

In turn, P.K. Feyerabend thinks that the monopoly which science achieved as a kind of knowledge in modern society as well as the monopoly towards which the particular theories strive and which is actually achieved by some of them is very harmful (according to him, an example may be quantum mechanics in its Copenhagen interpretation). Constructing alternative theories in a given discipline not only constitutes an efficient means of fighting the monopoly but also allows establishing facts that serve as raw tests (in Popper's sense) for the existing theoretical systems. In this way, according to Feyerabend, the strategy of theoretical pluralism twice contributes to overcoming dogmatism in science.

Most attention to the problem of theoretical pluralism was paid by Feyerabend in his early scientific activity, which preceded the period of epistemological anarchism. In his works of that period, Feyerabend stresses that only alternative theories in relation to the accepted theoretical system may be a real foundation of its severe criticism, understood as an attempt at refutation (see Feyerabend 1962, p. 30; Feyerabend 1963, p. 6). That is why methodological directives which are formulated demand that the theoretical systems suggested for the description of some field of inquiry be incompatible with all the hitherto existing systems concerning the discipline that have been developed so far, and even with (at least some) of their observational consequences.

Using the example of Brown's movement of molecules, which was the basis for falsifying the second principle of thermodynamics in its classical (non-statistical) formulation, Feyerabend shows that empirical data which falsify some theory may be discovered and – what is most important – interpreted as just falsifiers, in principle only on the grounds of a theory alternative to the tested one. On the grounds of the tested theory these empirical data may not have any interpretation at all.

In the case considered here, even if it were accepted that the phenomenon of Brown's movements might have been discovered without the statistical kinetic theory, then still it would not have been treated as a falsifier of the second principle without the knowledge of this kinetic theory (see Feyerabend 1963, pp. 23-24; Zamiara 1974, pp. 80-97).

The question considered by Feyerabend concerns an important issue of the possibility of conducting in science the so-called crucial experiments. Methodological analyses of this question show that actually the occurrence in science of the possibly numerous alternative theories considerably strengthens the crucial power of scientific experiments conducted for verification purposes (cf. e.g. Such 1982, pp. 113-126; see also Such 1975).

On the other hand, what is controversial is the negative attitude of Feyerabend to the well-known conception of the "unity of science," according to which science attempts to reduce particular theoretical systems of various fields of science to a certain universal, all-embracing physical theory. The reason is the same: the lack of alternative theoretical interpretations which made it impossible to undertake attempts at reliable testing of such a theory and so would lead to dogmatism of scientific knowledge (see Feyerabend 1962; Feyerabend 1963; see also Zamiara 1974, pp. 96-97). This standpoint is in disagreement with the unifying tendencies in modern physics and with an attempt of many outstanding physicists to construct "the Theory of Everything" (see Einstein, Planck, Heisenberg, Hawking, Weinberg, Weizsäcker).

V. The Problem of Continuity of Knowledge as Interpreted by Hegel and Kuhn

Of the well-known conceptions of the development of science, formulated in the modern philosophy of science, one that undoubtedly differs the most from the Hegelian theory of development of knowledge is T.S. Kuhn's theory of the development of normal science and scientific revolution (see Kuhn 1962).

Let us point out the main differences, adducing Hegel's conception of development of philosophy as representative of his theory of development of knowledge.

First, Hegel's theory is extremely cumulativist, while Kuhn's theory represents a rather determined anticumulativism.

Second, Hegel's theory is based on the idea of a most optimistically understood progress, although slow and of devious nature. By contrast, Kuhn's theory allows only very restricted cognitive progress in the periods of normal science, and in the whole course of the development of science, takes into consideration only the "heuristic" progress, whereby the new paradigm enables one to formulate (and possibly solve) more problems-puzzles than the previous one. Progress in the periods of normal science consists in accumulating the puzzles that have been solved, and in the periods of crisis and revolution the place of scientific progress is, according to Kuhn, actually taken by the scientific change which does not have any distinct signs of the cognitive progress.

Third, Hegel's theory precludes the possibility of going back in the development, and in this way, making errors, and what follows, he precludes the occurrence of the periods of revival which "rehash" what has already been dead. On the other hand, in Kuhn's theory the return to the formerly abandoned conception is always possible. This is connected with the fact that, as has been rightly observed by Lakatos, Kuhn's psychological epiphenomena of "the crisis" and "conversion" may accompany both the changes that are objectively progressive as well as those which are objectively degenerating, both revolutions and counterrevolutions which cause those two types of changes to be actually indistinguishable from each other.

Fourth, according to Hegel's theory of "removing" (*die Aufhebung*) each subsequent stage of development contains in itself as indispensable the moments which enrich it with the content of all the previous stages. According to Kuhn, on the other hand, the transition to a new paradigm means both gains and losses. First of all, it concerns the resources of problems which can be formulated within a given paradigm as well as the repertory of explanations which may be obtained on its grounds. The change of paradigm consists primarily in the revolutionary transition to a new theory which is, logically and empirically, incommensurable with the hitherto existing one. This means, among other things, the change of the set of concepts which reaches deep enough so that at least certain previous problems (and their solutions) lose their sense, and the new ones which cannot be formulated on the grounds of the old paradigm appear in their place. The new paradigm also changes scientists' views on what is good explanation and provides new ways of explaining which invalidate certain explanations that have so far been considered to be correct.

And finally, Hegel's theory represents internalism in the sense that the development of philosophy is determined exclusively by cognitive (rational, logical) factors while Kuhn's conception is externalistic: the main determinants of the development of science are of the socio-psychological (non-intellectual) nature.

As can be seen, Hegel's and Kuhn's theories, confronted above, represent two extremes: the extreme of the cognitive optimism, today viewed as naive, and another extreme of excessive scepticism. From the present point of view it may seem even stranger that the development of philosophy on which Hegel based his optimism does not show such distinct signs of progress as the development of science whose examination led Kuhn to such pessimistic conclusions. An adequate description of the development of human knowledge in various fields seems to lead to more moderate standpoints.

VI. Conclusion

The confrontation of Hegel's conception of the development of philosophy with three modern theories of the development of science shows both some strong points of Hegel's historicism and its certain weaknesses. Undoubtedly, a strong point is his perception that all true development, if it is to represent progress at the same time, must contain elements of the directed cumulation. Therefore, it must constitute a set of changes, ordered in some way, leading to perfecting of the organization.

Another valuable idea of Hegel's historicism is the conviction that although the development of any discipline of thinking is influenced both by internal and external factors, still every time the nature and peculiarity of this development is determined by internal factors. This is the so-called idiogenetic interpretation of development which, in contradistinction to allogenetic expression, assumes homogeneity (identity of the type) of developmental phenomena with the phenomena that determine them.

The third element of Hegel's theory of development which should be positively evaluated is his view that the development of knowledge consists in "pushing more and more into the shadow" of empirical elements and in "bringing into the foreground" of the theoretical elements of a logical, intellectual character. This penetrating view of the cognitive process through notions in Hegel goes hand in hand with the brave rejection of the rigid common-sense approach to cognition for the sake of dialectical interpretation, which is characterized by smoothness and mutual permeation of notions until the perception of the unity of opposites is achieved. Taking this path, which was undoubtedly the right one, Hegel went somewhat too far, since he did not take into consideration Kant's admonitions and came to approve and even praise the pure speculation in thinking, which did not take into account the facts that were provided by special sciences.

The weaknesses of Hegel's conception of development were partly the weaknesses of his epoch. Beyond doubt, they included an overly optimistic 19th century faith in progress which does not retreat from anything and does not take any faulty steps. That faith was accompanied by certain dogmatism, consisting mainly in the conviction that in any field there is only one way of real development and progress and that (human) spirit progresses in a manner which is indeed very slow and devious but at the same time it does not stumble or make mistakes. Hence the dogmatic solution of the paradox of "unity and multitude of philosophy," which is so gross in the light of what has been established by modern methodology.

Above all these advantages and drawbacks of Hegel's historicism rises his rationalist thought which has been the greatest achievement of his philosophy. Therefore, I would be inclined to consider Hegel's rationalism as his main message for our times, which are endangered by the disbelief in human mind.

Part IV
Problems of Verification of Knowledge

ARE THERE DEFINITIVELY FALSIFYING
PROCEDURES IN SCIENCE?*

The point of departure for the present discussion is the fact, well documented by the history of science (especially physics) that certain scientific theories are rejected – and, what is more, are rejected finally, definitively and once for all – on the basis of falsification in the strict sense of the words. At times, the same holds for the theories that had formerly been considered well confirmed experimentally and that have found numerous applications, both theoretical and practical. An example of a theory that has been falsified beyond any doubt is classical Newtonian mechanics.

The degree of reliability with which the falsification of a given theory may be considered real and definite is obviously connected with the degree and range in which the theory has been falsified. And so, the degree (depth) and range of falsification of Newtonian mechanics (determined by the special and general theory of relativity, quantum mechanics, quantum electrodynamics and other theories and facts that lie at their basis) is at present so considerable that those who do not recognize the falsification of this theory as ultimate are simply those who turn their eyes away from well-grounded results. All the more so that along with the development of physics this degree and range increase continually, i.e. Newtonian mechanics appears to be inapplicable (with a given degree of approximation) to the ever newer subranges of its alleged (originally assumed) applicability.

In view of the definitive falsification of certain theoretical systems that undoubtedly occurs in science, a question arises concerning procedures that lead to this falsification. The adherents (formerly numerous, few at present) of the *experimentum crucis* believe that a research procedure which leads to definitive falsification of theoretical systems (and according to some authors to verification as well) are crucial experiments. However, this standpoint cannot be maintained and even seems rather naive: the very fact of the possibility of different, even numberless theoretical interpretations of a given experimental result (or, in

* First published in Władysław Krajewski (ed.) 1982, *Polish Essays in the Philosophy of the Natural Sciences* (*Boston Studies in the Philosophy of Sciences*, vol. **68**). Dordrecht Boston London: D. Reidel Publishing Company, pp. 113-126. This text is reprinted here with the kind permission of Kluwer Academie Publishers.

other words, the possibility of the reconciliation of a given experimental fact with the unlimited number of alternative theoretical conceptions, "explaining" this fact) makes the *experimentum crucis* – which would be capable of falsification (*eo ipso*, definitive rejection) and, even more so, of verification (*ergo*, ultimate recognition) of the tested theoretical system – impossible.

Nevertheless, a certain rational nucleus seems to be inherent in the conception of the *experimentum crucis;* it should be revealed and maintained. Experiments and observations achieved in science sometimes differ even considerably with respect to the "crucial power" belonging to them in the process of the selection of theoretical conceptions being tested. However, that crucial or decisive power is a gradual imprecise feature; it belongs fully to none of the experiments. Therefore, the division of experiments into "crucial" and "other" should be replaced by ordering experiments with respect to their degree of decidability, thanks to which, instead of a dichotomous division of experiments, a continuum of decidability is obtained.

I believe that the procedure that is capable of definitive falsification of theoretical systems is not a single experiment nor any other purely empirical procedure (i.e. composed of either experiments or observations) but a certain more comprehensive mixed procedure: a theoretical-experimental one which might be called a "crucial situation."[1] The following observation leads to the concept of crucial situation. The crucial power of a given experiment in relation to a definite hypothesis or a theory is not its immanent feature. It depends on what other experiments the tested competitive theoretical systems have been confronted with, as well as on certain features of the systems themselves. At the same time, it appears (and this is extremely significant fact for our present considerations) that there exist the so-called "complementary experiments." These are experiments whose crucial power is "positively connected" to each other and thus ones which determine the fate of relevant competitive theoretical systems to a much greater degree than when their results are formulated jointly. Only such a cumulative result is confronted with theoretical systems.

Two physical experiments may serve as clear examples here. They were decisive for the special relativity theory[2]: the experiment carried out by Fizeau (1851) concerning the speed of the propagation of light in moving water, and the famous experiment conducted by Michelson, performed thirty years later (1881), whose aim was to establish the

---

[1] I introduced the concept of a crucial situation in Such (1975a).

[2] Indeed not in a historical-genetic sense because they were recognized as such only *ex post*: Einstein, as he himself stated afterwards, did not know the results of the second of the experiments (Michelson's experiment) before he formulated the special relativity theory.

movement of the Earth in relation to the ether (the first proved that the speed of light in water depends on the movement of water, the second that the speed of light does not depend on the movement of the Earth). Each of these experiments undoubtedly had considerable crucial power in relation to the then existing hypotheses assuming the existence of ether. The first one suggested the falsity of hypotheses that ether is dragged completely by the bodies moving in it, although it was easy to reconcile it with hypotheses of resting or partly dragged ether. The second one rather precluded hypotheses of ether at rest (in absolute space), although it was easy to reconcile it with conceptions that ether is completely dragged by moving bodies and that in this connection the ether "wind" does not exist.

The results of both experiments interpreted together led to the rejection of all hypotheses assuming the existence of the ether, either fixed or partly or wholly dragged by moving bodies. Consequently, the crucial power of each of those experiments, one with a negative result, the other with a positive result, in relation to hypotheses assuming the existence of the ether, thanks to the interpretation of results of one of them in the light of the other one, increased considerably (exceeding the crucial power of each experiment if considered separately). They laid a foundation for the conclusion that none of (the existing) hypotheses assuming the existence of the ether is in a position to cope with the problem situation which appeared in "the electrodynamics of moving bodies." Of a similarly complementary nature, as far as crucial power is concerned, was a property of the pair of experiments (or astronomical observations), one of which suggested that ether is dense, and the other that ether is rarefied, or of a pair of experiments (observations) one of which led to the conclusion that the ether is elastic, whereas the other led to just the opposite conclusion. There are many examples of mutually complementary scientific experiments. Their distinctive feature is that each of them (naturally, there may be more than two of them) clearly gains in importance in the process of testing theoretical knowledge. It is thanks to the comparison of its results with the results of the other experiment (or the other ones), which forms a pair (or triad, etc.) of mutually complementary experiments.

Which of the experiments which enter into the set of complementary experiments is included by specialists and historians of science in the class of "crucial experiments" often simply depends on the order in which they were made. One can, for instance, assume quite reliably that if the experiments conducted by Fizeau and Michelson were made in a reverse order, then, due to their mutually strengthening and significant crucial power, Fizeau's rather than Michelson's experiment would be considered crucial. However, this is not usually the case (that description is left to Michelson's experiment).

However, the increase in the crucial power of experiments is never so significant that they might deserve, even jointly, the name of the "crucial" experiment (or, correspondingly, of the set of crucial experiments) in the full sense of the word. Because of the possibility of various (though not so significant as far as their number and range are concerned) theoretical interpretations of their common result as well, they are not capable by themselves of falsification of any theoretical system.

Hence, in my view, one can draw the conclusion that the falsifying procedure may never be exclusively experimental in nature, and that the theoretical component, especially the action of interpreting experimental results on the basis of definite theories, may not be left out of account at any stage of the empirical testing of scientific knowledge. In particular, it appears that if a law is theoretical in nature (i.e. it is a component of a theory) and is confirmed to a certain extent, i.e. if it cannot be altogether refuted, then renouncing it or at least giving up its further application to a certain degree must depend on the discovery not only of cases which contradict it (to a certain degree) but also of more precise and more general laws concerning phenomena of the same kind. The new law should be of such nature that the cases confirming the old law should confirm it as well. In order to determine whether the newly-observed phenomenon actually constitutes a counter-example to the old law, one should interpret the phenomenon appropriately, i.e. in the light of some new, more perfect law, one more adequate to reality and better fulfilling its explanatory, prognostic and heuristic functions. The discovery of the displacement of the perihelion of Mercury, which appeared to be at variance with the Newtonian law of universal gravitation, may serve as an example. However, the fact that this nonconformity is not merely apparent but real was first established eighty-five years after that discovery because of the creation of the general relativity theory. Caution is advisable in this respect if only because of the well-known fact that actually no precise theory constructed in the natural sciences is in conformity at any stage with all the relevant facts established in the sphere of its applications. Here is one of the reasons why one cannot reject an old theory without constructing a new one: a former theory may conform to experiment, and the new phenomenon cannot be predicted and might not be explained on the ground of other reasons, e.g. due to neglect of some cause which produces this phenomenon according to this theory.

Considerations of this kind lead to the conclusion that the crucial procedure, a definitively falsifying one, should include – apart from an experimental component (composed of a certain number of mutually complementary experiments) – a theoretical one; it should be a mixed (theoretico-experimental) procedure. The presence of a theoretical component in each crucial situation is also conditioned by the fact that all theoretical considerations usually result in establishing magnitudes of crucial effects between competing theories., i.e. effects predicted by one competing theory and denied by another one. As long as the magnitudes of such effects are not determined, it is difficult to talk of the existence of a crucial situation. Therefore, until the answer to the question of, for instance, how large an effect of the diffraction of light should be expected for waves of a given length when the truth of the wave theory of light was not known, one could not speak of the existence of a crucial situation between the wave and corpuscular theories of light, nor in any case of a situation based on experiments with the diffraction of light. Similarly, as long as it was not possible to establish, within general relativity, certain effects pointing to the non-Euclidean character of the space-time continuum, it was not possible to speak of the occurrence of a crucial situation refuting physical theories based on the assumption of the Euclidean character of space and time. After all, in a crucial situation both (1) the magnitude of the expected effect, and (2) the degree of accuracy of appropriate measurements must be established. It should be possible to compare them and to determine whether the first one is sufficiently large, and the other sufficiently precise for the effect to be perceptible for the measurement. Therefore, at least one of the theories belonging to a given crucial situation should indicate not only the magnitude of the expected crucial effect but also whether in the existing – experimentally feasible – conditions a given effect may be observed and measured with appropriate accuracy.

A question arises as to what should enter into the composition of a given crucial situation as far as its theoretical component is concerned. It seems that the theoretical component, like the experimental component, is always multi-segmental: each time it includes a number of hypotheses and theories of a high degree of testability, of which some are the (tested) mutually competing hypotheses (theories), whereas the others help in deducing interesting predictions which are then confronted with the results of complementary experiments of great crucial power. A more precise answer to the question of the content of a theoretical component of the crucial situation is possible after answering the fundamental question, on what conditions is the definitive falsification of scientific hypotheses on the basis of actual experiments possible?

These conditions may be expressed as a single general sufficient condition in a statement that the definitive falsification of a given hypothesis or theory is possible under the stipulation that the results of certain experiments (or even of a single experiment) are of the kind that such an interpretation of all these results is impossible neither at present nor in the future, i.e. on the basis of not only existing but also of constructible and at the same time (relatively) true theories, which would make them compatible with the tested hypothesis (theory). For it would be impossible, then, to demonstrate that these results are not counter-examples for a respective hypothesis (theory).

However, at least since Duhem we have known that there are no experimental facts (or their sets) which could not be made compatible with an existing theory either by way of modification of the theory itself or by way of the appropriate modification – even if causal or conducted *ad hoc* – of the remainder of knowledge (e.g. auxiliary hypotheses) and, consequently, by appropriate interpretation of these results.

The above circumstance (perceived explicitly for the first time by Duhem), at times called the paradox of commonplace empiricism, prompted Einstein to support the criterion of "external confirmation," i.e. the conformity of the theory with experiments, with the criterion of "internal conformity" which Einstein called the "inner perfection" of a system. The paradox of commonplace empiricism is clear evidence that the criterion of conformity with experiment is not sufficient, in spite of appearances, to defend the empirical standpoint in science. Consequently, certain restrictions should be imposed on admissible modifications either of the tested theory (hypothesis) or of the remainder of theoretical knowledge, e.g. one which would preclude introduction to the recognized theories of so-called *ad hoc* hypotheses. There is nothing easier than to explain – by means of a new hypothesis, constructed especially for this purpose – why a given effect, which, according to the predictions of a theory should not occur, did actually take place or, conversely, which was to occur but actually did not take place.

A complex problem arises concerning the establishment of effective criteria which allow one to distinguish situations in which the modification of existing knowledge is prohibited as *ad hoc*, from the modification of situations when it is fully justified by the existing circumstances. One could certainly accept – as a postulate – the view that not every modification of the tested theory, forced by experiment, protects the theory against refutation but may only save a modified theory which is not identical with the initial theory. Thus, one might agree that the theory becomes falsified in each case in which it cannot be adjusted to experiment in the initial (unmodified) form. Therefore, it is falsified both when the need arises to replace it with the fundamentally different competitive theory and when it should be replaced with a theory which is its modification, i.e. by introduction of certain corrections to it. At the same time, a modified theory should be understood as any theory which was created from the initial one by way of the addition of new theorems to it, the rejection of some theorems included hitherto in the theory (sometimes a single one) or the change of any of them.

However, in view of the fact that was pointed out earlier that every theory may be saved (in its initial version) against the threat from experiment by way of the modification of the remainder of knowledge, the above assumptions and terminological statements will be good for nothing if appropriate restrictions of the admissible modifications of the existing theoretical knowledge are not imposed.

Therefore, following the direction pointed to by Einstein (and earlier by Leibniz) one can explain the postulate of "inner perfection" of a theory introduced by him by means of the concept of the so-called logical simplicity of a theory. As one measure of the logical simplicity of a theory, a number of mutually independent initial assumptions or postulates (axioms) of a theory are usually assumed, following Einstein: a theory which is logically simpler assumes less. However, one should not disregard another and, as it appears, no less significant component of the logical simplicity of a system. This is the informational (logical) content of a theory: the more logical consequences it contains, the more reason to consider it to be logically simpler. Therefore, the global determinant of the logical simplicity of a theory is a relation of its logical (informational) content to the number of assumptions: the greater the logical content of a theory and the smaller the number of its initial postulates, the logically simpler the theory. Thus its determinant is a formula: *logical simplicity = informational content / number of initial assumptions.*

Since the informational content of a theory is determined by the set of its implied theorems (being logical consequences of its postulates), one can also say that the logical simplicity of a theory is marked by the quantitative relation of the number of its implied theorems to the number of its initial theorems.

The absolute measure of the scale of logical simplicity thus conceived may not, obviously, be effectively designated for any actual scientific theory. This is for at least two reasons: first, we do not have an absolute measure of the degree of informational content of a theory, and second, we do not have at our disposal an absolute measure of the number of initial assumptions, or postulates of a theory. The first fact is connected with the circumstance that the information (logical) content of each theory is infinite in the sense that one may deductively infer an infinite number of theorem-consequences from the postulates of a theory. The second fact is in turn caused by the circumstance that the theories of the empirical sciences (not axiomatized and formalized in a precise manner) contain many premises that are adopted tacitly, or not even realized by their founders and adherents; those are premises which, for the time being, cannot be reconstructed as a whole by any logical reconstruction of a theory. Furthermore, the initial assumptions of a theory, like other sentences, may be joined and divided in many different ways, which prohibits the existence of any univocal method of establishing their number.

However, the above difficulties do not turn a given concept of logical simplicity into an entirely undefined one, especially if the criterion of the selection of a theory formulated on its basis were interpreted not as a criterion of an absolute but only of a relative (comparative) logical simplicity of two or more theories. For instance, when comparing two theories with respect to informational content, one may establish which of them is richer in information by way, say, of proving that from the first one all consequences (or their analogues) follow, which may be deduced from the second. Moreover, the first one contains consequences which cannot be deduced from the other (nor do they have in it their analogues). In such a way, one may prove, for example, that relativist mechanics is richer in information than classical mechanics.

Things are similar as far as the number of initial assumptions of a theory is concerned. At least in certain cases, the comparison of two theories in this respect leads to definite (univocal) results. For instance, when both theories contain certain mutual assumptions (or when certain assumptions of one theory have analogues in the assumptions of the other theory) and moreover one of them contains additional assumptions which do not have analogues in the other theory (but not vice versa), or when the second theory contains such assumptions, each of which constitutes a combination (generalization) of several separate assumptions of the first theory. Einstein's belief in a greater "inner perfection" of the general relativity theory as compared with Newton's theory of mechanics and of universal gravitation was based on the fully justifiable opinion that the former exceeds the latter both with respect to informational content and to the small number of initial postulates.

The logical simplicity of a theory appears to be negatively correlated with the mathematical simplicity of a theory,[3] but it is positively correlated with the degree of confirmation of a theory. The main reason why logically simpler theories are preferred is just that the simpler a theory is from a logical point of view, the smaller number of initial postulates it contains, and therefore the more credible it is as a whole. In order to accept it, it is enough to assume that the degree of the confirmation of the theory is a logical product of the degree of confirmation of its particular initial assumptions and that the initial assumptions of the logically simpler theory are no less credible than the initial assumptions of the theory which is logically more complex. If, for example, two theories differ in a number of premises (postulates) and do not differ in the degree of confirmation of

---

[3] Einstein explains this negative correlation of the two kinds of simplicity of knowledge by the fact that the simpler the theory, from the logical point of view, the longer and more complex the sequences of inference that lead from its initial postulates to observational consequences (e.g. predictions) and therefore, the more powerful and more complex the mathematical apparatus that must find application in this theory. Einstein's famous saying that our "scientific knowledge is so complex because it is so simple" should be interpreted as follows: our scientific knowledge is so complex mathematically because it is so simple logically.

any of them, then the more credible theory is the theory containing a smaller number of premises, or in cases when theories do not differ in informational content the theory which is logically simpler.[4]

It becomes clear, in view of the above mentioned relationship of logical simplicity with the degree of confirmation, why saving a theory that is incompatible with experiment, by adopting additional *ad hoc* causal hypotheses constitutes reliable evidence of its falsity. It is the case both when the tested theory itself is modified in this way and when the remainder of knowledge involved is modified in this way. The introduction of *ad hoc* hypotheses leads to logical complication of knowledge, to its loss of logical simplicity. It explains the need, indicated by Einstein and others, to combine the criterion of conformity with experiment with the criterion of logical simplicity. The rejection of the latter leads to conservatism in the field of thinking: at the cost of the logical complication of the initial system and of the introduction of additional *ad hoc* hypotheses, one may obtain increasingly satisfactory "conformity" of a theory with experiment. However, such a conformity with experiment will not be evidence of the truth of a theory. Without the criterion of simplicity we would stick to old theories and introduce (to them or to the remainder of pertinent knowledge), if a need arises, new *ad hoc* modifications and particularizations. Consequently, one should aim at conformity of a given theory with experiment but not at conformity at all costs: not at the cost of the excessive logical complication of knowledge.

Introduction of *ad hoc* modifications to knowledge deprives it of its original informational content and heuristic fertility: a cognitively fertile theory may be deprived of heuristic power, and even of its empirical character, through the application of the conventionalist strategy of saving theories threatened by experiments by means of the introduction of ever new, causal auxiliary hypotheses, as new facts, incompatible with the theory, are revealed.

Therefore, one should require that the auxiliary hypotheses introduced do not decrease informational content (and empirical content), or logical simplicity, nor consequently the degree of testability and heuristic fertility of the modified knowledge. Apart from the problem of the heuristic fertility of the tested knowledge, the problem of the heuristic fertility of theorems which serve as its modification also appears.

When considering the developmental aspect of science, one can introduce, on the basis of the concept of logical simplicity, the notion of a dynamic simplicity of knowledge (of a theory in particular). In contradistinction to logical simplicity, which is of a static character as it constitutes a property of knowledge (theory), belonging to it at a definite

---

[4] More thorough considerations on the subject are contained in Such (1975a, pp. 171-180).

stage, dynamic simplicity is characteristic not of a definite situation but of a (long-lasting) process of the development of scientific knowledge (theory).

This concept may be introduced in the following way (Mierkulov 1971). Let us take two theories T_1 and T_2, each of which explains, in its own manner, a group of facts E_1. Let us further assume that T_2 is true, although there is no particular reason to prefer it at present, since T_1 is as confirmed and as simple (or as complex) as T_2.

Let further investigations reveal a new group of facts E_2 which conform to T_2 but not to T_1. However, the advocates of T_1, saving their theory, introduce a new hypothesis H, such that $T_1 \wedge H$ is as confirmed as T_2. This process may last almost endlessly through the introduction of all possible hypotheses in order to remove contradictions which appear between the incessantly modified theory T_1 and experiment. If the introduction of the constantly new hypotheses H_1, H_2, etc. leads to an increasing logical complication of the initial theory T_1, then we say that this theory does not fulfil the criterion of dynamic simplicity. On the other hand, a given theory fulfils it when its logical simplicity grows along with the theory's development, or when it takes place at least on the same level. If theory T_2 fulfils this criterion and the competitive theory T_1 does not, then the dynamic simplicity constitutes evidence of the (relative) truth of T_2 as well as evidence of the falsity of T_1. At the same time, it will constitute the criterion of selection between T_1 and T_2, and the criterion pointing out that hypotheses H_1, H_2... which complicate system T_1 are *ad hoc*.

The relation of dynamic simplicity and heuristic fertility to truth is evident because the heuristic fertility of the theory in the sphere of experiment means its conformity with future experiments, whereas in the sphere of theory it means conformity with future theories which characterize these experiments and with the reality that they represent. The auxiliary *ad hoc* hypothesis is a hypothesis which complicates the theory logically and adjusts it to a given (known) experiment but it does not make it heuristically more fertile nor more predictive, i.e. it does not adjust it to future experiment. Therefore, an *ad hoc* hypothesis may only perform an "apologetic" role towards practice that has already been performed; it is helpless in relation to future experiment. In contradistinction to it, a heuristically fertile hypothesis performs the function of a reflector shedding light on future practice, the experiment of tomorrow. It explains why the introduction of some *ad hoc* hypothesis to a given theory is not of much use, and why it is an extremely provisional procedure requiring subsequent introduction of several further *ad hoc* hypotheses, hypothetical patches, which serve to piece up the existing theoretical system.

That is why in the case where the previous theory may be saved in the face of the negative verdict of the tribunal of experiment only by violation of the criterion of dynamic simplicity (i.e. by the loss of logical simplicity, either by itself or by the accompanying knowledge), one should proceed to the construction of a new theory, which would obviously be logically simpler and thus fulfil the criterion of dynamic simplicity.

It seems to me that physics and the methodology of physics obtained a significant, and practically reliable (and in any case effective, albeit partial) tool for establishing whether the transition from a given theory to a new one fulfils the criterion of dynamic simplicity or not. The tool is the principle of correspondence, formulated by N. Bohr. The relation of correspondence between two subsequent theories in the same field means that, simply speaking, the earlier theory, restricted to the range of its actual empirical applicability, constitutes nearly an approximate, specific case of a later theory. The latter describes reality not only more accurately but in a more general way as well (cf. Krajewski 1977). Therefore, it is both more informative and logically simpler on account of the smaller number of more general initial postulates, which it requires.

The necessity of introducing the criterion of dynamic simplicity, and therefore the criterion of a clearly marked time span for the consideration of definitively falsifying procedures indicates that the crucial situation (i.e. a definitively falsifying one) is not a static and invariable structure but a dynamic and timebound one.

The asymmetry between confirmation and falsification consists, among other things, in the fact that, thanks to the development of practice and the growth of precision in measuring and experimental techniques, the theoretical theorems, increasingly well confirmed until now, may at some point appear to be falsified by the new, more precisely established, or more differentiated facts. On the other hand, a theorem, once falsified, may not afterwards be adjusted to reality in its whole range. Quite on the contrary: new, more accurate empirical data will in the future reveal new areas of its ever greater nonconformity with the described states of affairs.

The best evidence that we are dealing with a crucial situation that falsifies a given, hitherto recognized theory, is simply that it undergoes a still further – one could say, more thorough – falsification with the new, more comprehensive empirical material. Further development of research in a given field, after the crucial situation comes about in it, is a history of such increasingly extensive falsification of this theory (i.e. increasingly narrowing the range of the actual applicability of a given theory), and, if necessary, a history of attempts at saving it by more and more radical modifications, through the introduction of a greater number of *ad hoc* hypotheses. This is connected with the fact that the hitherto prevalent theory, confirmed to a certain degree, does not describe reality accurately to the same degree in various "places" of its applicability. Thus it cannot be fully falsified, not in a single act, nor on a given level of development of a definite branch. One could even state that the theory, formerly recognized and confirmed to a certain degree, is submitted to subsequent falsification in its various aspects in the subsequent crucial situations which are formed in a given discipline as the latter develops. Thus, for instance, classical mechanics was submitted to more thorough falsification in the ever new sub-ranges of its presumed applicability, in situations which successively were decisive about the advantages of special relativity theory, general relativity theory and quantum mechanics, and which eliminated their most important rivals. The process of more thorough falsification certainly goes on and will continue to do so. And if today no sensible physicist doubts that classical mechanics has been ultimately falsified, then it is because during more than the last half century of conflict with the special relativity theory, its falsity in many aspects has been shown by means of increasingly enormous experimental and theoretical material. Therefore, even if classical mechanics liberated itself in a certain aspect and were "cleared" of certain objections, it is out of the question that it could ever regain the acceptance in science which it formerly enjoyed. On the contrary, there is no doubt that generally its situation will be worse as science develops, although it will surely never lose its significance completely, particularly as far as its practical applications in the traditional branches of technology are concerned. Each of the stages of falsification is connected with the new, victorious theory and with falsification of some other theories, which indicates that this is a process of a theoretical and experimental character and that certain crucial moments and situations are involved, which have been described here as crucial situations.

A crucial situation falsifies a given theory definitively because it constitutes the beginning of the increasingly intense crisis in which the falsified theory is sunk as a given discipline, and related disciplines develop. It becomes obvious at last that the way of interpreting and explaining facts itself which this theory represents is outdated.

It is evident from the considerations presented that a crucial situation is not so much a single act of falsification of a given theory as the process of intensification of contradiction between this theory and experiment and the remainder of knowledge, the process in the course of which the constantly new sub-ranges of applicability of a theory are falsified. This process might be called the increasingly intense crisis of a theory.

It means that the elements that are indispensable for the formation of a crisis do not emerge simultaneously but are accumulated gradually as long as the situation that occurs is not considered "sufficiently significant" and univocally decisive as to the disadvantage of a given theory. It takes place when it becomes obvious that the future results can only "crush" the theory and provisionally establish one of its rivals (no matter which one).

Translated by Sławomir Magala

ON THE SO-CALLED COMPLEMENTARY EXPERIMENTS. THE EXAMPLE OF FIZEAU'S AND MICHELSON'S EXPERIMENTS*

I. I would like to present one of the ideas contained in my book which came out in 1975 in Polish, entitled *"Czy istnieje experimentum crucis?"* [Is there an Experimentum Crucis?] (Such 1975a).

The idea has its source in the observation that in the process of choosing between competitive theories – which was the case of, for example, the special theory of relativity (STR) and Lorentz's electron theory – it is not so much individual experiments taken in isolation that are of great importance but pairs of experiments whose results are complementary to each other. They impose certain constraints on each other, thus excluding certain interpretations that are possible when the experiments are considered separately. I term experiments in such a pair complementary experiments (complementary to each other).

It turns out that the crucial power of a given experiment towards a particular theory is not its immanent feature but depends on which other experiments the tested competitive theoretical systems have been confronted with. There exist, at same time, complementary experiments, i.e. experiments whose crucial power is "positively coupled," experiments which thus determine the validity of respective competitive theoretical systems to much a greater degree when their results are taken together and only such a combined result is confronted with theoretical systems, than when each result is considered individually and separately compared with relevant systems.

Two physical experiments can serve as a clear example. The experiments are sometimes said to have been the basis for STR[1]. they are: Fizeau's experiment of 1851 concerning the speed of light propagation in moving water and the famous experiment conducted by Michelson thirty years later (1881), whose aim was to establish the movement of the Earth

* Translation of "O tak zwanych eksperymentach komplementarnych," published in D. Sobczyńska, P. Zeidler (eds.) (1994), *Nowy eksperymentalizm – Teoretycyzm – Reprezentacja* [New Experimentalism – Theoreticism – Representation]. Poznań: Wydawnictwo Naukowe IF UAM, pp. 123-132.

[1] Not in the historical-genetic sense, since they were recognized as such only *ex post*, as Einstein had not known the results of Michelson's experiment before formulating the STR – that is, at least, as he maintained later.

in relation to the ether. The former proved that the speed of light in water was dependent on the movement of water, the latter – that the speed of light was not dependent on the movement of the Earth. Fizeau's experiment is now a classical example of an experiment with a positive result (i.e. manifesting the expected effect) and Michelson's experiment is, in turn, is a classical example of an experiment with a negative result. Each of those experiments had undoubtedly considerable crucial power in relation to the then-existing hypotheses assuming the existence of the ether. The first one suggested the falseness of hypotheses assuming that the ether is entirely drawn by bodies moving in it, though it was easily reconcilable with hypotheses of the resting or partly drawn ether, whereas the second one rather excluded hypotheses of the ether resting (in absolute space), although it was easily reconcilable with conceptions assuming that the ether is entirely drawn by bodies moving in it and that consequently the etheric "wind" does not exist.

It is obvious that the results of both experiments interpreted together led to the rejection of all hypotheses assuming the existence of the ether, either resting or partly or entirely drawn by moving bodies. Consequently, the crucial power of each of them towards the hypotheses assuming the existence of the ether, due to formulating the results of one of them (whichever one) in the light of the other, increased significantly: much stronger than individually (and both "totally" if considered separately), they substantiated the conclusion that none of the (existing) hypotheses assuming the existence of the ether was in the position to cope with the problem situation which occurred in "the electrodynamics of the bodies in motion." Similarly complementary character – with respect to the crucial power – was displayed by the following: (1) a pair of experiments (or astronomical observations) of which one suggested that the ether was dense and the other that it was rare and (2) a pair of experiments (observations) of which one led to the conclusion that the ether was elastic, whereas the other one led to an opposite conclusion (as the more elastic a medium is, the quicker a wave moves, and electromagnetic waves move at enormous speed, hence the ether should be very elastic; on the other hand, the ether would have to be very penetrable and rare, so that planets could move in it without noticeable friction). In that situation, the requirements that the ether would have to fulfil (the properties it would have to have) turn out to be mutually contradictory, excluding each other.

Which of the two elements of a pair of complementary experiments should be recognized by specialists and science historians as entering the class of "crucial experiments" depends, at the same time, simply on the order in which they are carried out. One can, for instance, reliably suppose that if the experiments of Fizeau and Michelson had been carried out in the reverse order, then – due to their mutually strengthening and, at the same time, considerable crucial power – not Michelson's experiment but just the experiment of Fizeau would have been recognized as crucial, though it is not usually called crucial, the notion being reserved for Michelson's experiment.

The point is that both the experiments had great crucial power – but in relation to each other, i.e. each of them in consideration of the other. Moreover, the crucial power of both those experiments is comparable and even – at least at first sight – identical, in spite of the fact that one of them had a positive result whereas the other one had a negative result, and that it is usually the experiment with the negative result that has greater crucial power.

In general, it is much more difficult to reconcile with a given hypothesis many results of various experiments taken together, especially experiments so matched that they make up pairs of complementary experiments, i.e. experiments whose results complement each other, constituting a verifying whole of greater crucial power, than individual results of experiments considered separately, even if some of the latter are later recognized to be experiments determining the fate of a given hypothesis.

The illusion that Michelson's experiment alone determined the fate of STR and its rival theories – has its source in the fact that the context (the problem situation) in which the experiment was carried out is not noticed. The crisis caused in physics by the negative result of Michelson's experiment, the result that no one had expected, was actually as well the effect of the fact that results of some other complementary experiments, especially of Fizeau's experiment, had already been known earlier. If it had not been for earlier experiments impairing the conceptions of the ether entirely drawn by moving bodies, the negative result of the experiment liable (in a natural way) to agree only with the conceptions assuming its mobility, would not have caused any deeper crisis in physics.

The authors who emphasize crucial significance of a singular experiment for a given theory are right if they mean not the crucial power of those experiments taken separately in the process of verification but their heuristic significance, their stimulative power, i.e. the fact that more than once it is just the unexpected result of a given experiment that impels a theoretician to give up an existing theory and to formulate a new one, which turns out to be better. However, a source of a theory is possibly never a single new experimental result (no matter how unexpected). The success of a new theoretical conception usually lies in the ability to select certain facts, usually not numerous but crucial for a given problem situation, from information abundant in empirical data and the ability to subject these facts to deep and thorough theoretical analysis. This kind of approach is characteristic of many eminent contemporary physicists, and especially for Einstein and Bohr.

It is then to be expected that it is not one experiment that is crucial in nature (determining fates of theoretical systems, even if it were subsequently to acquire the status of *experimentum crucis*) but rather pairs of such experiments, whose results can be appropriately put together, constituting some sort of a positive feedback, increasing their crucial (falsifying) power. Therefore, it is pairs of experiments (of complementary features) and not singular experiments that come closer to the ideal of *experimentum crucis* which is non-achievable in practice.

II. I believe that the role of pairs of complementary experiments became extremely evident exactly in the problem situation which determined the victory of STR over its rival theories, above all over Lorentz's electron theory.

The experimental element of that situation was composed of a whole series (about thirteen) experiments, including at least seven experiments on light. They were experiments in effect of which: (1) the aberration of light was discovered, (2) Fizeau's convection coefficient was determined, (3) the negative result of numerous experiments of Michelson-Morley experiment type (conducted since 1881) was obtained; and further on, the following ones: (4) Kennedy-Thorndicke's experiments with the inferometer, with unequal arms, (5) with movable sources of light and mirrors, (6) De Sitter's and others astronomical observations concerning the double stars spectroscopy, (7) Michelson-Morley's experiments with sunlight. Besides the experiments with light, the other ones which supported the theory of relativity (against its rival theories) were the following ones: (8) the experiment revealing the dependence of molecular mass (e.g. of electrons) on speed and (9) the proportionality of mass and energy; next, (10) experiments concerning the radiation of moving charges, (11) experiments determining the slowing down of the disintegration of mesons moving at great speed and, finally, the negative results of (12) Trouton-Noble's experiments (aiming at the discovery of the twisting moment, affecting the condenser) and 13) of experiments on unipolar induction with permanent magnet.

Besides the theory itself, the theoretical element of the problem situation, which determined the verdict in favour of STR was, among others, composed of Lorentz's and Fitzgerald's theories in various versions, the wave and corpuscular theories of light in their diversity, namely the wave theory (1) with the resting ether, not assuming the contraction of bodies in motion, (2) with the resting ether, assuming the contraction of bodies in motion, (3) with the ether moving together with a body (carried by it), and three corpuscular theories (emissive; among others Ritz's emissive theory) assuming respectively that the speed of light after a reflection equals c/n_1, c/n_2 and c/n_3 (where c is the speed of light in a vacuum and n_1, n_2, n_3 are the refractive indices towards successively (1) the source, (2) the mirror, (3) the image).

Thus, the problem situation which determined the fate of STR was composed of at least thirteen experiments of great crucial power and at least nine theories (including STR).

It turns out, at the same time, that on the one hand none of those theories is inconsistent with all the mentioned experiments and observations and that, on the other hand, STR is the only one of those theories which agrees "from the start" with all the thirteen experiments, without the need for any *ad hoc* modifications,[2] which would complicate it from the logical point of view. Also, none of those experiments – taken separately – can be treated as the *experimentum crucis* deciding on the choice between STR, on the one hand, and all its rival theories, on the other, or at last, falsifying one of the latter ones. However, mutual constraints which these experiments impose on their possible interpretations are of such a type that all of them considered together cannot be made compatible with any one of the discussed theories, except the STR and the modified Lorentz's theory. That does not mean, of course, the definite verification of the STR or of the modified Lorentz's theory but it means only the falsification of all the other theories (amounting to seven).

One ought to agree that a singular experiment even as significant as Michelson-Morley's experiment, does not constitute the *experimentum crucis* (the falsifying one, to say nothing of the verifying one) in the traditional, methodological sense of the word. The conclusions which Michelson himself inferred from his experiment can serve as an illustration, for he was far from supposing that the negative result of his experiment refuted the classical mechanics or at least the conception of the existence of the ether. He only supposed that it slightly modified the theory of the ether assumed by Maxwell leading to the rebutting of the hypothesis of the friction of the ether and of the existence of the etheric wind. To confirm his hypothesis of the carrying of the ether by a moving body, he conducted his experiments, among others, at a great height, between peaks of mountains. And although the negative result remained

---

[2] It should be noted that the above-mentioned facts did not lead directly to the falsification of the modified – under the influence of some of them – electron theory of Lorentz (containing hypotheses of Fitzgerald-Lorentz's abbreviation, which complicated it and were undoubtedly *ad hoc* in nature), although STR contains the simplest, the most elegant and natural (of all known) interpretation of those facts. Nevertheless, also Lorentz's theory in its version adapted to the experiments under discussion can be considered falsified: it was refuted by further development of theoretical and experimental physics, which so much rested on STR that without it would not be conceivable. That type of falsification could be called indirect falsification, which, to a great extent, is theoretical in nature.

unchanged, Michelson still believed that the existence of the ether was possible to be proved – to that end the experiment needed only to be repeated at still greater height. It is obvious that Michelson could interpret the result of his experiment in that traditional way only because he did not take into account experiments of other kind collected in growing numbers and still multiplied, which were complementary towards his experiment and, by the same token, jointly contradicted the theories assuming the existence of the ether. In particular, he did not take into consideration the Fizeau's experiment, carried out thirty years earlier, which excluded the conception of the complete drawing of the ether by a moving body (water).

Obviously, Michelson's experiment alone not only did not refute the hypothesis of the existence of the ether, which simply stated that the ether existed (since universal existential propositions cannot be, as is known, falsified at all) but it did not falsify any of the hypotheses or theories which rivalled one another at that time, most of which assumed the existence of the ether – moving, with bodies or resting in absolute space, although none of them reduced to that existential assumption (if they did, they could not be falsified).

Michelson's experiment, as any other experiment, taken in isolation, can be interpreted in many different ways. Just in relation to this possibility of extremely different interpretations of results of particular experiments, considered in isolation, there remains the problem of complementary experiments, "complementing each other" in such a way that the results of one of them exclude certain significant interpretations of the other one. Therefore, it should be emphasized that in the process of choosing a theory, the aim is not to put together all relevant empirical facts, i.e. assignations concerning a given hypothesis, law or theory. The aim in the process is essentially to set together such experimental results which jointly lead to eliminating at least one serious hypothesis from a set of rival hypotheses and which, in that sense, complement each other, i.e. are mutually complementary. In the process of choosing a theory, the way of putting together facts that had been gathered previously, which led to the well-known Bacon's tables, or putting together data in the form of points in diagrams, to subsequently joining these data with the simplest curves, determine in the experimental way functional relations between phenomena that have the character of experimental laws, is then not of great importance. Gathering data of that type and putting them together aims at formulating empirical factual laws in the inductive or quasi-inductive procedure, whereas setting together complementary experimental results of great crucial power aims at eliminating certain theoretical hypotheses that are quite often idealizing in nature and thus testing highly theoretical knowledge.

Such was also the meaning of putting together the results of Fizeau's and Michelson's experiments. The former, determining the speed of light in moving water, interpreted on grounds of the then prevailing wave theory of light assuming the existence of the ether, made it impossible (at least in practice) to formulate the conclusion that the Earth does not move in the ether that the so-called etheric wind does not exist. The latter, in turn, within the same theory of light, made it impossible to formulate the conclusion that the Earth does move in the ether (i.e. that the etheric wind exists). It would seem then that the situation was a deadlock (there was really no way out of it, if one assumed Fresnel-Youngs wave theory of light or any other theory assuming that light is the undulation of the ether): according to one of the experiments the Earth moves in the ether, whereas according to the other one it does not. Einstein was the first to understand what the matter was: both the experiments together excluded the possibility of interpreting their results as consistent with the assumption of the existence of the mechanical ether in the circumterrestial area (or, to be more precise, with the assumption that light is a wave propagating through the ether) and, thereby, precluded the traditional mechanical approach of the issue of the electrodynamics of bodies in motion.[3] Thus, they led to the conclusion that physical theories assuming the existence of the ether (including Lorentz's initial theory) were incorrect.[4] The same applied to a number of other mentioned experiments and astronomical observations. Therefore, only ascribing contradictory (or at least mutually exclusive) features to the ether allowed an explanation of the entire range of phenomena of the type under study, revealed by various "complementary" experiments and observations: these phenomena and experiments (or observations, respectively) carried out on them did not lend themselves – without additional, complicating hypotheses – to a noncontradictory interpretation within the theory of the ether. Within the existing theories of the ether, that led to the following alternative: either a contradiction in the theory itself or a contradiction of the theory in relation to results of certain experiments. The situation was therefore clear: at least some previous theories assuming the existence of the ether not only constituted unnecessary hypotheses but yielded to falsification.

A correct interpretation of a given experimental result should therefore be carried out in the light of the results of all other relevant experiments exerting influence on its interpretation, especially the results of experiments complementary to it. This is because experiments, although in many cases not without considerable crucial power even when considered

---

[3] I do not wish to embark on the discussion of whether Einstein had really known the results of both the discussed experiments before formulating STR, or whether he had only assumed (anticipated) them in his considerations.

[4] That does not mean that they proved the non-existence of the ether (nowhere and never). They led to the conclusion that the assumption of the existence of the ether was an unnecessary hypothesis since it did not explain the occurrence of those phenomena (e.g. light effects) for the explanation of which the assumption had been accepted.

in isolation, usually gain in significance when they are taken together in mutual relation and given a uniform theoretical interpretation.

Experiments and their results can be grouped, according to the scientist's intentions, in various tables and sets corresponding to different patterns, e.g. Bacon's tables or Mill's canons. In this paper, I have intended to emphasize the importance of putting together experimental results in pairs (possibly also in sets of three, four, etc.), in consideration of their possible complementary properties.

In this connection, methodological directions can be formulated for the scientists that, in the process of designing and conducting verifying experiments, they should pay more attention to carrying out just the experiments that are complementary to each other, i.e. such experiments that the result of one of them exerts direct influence on the way of interpreting the other one, whereby both the relevant experiments provide mutually reinforcing results, i.e. arguments against a hypothesis or a theory. An additional suggestion is that they should confront obtained results of experiments not only with theories (through their observational consequences) but that they should also confront one group of experimental results with another one, namely that they should carry out the confrontation of facts.

That leads to the holistic approach to experimentation. Similarly, as we do not confront particular theories taken in isolation with an experiment but instead, in the process of their experimental verification, we compare them one with another, we should not confront theories exclusively with individual experimental facts but also, or perhaps above all, with sets of such facts, with a certain whole of a definite order and structure, a whole formed by means of interpreting each of them in the light of the others, hence a whole that is uniformly theoretically interpreted.

TESTABILITY OF KNOWLEDGE AT VARIOUS LEVELS OF ITS DEVELOPMENT*

In the course of its development, scientific knowledge achieves an ever higher level of empirical testability. That is why, in empirical sciences, testability may be one of the indicators of the level of development achieved by one or the other field of science. However, the general increase in the degree of testability may not be observed in every respect. From many points of view, scientific knowledge becomes more and more difficult to test in the course of its development, be it in a positive sense, whereby verification and confirmation are concerned, or in a negative sense, pertaining to falsification and disconfirmation procedures. Therefore, the problem requires thorough consideration and appropriate relativization.

The aim of this essay is just to establish what advantage, on the one hand, and what losses, on the other hand, are brought by the progress in science as far as testability of knowledge is concerned.

Let us start with the benefits, as these are easily visible to everyone and there are no doubts about them as they outweigh the disadvantages. They stem both from the theoretical maturity which science gains in the course of its development, and from the ever greater technical skilfulness of science, including such procedures as observation, measurement and experiment. Let us look first at what is obtained by science in the field of testability thanks to its theoretical maturity.

Any discipline achieves its theoretical maturity in the course of realizing two closely related processes: in the course of the transition from formulating particular empirical generalizations, e.g. empirical laws that are registrational in nature (experimental laws), for the construction of theoretical systems, composed of those generalizations, or as binding them in a whole and as superconstructing over the knowledge obtained of the ever newer levels consisting of statements (e.g. laws) of a higher and higher degree of theoretization.

As can be easily noticed, the two processes mentioned above are as a matter of fact two sides of the same process of constructing more and more

---

* Translation of "Sprawdzalność wiedzy na różnych szczeblach jej rozwoju" published in: J. Such (ed.) (1980), *O swoistości uzasadniania wiedzy w różnych naukach* [On the Peculiarity of Justification of Knowledge in Various Sciences]. Poznań: Wydawnictwo Naukowe UAM, pp. 7-18.

capacious theoretical systems, whose increase "in breadth" requires, because of their (approximately) deductive character, i.e. because of the fact that these are sets of statements logically ordered by means of the relation of consequence, extension towards superstructuring over them ever higher levels consisting of statements of a higher degree of generality, informational content and theoretization.

As it is, from both of these two aspects of the process of theoretization of knowledge there stem definite advantages for the testing of knowledge. Combining particular statements (e.g. laws) into logically ordered systems (called theories) results in the fact that while testing one of them we are testing to some extent all the others; therefore, we are testing the whole system. According to Ajdukiewicz's visual metaphor, such a system resembles a bridge all the spans of which support each other. In this case, the empirical consequences of the system tested directly may be considered as pillars which support directly the adjacent spans but indirectly all the other spans of a bridge. As a result, a given statement (law) included within a system (theory) become capable of more versatile confirmation since it can be indirectly tested, referring to cases which do not directly fall within its range but which enter into the range of the applicability of a theory (i.e. its other statements) that is, it obtains, apart from the possible direct confirmation an indirect one. It is particularly important for these statements, which for one kind of reasons or another, cannot be tested (at a given stage) directly, i.e. through the study of cases which fall under them. One example of such a statement is the statement of the special relativity theory which describes Lorentz's contraction, which has not been tested until now but has been confirmed thanks to the testing of other consequences of the theory.

In turn, building over the given theories of statements of a higher order of theoretization and informational content, the former of which result deductively, causes the initial statements to obtain, apart from testing by means of their consequences confronted with empirical facts, an indirect justification that is theoretical in nature, consisting in referring to their logical reason (or a set of statements from which they result deductively). Obviously, a given theorem obtains theoretical justification on the part of statements of which it results logically, only in this measure in which the latter have been confirmed either by cases which fall under them (and not falling under the statement being justified), or in some other way. In this manner, the above-mentioned process of indirect testing (i.e. testing by means of statements at the same level of theoretization), links (and actually unites) the process of theoretical justification at hand, just as both procedures of theoretizing knowledge are combined closely into a uniform process of constructing theoretical systems.

Apart from these direct advantages, the process of theoretizing knowledge also involves a number of advantages that are indirectly conducive to the testing of knowledge. Logical and mathematical ordering, obtained in the process of constructing a theoretical system of an (approximately) deductive character allows one to realize the actual content of this knowledge, i.e. the actual resources of statements contained in the system, usually further raises the degree of exactness of knowledge and considerably facilitates deducing logical consequences from it, without which empirical testing of knowledge is not possible at all. It facilitates it both from the point of view of quantity where what is meant is the number of logical consequences deduced (observational ones, e.g. including prognostic sentences), as well as the quality is concerned, where the preciseness of consequences is concerned.

Obviously, the testing of separate theoretical systems is not the end; in the developed empirical disciplines many theoretical systems (many theories) are often involved in the process of testing which are connected with one another in different ways (e.g. the observational consequences, which are most interesting in the course of testing, prognoses including, are often deduced not from one but from many theoretical systems).

Still more evident are the advantages that are afforded by the testing procedure with respect to the progress in observational, measuring and experimental methods and techniques. First of all, it allows reaching ever more varied and subtle phenomena and studying them more and more exactly. As a result, we obtain a quantitative and qualitative increase in empirical facts that are at a scientist's disposal, which allows ever more precise and ever more varied confrontation of theoretical systems with empirical facts. Obviously, such a confrontation always occurs by means of observational consequences derived from the former. Therefore, we can see that the progress of knowledge on the theoretical plane as well as in experimental techniques leads to an increase in the possibilities of testing thanks, since it allows the use of ever more numerous and ever more precise and varied testing procedures.

The above picture of the development of science, outlined from the point of view of verification possibilities would, however, be overly optimistic if one did not account for the fact that cognitive progress inevitably leads to obtaining results of research whose testing sometimes becomes more and more difficult, and sometimes simply problematical (at least at a given stage). Let us begin the list of inevitable costs concerning the problems of testability and connected with progress in science with an observation that superstructuring the existing ones with ever newer levels of knowledge of the higher and higher degree of theoretization has an important drawback of complicating the process of the falsification of knowledge. The more theoretical the knowledge, the larger the units of that knowledge which can be submitted to independent negative testing (falsification).

Methodologists know and almost universally approve of the fact of the

asymmetry of positive testing (verification) and negative testing
(falsification) of strictly general statements and of their sets (scientific
laws and theories including), consisting in the fact that – if we consider
this problem from the purely logical point of view – one contradicting case
(fact) with a statement of this kind (consisting for it a counter example)
leads to its falsification while no (limited) number of positive cases, which
are in agreement with a statement of this kind is capable of fully verifying,
or confirming in the whole scope of its presumed applicability due to the
strict generality and an open character of the statement.

However, pragmatic considerations cause this asymmetry not to be so
distinct as one might think on the purely logical grounds. The thing is that
interesting prognoses, which are the consequences of the theoretical
statements tested, are usually derived – as has already been mentioned –
not from particular statements (e.g. laws) but the whole sets of them
(theories), and sometimes even from sets of theories. This means that in a
case when consequences of this kind (prognoses) are not tested what is
falsified is not a given particular statement that we want to test but the
whole set of statements (respectively, theories) from which, cumulatively,
a prognosis has been derived deductively (we disregard the problem of the
so-called initial conditions). Thus, the logical scheme of falsification
assumes the following form: $(T_1 \wedge H_1 \wedge \ldots \wedge H_n \to O) \wedge \neg O \to \neg(T_1 \wedge H_1$
$\wedge \ldots \wedge H_n)$, where T_1 is the statement (law) being tested or a set of
statements (theory), $H_1 \ldots H_n$ are the so-called accompanying hypotheses,
which together with T_1 constitute a sufficient condition from which a
prognostic sentence O is derived. It can be seen from the scheme that in
this case the whole conjunction $(T_1 \wedge H_1 \wedge \ldots \wedge H_n)$ is falsified, which
does not prejudge the question of which member or which members of the
conjunction should be rejected.

According to one of de Morgan's laws, negation of a conjunction
equals logically an alternative of negated members of this conjunction, so
we have: $(T_1 \wedge H_1 \wedge \ldots \wedge H_n \to O) \wedge \neg O \to \neg T_1 \vee \neg H_1 \vee \ldots \vee \neg H_n$. This
means that only considering the truthfulness of all the accompanying
hypotheses $H_1 \ldots H_2$ can one recognize the T_1 theorem being tested as a
falsified one. Certainly, the same concerns the other members of the
alternative: only in a case when the recognition of truthfulness of all
members but one are can the last one be considered to be falsified.
Falsification of a conjunction actually means a definite disconfirmation
(weakening) of each of its members, and does not mean the falsification of
each (none) of them; this is because without the application of further
testing procedures we are not able to state which member of a conjunction
(or which ones) is (are) responsible for the false prognosis.

Analyses conducted by Duhem and other methodologists show that the
more knowledge is tested, the more theoretical its nature, the larger areas
of the related knowledge (represented by accompanying hypotheses) must

be involved in the process of its testing (see Duhem 1911). Thus, the higher the level of theoretization that the tested statement represents, the more accompanying hypotheses generally take part in the process of its testing, and as a consequence the more difficult it is to falsify them unequivocally. Hence, in the course of development of a given discipline and as it achieves an ever higher degree of theoretization, units of knowledge which are submitted to independent testing (in any case, falsification) become ever more capacious, which makes the above-mentioned asymmetry of positive and negative testing less and less distinct. Problems with empirical testing, observed by conventionalism, which found their expression in the well-known "holistic" thesis of Duhem-Quine that while testing a given theory we are actually testing "the whole body of science," and even "the whole human knowledge in general" can be really seen, and even increase in the course of the development of science. Yet they do not lead to holism in such an extreme form as presented by Duhem and as claimed at present by Quine, but actually confirm the standpoint of moderate holism, according to which the most natural units of theoretical knowledge which can be submitted to independent testing (falsification) are in general scientific theories.[1] Only in the case of observational statements of a low degree of theoretization can one speak of testing statements individually, without referring to other accompanying statements.

By the way, difficulties of the same kind were perceived by Duhem, resulting from the development of experimental and measuring techniques and their application on the larger and larger scale in the process of testing scientific knowledge. As Duhem showed again, testing by observation, equipped with technical devices, measurement and experiment inescapably involves in the process of testing a given hypothesis also theories of the functioning of the research equipment as a result of which units of knowledge which can be submitted to independent testing (falsification) must each time be extended not only with the accompanying hypotheses but also with appropriate theories describing the action of experimental devices. The role of those devices increases, especially in those fields which, like physics, penetrate ever more distant levels of the structure of the matter. That is why the interpretation of experimental results concerning either the cosmic level or the microlevel becomes a more and more tedious, complex and more risky procedure. All this makes testing the theoretical knowledge concerning levels of reality other than the level that is directly available to us (macrolevel) an extremely complex activity, which has to overcome a number of new difficulties as compared with testing the earlier knowledge, which concerns merely the macrolevel.

---

[1] This standpoint is presented in detail in Such (1975a), chapter VI: Atomistic Empiricism or Holistic Empiricism?

Increase in the degree of theoretization of knowledge complicates the process of its testing in yet another respect. Namely, it leads to the increase in the degree of the idealization of knowledge: the number of idealizing assumptions with which statements entering the composition of the newly constructed theories should be provided generally increases.[2] Consequently, the process of testing laws and theories – through gradual concretization – includes more and more indirect links, i.e. it becomes prolonged and complicated. This makes it difficult to confirm or to falsify the laws and theories of a high level of idealization.

Generally speaking, the costs that scientists must pay with respect to testing of knowledge for the progress in science are connected with prolonging the chains of deductions, leading from initial statements (postulates) of a theory to its observational consequences (Einstein once drew the attention of the theoreticians of science to this). This also involves an increase in the degree of mathematical complexity of newly constructed scientific theories when compared with the previous ones.

On the other hand, the profits pertaining to testing and stemming from the progress of scientific knowledge, are connected with the fact that new theories achieve an greater logical simplicity.[3]

Logical and mathematical simplicity are negatively correlated with each other, so that scientific knowledge, while realizing the postulate (goal) of logical simplicity, must construct theories which are more and more complex in mathematical terms. That is, they must deviate from mathematical simplicity, whereas both logical and mathematical simplicity are advantages of a theory from the point of view of their testing possibilities. Thus, it is small wonder that the profits connected with possibilities of testing and stemming from the growing logical simplicity of knowledge inevitably go hand in hand with the disadvantages resulting from the growing mathematical complexity of knowledge.

However, generally speaking, there is no doubt that together with the

---

[2] Arguably, a certain fundamental exception in this respect is quantum mechanics as well as quantum electrodynamics and other theories of microphenomena in which certain hitherto idealizing assumptions which play a fundamental role in classical physics (macrophysics) – such as an assumption of an absolute isolation of the investigated system, or the assumption of the pointness of the studied objects – seem to lose their sense completely, and in this connection some physicists (e.g., von Neumann) even assume that the laws of microphysics – in contradistinction to macrophysical laws – are in principle concrete (factual) rather than idealizational in nature.

[3] Logical simplicity of a theory is determined by the relation of its informational content (the number of derivative statements) to the number of its postulates (initial statements): the greater the informational content and the lower the number of its postulates, the logically simpler the theory. Logical simplicity, understood in this way, may be considered as an explication of the value of theoretical knowledge which Einstein called an "inner perfection" of a theory.

progress in the development of science we can also observe a general progress in the testability of scientific knowledge, and that the requirements for the degree of testability of scientific theories grow in time. Consequently, new theories appear to be better tested than the hitherto existing ones, i.e. they obtain a higher level of adequacy to reality and a higher degree of confirmation on the one hand, and on the other, can be submitted to stricter attempts at falsification.

As an example illustrating the above findings, let us consider the development of modern physics, especially mechanics. It began with the formulation, in the late 16th and in the first half of the 17th century, of separate experimental laws at a low level of theoretization. They were formulated by Kepler, Huyghens, Galileo and others. The latter constructed the first theory of mechanics, of an indeed limited range of applicability and not without some essential flaws (even internal incompatibility). Still, it was a theory consisting of a number of logically interconnected laws (among other things, the law of inertia, the law of the free fall of objects, the principle of the composition of forces, and the laws of ballistic movement).[4] Already in the system of Galileo's mechanics, laws formed a certain hierarchy: the less general ones result from the more general. However, only the system of classical mechanics, constructed by Newton, together with his law (theory) of universal gravitation was distinctly a new level of knowledge, with the degree of theoretization significantly exceeding the level represented by empirical laws which could be approximately derived from it: Kepler's laws of planetary motion, the laws of the fall of bodies and Galileo's law of ballistic motion, Huygens' law swinging motion and a number of other laws.

Thanks to those empirical laws, Newton's classical mechanics (together with the theory of gravitation) had achieved an unusual empirical grounding in science, and, in turn, those laws themselves obtained a theoretical justification, i.e. were confirmed also in an indirect way (that is, not through their consequences but through testing, independent of them, of the premises from which they approximately resulted logically). Newton did not have to bother to submit the new theory of mechanics to tedious testing procedures, which referred to particular experiments (single empirical facts) since, thanks to his predecessors, he had at his disposal a number of empirical laws, which described with sufficient approximation the facts that were indispensable for testing his theories.[5]

---

[4] Cf. Matuszewski (1980) in which the author rejects the view which is commonly held by physicists and methodologists that Galileo's achievement, like of the other modern physicists before Newton, represented only a formulation of several laws which were independent of one another, and not a construction of a uniform theory of mechanics.

[5] The only independent fact which Newton refers to when testing his law of universal gravitation is the motion of the Moon. All the other facts that were known at that time and concerning these questions could be classed as the above mentioned empirical laws,

The development of technical devices of research, achieved on the basis of Newton's mechanics, also contributed significantly on its part to the extension of the verification base of this theory and to a better direct and indirect justification of its principles and statements. However, it appeared that neither Newton's law of universal gravitation can be tested without the laws of classical mechanics, nor the laws of classical mechanics can be submitted to exact testing (in astronomical conditions) independently of the law of gravitation. The same applies to a number of other principles and laws of Newton's mechanics of a high degree of theoretization, which can be tested well only within the whole system.[6] The level of theoretization of knowledge was raised as far as mechanics and theory of gravitation are concerned and so there was an increase in the number of (the least) units of knowledge which could be submitted to independent testing, i.e. testing independent of the other knowledge.

The level of idealization of the fundamental statements of mechanics and of their direct consequences was also increased, which obviously considerably prolonged concluding and other testing procedures. One example is the law of free fall of bodies which, as can be seen from its reconstruction, contained only one idealizing assumption in Galileo's formulation while in Newton's formulation it contained as many as thirteen such assumptions.[7]

As we know, Newton's classical mechanics was then generalized and assumed a much more abstract and formal form in Lagrange's system, in that of Hamilton, and in the systems of other scientists. Its mathematical apparatus became more complex and achieved a higher level of theoretization, becoming much similar to the axiomatized and formalized mathematical systems.[8] With regard to testing, this was followed by results – both positive and negative – similar to the rise of Newton's mechanics.

An even higher level of abstraction was achieved by mechanics thanks

---

formulated by Newton's predecessors. Also, the independence of this fact of the laws discussed is relative, since the motion of the Moon occurs (obviously approximately) according to Kepler's laws if we interpret them more broadly than it was done by Kepler himself. As we know, he applied them only for the motion of planets.

[6] It is no coincidence that one of the classical examples given by Duhem in favour of the holistic point of view regarding testing is an example of Newton's law of gravitation, which, as Duhem showed, cannot be tested exactly under astronomical conditions without reference to the laws of mechanics, to the laws of optics, and even to certain laws of physics and chemistry, concerning the functioning of respective observational and measuring instruments (e.g. telescopes or photo cameras).

[7] Logical reconstruction of this law in both formulations was conducted by myself in Such (1978).

[8] That is why one should not wonder that theoretical mechanics in its most abstract sections and interpretations is often included within mathematical disciplines and lectured at many universities not by physicists but by mathematicians.

to its "correspondence generalization" onto any fast motions, which was performed in the special relativity theory and in particular thanks to the repeated, and even stricter than Newton's, combination with the gravitation theory, conducted by Einstein in the general theory of relativity. In the form of both these theories, representing the so-called relativist mechanics, mechanics is understood as a theory of motion ("the science of the motion of earthly and celestial masses," as it was described e.g. by Engels) was combined with the theory of space-time, which significantly raised the level of theoretization and the degree of abstraction (including idealization). Again, this simplified the structure of the system of mechanics (because the new theories appeared to be more capacious from the informational point of view than the old ones, and also thanks to deriving from the fundamental postulates of the former of a number of statements which in the earlier pre-relativist theories of mechanics were to be adopted as independent premises) and it complicated its mathematical form, which involved appropriate results as far as the problems of testing are concerned. One of them was the crossing out of the previous boundaries separating mechanics from other fields of physics, especially the line dividing mechanics from the gravitation theory and the physical theory of space-time (physical geometry) thanks to the derivation of equations of motion of the universal relativity theory from the equations of the (gravitation) field of this theory. This considerably extended the range of applicability of relativist mechanics as compared with the pre-relativist mechanics; the former appeared to be applicable from elementary particles up to the most distant areas of the Universe, accessible to astronomy.

According to Einstein and in line with actual state of affairs, both the physical part F and the geometrical part G enter relativist mechanics, so that neither the geometry of the physical space-time G can be tested independently of the physical part F nor the latter (which is a contemporary equivalent of classical mechanics and classical gravitation theory) can be tested without the former (G). After all, the very division of relativist mechanics, represented by the general relativity theory, into two above parts $F + G$ is very conventional and cannot be conducted unequivocally.

Finally, the rise of quantum mechanics reduced certain further barriers separating mechanics from other fields of physics, and also mechanics and physics in general from chemistry, and even to some extent from biology; the fundamental statements of chemistry, and also a number of biophysical and biochemical statements in fact appeared to be consequences of statements of quantum mechanics, which found their application in specific conditions under which chemical and biological phenomena occur.[9] On the one hand, it extended considerably the scope and possibilities of testing the laws of mechanics (and physics in general) since even the study of chemical and biological phenomena supplies material for their testing. On the other hand, however, it also resulted in units of physical knowledge submitted to independent testing becoming more extensive; at present, even the whole knowledge of mechanics (about the mechanical motion in the classical sense) does not constitute such a unit since its testing (and even its formulation and interpretation alone) involves branches of physics that are different from the point of view of the classical tradition, and sometimes even those of chemistry and biology.

The above increasingly comprehensive structuralization of physical knowledge, removing old divisions, constitutes a significant objective fact of the development of knowledge. This obviously leads to the necessity of a more and more comprehensive interpretation also of testing procedures (and of a number of other procedures, e.g. explanation or prediction in which we are not directly interested here), which produces, as could be seen, both positive and negative results, with the considerable prevalence, as can be supposed, of the former.

Similar tendencies occur in other disciplines of empirical sciences and they certainly bring similar results as far as problems of testing are concerned. The only important difference seems to be that other empirical sciences lag (a little or much) behind physics that is why both profits and losses resulting from progress are less visible in their domains. Therefore, on the one hand, knowledge which they represent is less reliable and less adequate to the reality since it is worse tested, but, on the other units of knowledge which can be submitted to individual testing are smaller on their grounds, and the degree of idealization is lower.

The proof of the degree of theoretical maturity of a given discipline are both the internal relations between its specialized branches as well as external connections with other disciplines. With the exception that interdisciplinary connections depend not only on the extent of maturity achieved by it but also on the degree of development of related disciplines. And so, for instance, close theoretical and experimental connections of physics with chemistry, and then of chemistry with biology and of physics and biology became possible not only thanks to the development of physics but also to the development of chemistry and biology.

As far as the degree of maturity of any given scientific discipline from the experimental point of view is concerned, the history of many sciences shows that it is closely connected with the degree of development of the discipline from the theoretical point of view: the ever deeper experimental penetration into the structure of an investigated object requires its ever

---

[9] For the relationship of biophysical and biochemical laws to the laws of physics and chemistry see Pakszys (1980).

deeper theoretical analysis, specifically a mental one (connected with its more and more adequate logical reconstruction). The situation may also be reversed. It seems that only those experimental sciences achieve theoretical maturity which are either directly or indirectly based on strict experimental and measuring methods. If in a given field experiments cannot be used directly on a large scale then it can only achieve theoretical maturity due to being based on experimental and measuring studies, developed in some related discipline (as is well known, non-experimental astronomy was based on experimental physics in such a manner).

Certain important differences in testing procedures, which occur between natural sciences on the one hand and social sciences on the other are connected with the fact that in those large sections of empirical sciences two somewhat different although similar procedures of modelling the reality are generally used. In natural sciences, it was the procedure of idealization and factualization, which was initiated already in ancient times (in the 3rd century A.D. by Archimedes, Eratosthenes and others), and applied on a large scale by Galileo, Newton and other founders of modern natural sciences, whereas in social sciences it was the procedure of abstraction and gradual concretization, initiated by Smith and Ricardo and applied on a large scale by Marx in *Das Kapital*. The former is connected with laboratory experimentation and instrument-aided observation and strict quantitative measurement, based on measurement scales of great power, while the latter is mainly based, according to Marx's description, on "the power of abstraction and imagination" of the scientist and on the qualitative measurement, using measurement scales of little power.

The so far non-experimental character of social sciences, especially of the sciences of man (the humanities) causes units of theoretical knowledge which can be submitted to independent testing to be smaller in that field than the analogous units in natural sciences, especially in physics. Also, they are smaller not only because the relevant theories of functioning of experimental and measurement devices do not come into play in this case, but also because of their lower level of theoretization and abstraction. From this point of view, social sciences dominate over natural sciences as far as possibilities of verification are concerned. This is worth emphasizing, since when making comparative analyses from this or other respect of the two great branches of empirical sciences, one can usually see only advantages of the latter (i.e. of natural sciences) and the drawbacks of the former (social sciences). Nevertheless, the degree of the testability of knowledge in social sciences is generally lower than in natural sciences. Furthermore, the former encompass some hypotheses which are not empirically testable at all.

However, it should be remembered that the disadvantages resulting from the principally non-experimental character of social sciences may be, and surely actually are, to a certain (and considerable) degree offset by the role which is performed in them by practice. This constitutes a criterion of truth, which allows the separation of more true hypotheses from the less true ones. Indeed, experimentation is also usually included within practice; however, the scientific experiment does not come within the basic practice, which is productive practice (i.e. productive activity). Only experiments which are set not on exploration of new phenomena and testing of the previously framed hypotheses (such is the character of heuristic and testing experiments, which are conducted within science) but those whose direct aim is productive activity (the production of material goods) may be subsumed under basic (productive) practice.

Also, the kind of practice on which social sciences are directly based does not belong to production practice; it is an activity directed towards the formation of social life and social institutions, hence an activity of a political, cultural, pedagogical nature, etc. Apart from that, social science may adduce – like natural sciences – experimentation whose aim is not so much recognizing the phenomenon but rather transforming the reality, in this case the social reality, both on the micro- and macroscale.

To conclude, let us note that the nature of the development of science presented here, which is ambivalent from the point of view of the possibilities of testing knowledge cannot make one wonder, if we take into consideration that in every field one has to pay a definite price for progress. Thus, if we considered not the justificational but, for instance, informational aspect of scientific knowledge then it would be easy to notice that the increase in the possibilities of science in this respect, achieved thanks to the application of ever more powerful technical means of exploration, on the one hand, and of theoretical means (mathematical ones including), on the other, also leads to new difficulties. This may best be seen in the example of quantum mechanics and microphysics in general, where difficulties in obtaining information on the state and future behaviour of microobjects, caused both by means and conditions of observation and measurement "disturbing" reception of information on the state of a microobject as well as by the "mixed" nature of wave function (i.e. one that is objective-subjective on the one hand, and objective-linguistic and metalinguistic on the other) describing this state are more than sufficiently visible and do not allow obtaining information on the objective state of microobjects in the "pure" form.

However, it does not interfere with microphysics to obtain ever newer, ever more precise and complete, and deeper empirical data and theoretical information on a microobject. It is just that one should abandon the classical ideal of a fully objective description of the objects under study, which would be independent of the conditions and means of cognition.

ATOMISTIC EMPIRICISM OR HOLISTIC EMPIRICISM?*

I. Atomistic Empiricism

Traditional empiricism was atomistic: it assumed that scientific knowledge may be tested through direct comparison of the "atoms" of knowledge, i.e. particular sentences, with the "atoms" of experience, that is separate results of observation and experiments, or through the confrontation of small units (if only sensible) of knowledge with small units ("elementary" units) of experience.

"Atomistic" tendencies in empiricism can be best observed in positivism. Especially the so-called doctrine of logical atomism in which the notions of atomic sentence and molecular sentence come to the fore, proclaimed mainly by Russell and Wittgenstein, represents decisively the viewpoint of atomistic empiricism.

One of the fundamental dogmas of neopositivism is the thesis on the possibility of empirical confirmation or refutation of every statement, taken separately, by means of particular elements of experience.[1] In principle, the thesis is also shared by a critic of some of other dogmas of logical empiricism – K.R. Popper.

II. The Concept of Holistic Empiricism

It appears that the systemic character of knowledge makes it impossible to juxtapose separate theoretical statements with experimental data; what is more, theoretical hypotheses (laws) are not only and cannot be on the whole separately tested but mostly are not (and cannot be) separated

---

* Translation of "Empiryzm atomistyczny czy empiryzm holistyczny?" *Studia Filozoficzne*, **12**, 1975, 149-160.

[1] W. Mejbaum writes that empiricism of neopositivists is based on a conviction about the divisibility of experimental material into particular building blocks: impressions in the subjectivist version, and facts in the objectivist version. We subordinate sentences to those events, and a set of those sentences may be considered to be a starting point of a programme of logical reconstruction of knowledge. The programme of reconstruction itself consists in the search for rules which would allow, on the basis of elementary sentences, to justify the whole of human knowledge which we want to consider as basic (Mejbaum 1960, p. 345).

proposed.[2] For example, the Galileo-Newton principle of inertia is so much in contradiction with direct observation made in everyday life that it could not be formulated and adopted (accepted) in a different way than only in connection with those other hypotheses, which later entered into the composition of Newtonian dynamic and his theory of gravitation. It was only thanks to drawing from it, in connection with those other hypotheses (or thanks to the deduction from the system into which it was incorporated) of empirical conclusions and their confrontation with the results of experiments, it was carefully tested, or to put it more precisely, the theoretical system of statements encompassing it was tested.[3] It also appears that such a "complex" indirect testing through the deduction of common consequences is generally more precise and more certain than the testing of a law taken separately. Indeed, the principle of inertia may be tested in isolation from other statements of the system, as it was done by Galileo when observing the behaviour of bodies moving on an inclined plane, or as it was done by those who immediately followed him when they studied the behaviour of bodies in a relative vacuum and the like. However, as shown by Newton, testing of this kind has never achieved, even in approximation, the degree of preciseness that is accessible when testing it together with other principles of dynamics (and Newton's theory of gravitation), e.g. when applied to the motion of planets or the moving missile (in which case the movement of an object "is divided" into its vector components, i.e. projects on the co-ordinate axis). Apart from that, thanks to the derivation from a given law, together with other hypotheses, of conclusions and predictions in various fields, we gain a possibility of confronting that set of initial premises (thus also of the law being tested) with experience at "various points" of its application to solving various problems, so that also "the line of junction" of this law with experience becomes many-sided. As a result, we obtain testing that is more precise, versatile and reliable than in the case of testing of a law taken in isolation. This is connected with the fact that a given experimental situation, which determines the fate of a given theoretical system, usually does not consist

---

[2] Also with this respect "the context of discovery" and "the context of justification" reveal their similarity in spite of neo-positivism. On other aspects of relationships between "the context of discovery" and "the context of justification," see Lakatos (1970).

[3] The same concerns Galileo's law of fall of bodies. An experiment consisting in throwing objects from the Leaning Tower of Pisa, which was as if to begin it, has probably never been conducted (see, e.g., on this question: Butterfield 1958, Part V). This law was formulated as an element of a new physical conception, which at the same time allowed its experimental testing. It is worth noting that facts of this kind, taken from the history of science, speak in favour of both holistic empiricism (holism) and the method of idealization and hypotheticism: those laws could not be the inductive results of generalizations of empirical data.

of one *experimentum crucis* of a great solving power but of many such experiments.

The necessity of referring a given hypothesis in the process of testing not to one but to many experiments which are solving to the greatest extent possible results from one more fact, indicated by conventionalism and developed by holistic empiricism and hypotheticism: a given experimental result may be differently (although by no means totally optionally) interpreted. This is not without influence on its interpretation as a confirming fact or possibly a falsifying one (counterexample) of a given theory. In principle, this is made possible, as shown by conventionalists, by mutually matching any result of an individual observation with a given theory (with some possible insignificant modification of this theory or of any of the accompanying hypotheses). However, the set of crucial results which are part of a given solving situation, taken from various fields and supplementing one another (complementary ones), often makes this kind of "adjusting" impossible or in any case difficult to conduct due to the fact that the results of those experiments must be in agreement with one another, and their interpretations, as expressed according to uniform theoretical principles, must be coherent.

Hence, no result of a single experiment is capable of verifying or falsifying any scientific statement, especially a theoretical one. Testing in science may, therefore, consist exclusively in "comparing" certain theoretical wholes (systems) with wholly (globally) interpreted experience, accumulated in a given field and related fields, and in the confrontation of certain elements inside these structures. Conventionalists were the first to indicate the complexity, indirectness and the holistic nature of the relation between theory and experience. This indirectness does not consist only in the fact that it is not the theory itself as such that is compared with experience but empirical conclusions which are drawn from it or predictions, but also in the fact that, in this respect, one cannot do without the mediation of technology, i.e. scientific research instruments, or without the mediation of other assumptions and theoretical systems that are external with respect to the theory tested.

All this indicates that, on the one hand, no result of any single experiment can have crucial power, i.e. cannot prejudge by itself the fate of a given theoretical system, and on the other that a situation crucial to the fate of a theory must be experimental-theoretical (and not only experimental) in nature. It can be said that the crucial situation of this kind is a number of (to some extent) crucial experiments, interpreted cumulatively (more or less coherently) by various theoretical systems, competing with one another. That theoretical system is going to win, which affords the most coherent interpretation of all results of all crucial experiments which come into play, and of experimental laws, theoretical principles including, which are propounded on their basis. Both the new hypotheses and laws as well as experiments conducted for testing them, accompany one another "from the beginning to the end," i.e. beginning at the stage of formulating and ending at the stage of justification, and generally they are neither offered or interpreted, or justified separately. The standpoint defended here, emphasizing "the holistic character" of experiment and theoretical knowledge, may be called (based on the name "experimental holism" for signifying a standpoint stressing the holistic character of experiment, considered both as a "source" and as a "test" of knowledge) a theoretical-experimental holism.

The theoretical-experimental holism emphasizes the importance of a theory as compared with particular laws and the significance of a set of various (mutually complementary) experiments. And it still more strongly stresses the significance of "mixed" structures, theoretical-experimental ones, of which a situation which is crucial for the fate of the theoretical systems competing with one another comes to the fore. In the light of this standpoint, even a theory in its classical understanding as a purely theoretical system does not constitute an independent unit of cognition. Such a unit is created only by a theory together with experimental laws and facts, which it explains (those which occur within the area of its applications), and observational predictions, which resulted from it (and from the initial conditions). To put it differently, it is created by a theory together with applications which it has been possible to obtain for it. As Kuhn emphasizes, such an interpretation of a theory agrees with a manner in which a theory is assimilated by new votaries of knowledge, especially by students. For example, in physics textbooks as much (but no more) attention is paid to theories in the classical sense as to their applications, both theoretical (to other theories or in the process of solving new problems) as to practical (to planning and carrying out new experiments and to achieving utilitarian results). According to Kuhn, we have to realize clearly that the character of scientific knowledge is of the kind that scientists never adopt concepts of a theory and laws as such in an abstract way. They are in touch with those intellectual tools through their application within a wider, historically shaped structure of education. A new theory is always supplied together with its applications to a certain concrete range of natural phenomena. Without it, it could not even pretend to be recognized. The process of assimilation of a theory depends on the study of its applications, together with the practice of solving problems using paper and pencil as well as instruments in a laboratory (Kuhn 1962, part 5).

Changes occurring in the structure of modern science go in the direction of logical and methodological integration, i.e. in the direction of strengthening ties between particular statements and theoretical systems, between experiments and observations that are made, and finally, between theories and their applications which confirm them by means of experiments. That is why just the holistic view of science and the scientific procedure of obtaining and justifying knowledge becomes important as science develops, and it becomes more and more prolific. As far as the justification context is concerned, probably the most important result of such a holistic view is contained in the thesis that, due to the non-existence of absolutely certain, irrevocable, basic statements of science (and in general of the scientific results, both theoretical and technical-empirical), the best justified general knowledge is obtained thanks to the "alliance" of theory with experience. And this is thanks to the mutual correction of theoretical components of knowledge, on the one hand, and of its experimental components, on the other. As a result of this mutual control of theoretical and experimental results, the testing of knowledge occurs on all its levels and as if on all its planes, at all the border points, not exclusively on the purely experimental level, i.e. at the interface of observational statements with experience and the place of junction of all the other statements with observational statements. All the levels of knowledge are at the same time, because of non-existence of a sharp division in science between theory and practice, levels of experience but they differ in the degree in which a given knowledge is "a theory," and in what a degree "an experience" (cf. Cackowski 1972). Due to this "mixed," theoretical-experimental nature of all knowledge (we think that to some extent even of logical-mathematical knowledge), even the most general, fundamental scientific principles serve testing new scientific results, experimental one including. And thus, for instance, also the principle of causality constitutes a certain kind of measure according to which concrete results of empirical sciences are evaluated. However, this does not mean that the theoretical statements and observational sentences are fully equal as far as testing procedures are concerned, i.e. that in the case of contradiction between the former with the latter equally are admissible two possibilities: we either reject a general theoretical statement, or individual observational statement. There is a certain "epistemological order" of statements which are formulated in science. B. Russell writes on this order determining a certain sequence of recognizing or refuting statements that it is determined by which of the two statements – from the point of view of their credibility (epistemological certainty and at the same time psychological certainty) – is more of "a conclusion than a premise" with respect to the remaining one.[4] And so the epistemological premises ("basic

---

[4] See Russell (1956, p. 132). See also Gordon (1966), chapter II. Apart from "the epistemological order" (it can also be called "justification order"), Russell also distinguishes "the logical order" and "the discovery order," emphasizing that the epistemological order is not identical with the logical order (where what is meant is ordering in a deductive system by means of a relation of consequence of statements that have already been accepted) nor with the discovery order (determined by the sequence in

judgements") of testing in empirical sciences are mainly individual observational statements, and not general theoretical statements (e.g. laws), as the former serve justification of the latter to a greater degree than do the former of the latter.[5] Obviously, what is meant here in principle are the premises of reductionist concluding, and not of the deductive one. The epistemological order discussed here can be seen in the fact that laws and general hypotheses are reductionist conclusions of justifying reasonings whose "ultimate" (although not irreversible) epistemological premises are observational sentences, i.e. statements on the observed individual facts or their small sets.

This privileged position, based on the differentiation of epistemological premises and their conclusions, in the process of testing individual observational statements, is absent in the rationalistic coherentism (professed by some conventionalists and rationalists) which demands the removal of premises and conclusions, and declares that knowledge is determined by the agreement of the whole set of beliefs (Russell 1956, p. 172). The whole system of view *en bloc* should, according to this position, be recognized (or not) at the very beginning, and not according to the concrete sequence to assume first truthful premises (and rules of concluding), and only then conclusions based on them.

However, actually there is no doubt that in the case of contradiction between general and individual sentences, the general theoretical sentences are rejected, if only the observational individual sentences are well-established, i.e. directly in agreement with experience, in agreement with each other and with the recognised general knowledge (with the exception of the theoretical system which encompasses those general sentences that are in contradiction with individual ones). In this sense, experience and practice are pre-eminent over theory in the process of testing of knowledge. In this sense, the main link of the process of verification are observational sentences rather than theoretical ones, which does not at all mean that any observational sentences may or should be recognized independently of any theoretical sentences.

All this indicates that in the course of development of scientific knowledge and methodology changes have occurred not only as far as the view on what kind and degree of certainty belongs to the knowledge we

---

which statements of a system were discovered), although it has relationships with both (Russell 1956, p. 16).

[5] Russell understands an "epistemological premise," or "the basic judgement" as such a premise the recognition of which is not based on some other conviction or belief, and therefore, is independent of any understanding. The class of convictions that are not based on other convictions is, according to him, constituted by those convictions which derive straight from experience, i.e. are based directly on sensual data (cf. Gordon 1966, p. 146). Following Ajdukiewicz, we can call sentences of this kind directly justified statements.

obtained (logical, epistemological or psychological certainty, absolute or relative certainty) but also on to what elements of knowledge this (distinguished) certainty belongs.[6] Antique scholars were convinced of an absolute character of mental truths, i.e. truths obtained by means of an effort of pure intellect, without an addition of sensual cognition and practice. They (especially Plato) subjected technology to science, and the practical aspect of cognition to its theoretical aspect. In the Middle Ages this question was also generally seen in this way. But as early as during the Renaissance action began to be granted a role that was superior to that of reasoning. Finally, neo-positivists began to see in the empirical component of knowledge – which was in addition subjectivistically deformed in their system – an absolute and irrevocable basis for the whole of knowledge. However, almost at the same time conventionalism started to promote an almost opposite conception: only theoretical knowledge based on conventions may be irreversible which is sufficiently simple and useful not to be ousted by any of its competitors (according to Poincaré, e.g. Euclidean geometry represents such knowledge). As it is, no "pure" observational data – independent of whether such data exist or not – can ever suggest, and even to a lesser degree, justify any statements.

Today, there is little doubt that golden-mean solutions (offered by hypotheticists and holistic empiricists) are appropriate in this area: no element of knowledge contained in empirical sciences is absolutely exact and certain; each one of them obtains relative reliability, exclusively thanks to relations which connect it with other elements of knowledge from various levels and with experience. This solution reveals – in contradistinction to almost all the hitherto conceptions – its holistic, and at the same time, antidogmatic nature. Previously, it was thought that "the durability may be ensured to science by a single theory, a single law, single observation, i.e. particular components of two basic fields which form science" (Geymonat 1966, p. 142). At present, it is obvious that – with respect to the non-existence of any absolutely constant and certain elements of knowledge – particular components of knowledge owe their relative constancy to the fact that they are composed of harmonized greater theoretical-experimental wholes. It is just such a standpoint of theoretical-experimental holism, which takes over rational elements inherent in coherentionism and conventionalism. According to this standpoint, experimental testing of a theory should not consist in searching for isolated facts and confronting them with particular statements of a theory, taken separately but in confrontation of a theory as a certain whole, as a system with facts which are combined with one another, with the

---

[6] Geymonat writes that summing up what has been submitted to change in the relation to old conceptions is not only the nature of certainty but also the subject of this certainty (Geymonat 1966, p. 142).

totality of the established empirical facts, with the whole experimental material accumulated in a given field. As it is, facts like phenomena do not exist in a vacuum but create related structural wholes. In this kind of "whole" confrontation, apart from relations between theory and experience, an essential role is played by mutual relationships of facts with one another, on the one hand (hence the role of comparative analysis of experimental results) and connections between statements of a theory, on the other. No wonder then that a logically simple and internally compact theory "wins" the comparison with a theory devoid of those values, even if it does not exceed the latter as far as its compatibility with experience is concerned.

III. Theoretical Holism and Experimental Holism

Confrontation of theoretical statements with experience assumes, however, a previous confrontation of results of particular experiments with one another (it also assumes – which is a more obvious and less controversial thing – an appropriate ordering of those statements into a system, or into a theory). Facts taken in isolation are not even practically reliable; what appears reliable is rather certain wholes encompassing many facts in their mutual connection, interpreted in the light of a proper theory. An empirical fact serving to test a theory should be confronted with the remaining respective facts and with the whole knowledge that is at one's disposal. As it is only such a comprehensive confrontation that can ensure appropriate interpretation of a fact, i.e. interpretation corresponding to the state of knowledge achieved. After all, it is not so that, at least in general, first isolated facts are established and only then the ties joining them are searched for. Facts are recognized together with connections which join them, and this requires proper interpretation of facts and their confrontation with one another. There is no place for the Lockean-Humean dualism of "empirical facts" and "necessary relations," assuming the existence of the world as an aggregate of facts and existence of necessary logical relations and analytical statements which describe them and which are empty with respect to the empirical world.

However, the process of testing assumes not only an interpretation of facts, which is always partly theoretical. No less important for the results of testing is an empirical interpretation of a theory. The process of testing, then, consists of interpretation procedures of two kinds: theoretical interpretation of the results of experiments and an empirical interpretation of the theoretical system tested. Holistic empiricism adopts a view according to which evaluation of a theory and of its empirical interpretation, revealing relations of a theory with experience (with empirical facts) is impossible without an evaluation of a class of facts and their (trial) interpretation in the light of competitive theories. Both these evaluations constitute a whole of two component parts, which are non-autonomous with respect to each other.[7] Confirmation of a given theory means choice – on the basis of certain criteria which have an objective value – of this interpretation of the set of obtained facts, which is compatible with this theory or even conducted on its basis. In turn, falsification of a theory or its modification involves the necessity of the interpretation of facts, which are initially interpreted as confirming that theory, and which at present must be expressed in the light of some new (or the hitherto but modified) theory. In this way, in the process of testing laws and theories there occurs a mutual adaptation of various component elements of science: theoretical and observational ones. Holistic empiricism interprets completely both statements of science and their relationship to experience, as well as experience itself, and even the reality that is "the concealed" behind it. Thus its components are: theoretical holism, emphasizing that the basic units of knowledge which are submitted to individual testing are theoretical systems (theories and their sets) and experimental holism which indicates that the particular experimental situations which solve the fate of these theoretical systems consist of numerous experiments which complement one another in the sense that they impose mutual restrictions on the interpretation of their results. Thus they preclude (falsify) those systems of knowledge which cannot be reconciled with interpretations of experimental results that are possible to accept.

   In favour of the standpoint of holistic empiricism speaks, among other things, the fact that in science one usually does not abandon particular statements and laws but (at once) the whole theoretical systems, and also the fact that even when we reject some of the statements of a system (modifying it) every time it is necessary to introduce modifications (changes in meaning and re-evaluations) of many (or all) the other statements, namely with respect to mutual logical and epistemological relations occurring between various components of a theory. Some holistic empiricists (e.g. Quine) even think that it is necessary to submit to the whole confrontation with not only systems of empirical sciences but also the logical-mathematical systems, creating with the former larger scientific systems empirically tested as a whole.

---

[7] A completely different approach in this matter is that of instrumentalism, for which the problem of truthfulness of a theory is resolved to the question of truthfulness of observational sentences (i.e. prognoses) derived from a theory. On such an approach, it is important not only whether each of the prognoses is true, and there is no sense asking whether everything that a theory "claims" in addition is true or not. It is worth noting that what is concerned here is the so-called partial instrumentalism: as it is, according to another – more radical kind of instrumentalism ("holistic instrumentalism," for instance, at present represented by Quine), also observational statements are devoid of logical value.

According to this standpoint, to put it exactly, confrontation is possible with experience only of the whole of our scientific knowledge, encompassing equally the statements of empirical sciences and the mathematical–logical hypotheses as well as ontological assumptions. Holistic empiricism, conceived of in such a way, although it sometimes formulates its theses in an exaggerated way, still rightly emphasizes not only the multiaspectual nature of relations occurring at a given level of theoretization of knowledge but indicates the multitude of the levels of these relations (their "multilevel character"), and also relations between and among various levels.[8]

The fact that we do not usually test empirically single statements or even single theories but whole (more extensive) theoretical systems is confirmed not only by the fact that interesting scientific prognoses are derived generally thanks to the application of several theories but also by the fact that the more complicated explanations are based on premises usually taken from many different theories. Contemporary studies in advanced disciplines engage various complicated technical research tools, and thus various theories of functioning of these tools, and scientific theories are so closely related with one another that even a relatively simple explanation refers to all the theories at least in the field of a given discipline. The same pertains to testing. It is obvious that if we take into consideration that testing is a procedure which also refers to technical research tools (e.g. experimental devices, measuring instruments), and what follows to the theory of their functioning, and is an activity which is based on predicting without which it is generally impossible, and predicting is closely related with explaining. The thesis that general statements (laws and theories including) which are the proper candidates for becoming theoretical premises of explanation reasonings are also equally good for general premises of predicting reasonings and vice versa, has numerous confirmations, although does not hold universally. In any case, the ability of statements to predict correctly may be considered – at least within exact sciences – as a test (the necessary condition, although not a sufficient one) of the value of scientific explanations provided by a given theory: it is an ability of predicting appropriate phenomena (i.e. phenomena of the same type as the phenomena which are explained)

---

[8] When discussing attempts at introduction of a new logic into microphysics ("the logic of microphysics," logic of quantum mechanics), A. Zinoviev writes: "The case of microphysics is not the only one of its kind. An attempt at giving great scientific discoveries revolutionary a character not only in people's views on one or another fragment of reality (now you cannot entrace anyone with it!) but also in the same logical foundations of science, constitutes one of the characteristic features of a cultural environment in which contemporary science develops. It is just from this side that the analysis of the situation with 'the logic of the microworld' is instructive" (Zinoviev 1970).

within the framework of that theory and known facts.[9] If, on the other hand, on the basis of sufficient facts for the explanation of a given phenomenon a theory (law) is not capable of predicting this phenomenon, then one can consider it not as a fully valuable theory (law) of strict empirical sciences but as a theory which is – at least partly – speculative, and explanation conducted on its basis – as a rather apparent explanation than the scientific one. What is concerned here is obviously an ability of a theory for real scientific prediction (understood as deriving knowledge about the future from knowledge about the present and the past), being a test, of correctness of explanation, and not e.g. about prophecy or clairvoyance (understood as achieving direct knowledge about the future without the participation of knowledge about what occurs at present or what happened in the past). It can happen that a scientist formulates an astonishing prognosis which is actually confirmed in future, and even in spite of the fact that his/her explanation was unacceptable. However, then it appears that it was actually not a scientific prediction but a lucky prophesy.[10] Thus, the holistic nature of scientific explanation and prognosing is an additional argument for the holistic interpretation of testing procedures, for the experimental-theoretical holism.

---

[9] This test, understood as a necessary condition (and not as a criterion by which one usually understands the sufficient conditions) of correct explanation, is commonly recognized only on the grounds of the so-called exact natural sciences (physics, chemistry and astronomy). As applied to social sciences or let us say to biological sciences (the latter are also sometimes considered now as exact sciences; however, it is right surely only with reference to certain modern fields of biology such as molecular biology or genetics), it seems to be exaggerated. Within those sciences there exist qualitative theories which have no mathematical formulation (e.g. Darwin's theory of evolution of species) which provide, on the basis of the established set of facts, many valuable explanations but do not allow – at least on the basis of this set of facts – any noteworthy predictions concerning phenomena of the same class as the phenomena being explained. Many times, an indispensable condition of a better (sufficient) testing of such theories is making them more precise so that it is possible to derive from them unequivocal prognoses of a quantitative character.

[10] Nevertheless, we can also predict phenomena where explanation is unattainable for us for the time being. The fact that we can sometimes actually predict phenomena in a given field without any ability of explaining them is indicated by the fact that sometimes our predictions are based on observational general statements (phenomenalist generalizations) of small or no explaining power or on analogies and the general principle of determinism: we can conclude from the fact that some definite conditions occur again that there will appear again phenomena determined by those conditions, which does not yet mean that we can correctly explain these phenomena. It is a proof that abilities of prognosing should by no means be considered as a sufficient condition for correct explanation.

IV. Conclusion

To conclude, it is worth emphasizing that the recognition of one of the holisms presented does not always go hand in hand with being in favour of the other one. Theoretical holism has many defenders among the representatives of various disciplines today. However, experimental holism does not have determined proponents, which is not a very well-known standpoint and, one could say, not very precise.[11] Oftentimes even those who favour extreme theoretical holism base themselves on experimental atomism. Duhem and a number of instrumentalists who are his contemporary advocates may be given here as an example. On the other hand, Einstein's numerous utterances may be interpreted in the spirit of experimental and at the same time theoretical holism. Undoubtedly, he applied holistic approach in his research practice as well. Let the following opinion serve as an example. "Thinking allows one to build a system: the content of the results of experiments and connections between them are presented by means of consequences obtained from theory. It is just in the possibility of such a presentation are contained the value and justification both of the whole system and of the concepts and principles that are at its foundations" (Einstein 1965, p. 62).

---

[11] A closer characterization of this standpoint and arguments in its favour are provided in Such (1975a).

THE NOTION OF AN *AD HOC* HYPOTHESIS*

In research practice, both at the stage of formulating problems and that of framing hypotheses as well as at the stage of their testing, scientists sometimes happen to use various procedures such as, for example, *ad hoc* explanation, *ad hoc* reduction of some theories to some others or saving theories, threatened by experiment, through the introduction of *ad hoc* modifications.

The component of all these *ad hoc* procedures, which are usually negatively evaluated both by scientists and methodologists is framing *ad hoc* hypotheses (literally, those serving this special purpose).[1] Hence the importance of this concept, which is after all applied very often in the process of analysis, and especially evaluation of new theoretical conceptions, both with respect tot the methods used in the course of their formulation as well as to the cognitive results achieved. Hence the need to explicate it.

Analyses of the concept, offered so far by C.G. Hempel or A. Grünbaum show, however, that giving a general definition of an *ad hoc* hypothesis in purely logical terms – thus without referring to some or the other circumstances of pragmatic nature – is either not possible at all, or unattainable at the present stage of methodological considerations (cf. e.g. Hempel 1968, Chapter III; Grünbaum 1963).

This is why this study introduces three concepts of an *ad hoc* hypothesis which correspond to the three main (and perhaps the only) kinds of *ad hoc* hypotheses that are encountered in science. I hope at the same time that in future it will perhaps be possible to supply a definition that would encompass in principle all those hypotheses that scientists recognize as *ad hoc*. Besides, also the definitions introduced below of the three kinds of *ad hoc* hypotheses are not free from reference to certain pragmatic circumstances, since they are not given exclusively in logical categories.

The first two concepts concern the so-called auxiliary *ad hoc* hypothesis, and the third one concerns the concept of the independent *ad hoc* hypothesis.

Let us begin with the first kind of an *ad hoc* auxiliary hypothesis.

---

* Translation of "Pojęcie hipotezy *ad hoc*" *Studia Filozoficzne*, **9**, 1975, 95-110.

[1] According to the dictionary definition *ad hoc* (a Latin word) means "for this (special purpose); with respect to this (subject or thing)."

I. Auxiliary ad hoc *Hypothesis of the First Kind*

The main intuition at the basis of the concept of an auxiliary *ad hoc* hypothesis of the first kind is as follows: an auxiliary *ad hoc* hypothesis *H* is a hypothesis that cannot be submitted to testing independently of an experimental result *E* that contradicts a certain theory *T*. This intuition relativizes the concept of an *ad hoc* hypothesis to the theory for the "saving" of which – against the danger on the part of experiment – the hypothesis has been formulated. Making this intuition more precise and explicating it should go along the line of explanation of what is behind the concept of the "possibility of independent testing." However, this concept appears to be highly complicated (see Hempel 1968). This is indicated by the fact that from any auxiliary hypothesis *H*, introduced in order to explain result *E* which contradicts theory *T*, it is possible to derive testable consequences that may be tested only together with one or several initial assumptions (postulates) of theory *T*. To put it differently, in the process of defining the concept of the "possibility of independent testing" and, consequently, of the concept of an *ad hoc* hypothesis one should take into consideration the circumstance that the tested consequence of hypothesis *H* has a contextual character (cf. Mierkulov 1971, p. 87). Due to the contextual character of the tested consequence of hypothesis *H* it seems hopeless to give in purely logical terms the definition of an (auxiliary) *ad hoc* hypothesis of the kind discussed. Whether a given hypothesis belongs to the class of the *ad hoc* hypotheses or not will depend on the concrete theoretical system to which this hypothesis is relativized (referred to), and which should be investigated beforehand in order to decide whether this hypothesis is an *ad hoc* hypothesis[2]; but as it appears it depends also on other circumstances – theoretical and experimental – under which this hypothesis is formulated.

Recognizing the above fact, we assume – following Hempel – the following and only approximate and not extended onto all cases of auxiliary *ad hoc* hypotheses known to physicists[3] – as this is demonstrated by A. Grünbaum's analyses – definition of an auxiliary *ad hoc* hypothesis of the first kind: hypothesis *H* which allows theory *T* to explain in conjunction with *H* an empirical result *E* negating theory *T* is an *ad hoc*

---

[2] In particular, it may appear that a given (the same) hypothesis in conjunction with one theory has an "independent testing," and in conjunction with the other does not; thus, while exhibiting the features of an *ad hoc* hypothesis in the second case, it does not in the first.

[3] Grünbaum in particular shows that this definition does not contain the case of an *ad hoc* hypothesis which occurs in the modified electron theory of Lorentz (what is meant here is Lorentz-Fitzgerald's *ad hoc* hypothesis concerning the shortening of bodies in motion and Lorentz-Larmor's *ad hoc* hypothesis assuming the retardation of time). See Grünbaum (1963).

hypothesis if no testable consequences are derivable from it which would essentially or considerably differ from the negating result E.[4]

The typical *ad hoc* hypotheses of the kind discussed here are hypotheses consisting in the modification of some hitherto well and comprehensively tested laws by the establishing of a certain experimental result of a given (one) kind which on the grounds of the hitherto knowledge cannot be easily adjusted to this law.[5] Of such an *ad hoc* nature were the various modifications of Newton's law of gravitation, suggested after observing certain anomalies in the movements of Uranus which were then correctly explained within the framework of that law (thanks to the discovery of Neptune). One of the proofs of their *ad hoc* and hence not very heuristic nature is provided by the fact (apart from the fact that these modifications had not been confirmed in other astronomical phenomena) that in order to explain these anomalies of the movement of Uranus, Newton's law of gravitation would have to be given an extremely complex form while the competitive hypothesis (an assumption that there exists a new, so far unknown planet of inappropriate magnitude and location), was, based on existing knowledge, an incomparably simpler and heuristically more fertile hypothesis. No wonder that when Leverrier showed that the assumption of existence of a new planet explains well the observed deviation of the movement of Uranus it had been – as the simplest of hypotheses possible to adopt – commonly recognized earlier than the new planet was discovered. Thus, the best evidence of its fertility was just the fact that a similar way out was found when anomalies were attested to this time in the movement of Neptune: Pluto was discovered, which is the outermost planet of our system.

However, it should be emphasized that an attempt at an analogous explanation of anomalies in the movement of Mercury (the precession of its perihelion) through the introduction into the solar system of additional masses, undertaken by Leverrier[6] and others, had not been successful and

---

[4] The following definition, also coming from Hempel, is somewhat simpler but at the same time worse: if from theory T results consequence E falsified by experiment then hypothesis H is an *ad hoc* hypothesis in the context of theory T only if the same testable consequences are derivable from conjunction $T \wedge H$ as from T, with the exception of the negating result E.

[5] The fact of frequent formulation of *ad hoc* hypotheses just in such circumstances is an additional strong argument for the thesis of "experimental holism," claiming that the situation determining the fate of a given theory cannot consist of a crucial experiment of one type but should contain experiments of many kinds which appropriately are complementary to one another. The concept of a crucial situation, consisting of complementary experiments to one another and with competing theories (theories competing with one another) was introduced in Such (1975a).

[6] In 1859, Leverrier stated that the shift of 38" per hundred years in the Mercury perihelion may be explained, assuming the existence of a planet closer to the Sun than

was ultimately considered to be an *ad hoc* attempt (one can, anyway, doubt whether it had any reason). Since, from the logical point of view, the hypothesis of the existence of mass close to the Sun (Vulcan) which was to explain anomalies in the movement of Mercury is analogical to Leverrier's hypothesis on the existence of an additional planet (Neptune) which explains anomalies in the movement of Uranus, and also to the hypothesis on the existence of Pluto which explains anomalies in the movement of Neptune, and what is most important, all these hypotheses are in the same relation to the maternal theory (Newtonian theory of gravitation), thus giving in purely logical terms not only of a general definition of the *ad hoc* hypothesis but even of the *ad hoc* hypothesis of the kind discussed here seems to be really impossible – at least at the present state of methodological investigations. In the case considered one could only discern certain pragmatic differences that are close to impossible to define clearly. One of them would consist in the fact that the assumption of the existence of larger masses near the Sun, which so far no one has been able to observe, is more risky than the hypothesis on the existence of Neptune or Pluto, i.e. planets that are more distant from us. However, one can doubt whether differences of this kind allow us to conclude that the hypothesis on the existence of mass near the Sun is an *ad hoc* hypothesis if analogical hypotheses (one of which was framed by the same scientist who also framed the first hypothesis) would lead to the discovery of Pluto and Neptune. Maybe one should consider that this was a correct hypothesis which, however, appeared to be false. On this example we can see that hypotheses considered to be *ad hoc* in nature sometimes also lead to predicting facts of a new kind, and whether a given hypothesis will ultimately be recognized by specialists in a given discipline as an *ad hoc* hypothesis largely depends on whether its prediction concerning a given case (in this instance assuming the existence of a new planet) comes true.

The matter is not hopeless, however, if we consider that hypotheses on the existence of Neptune and Pluto had further observational consequences. Their confirmation made it possible to justify them "in an independent way" (these included predictions that were tested somewhat later concerning disturbances in the movements of other planets and of their moons). On the other hand, no further predictions resulting from the hypothesis on the existence of Vulcan found their confirmation, and in this connection, the hypothesis was not independently confirmed either at a later time. The problem is complicated enough that only in the light of further studies, thus *ex post*, is it possible to establish whether a given hypothesis has further observational consequences by means of which it

Mercury. When in the same year a French physician Lacerbaut noticed a passing of a dark body in front of the Sun, Leverrier claimed that this was the planet that was looked for, calculated its orbit, and called it Vulcan.

obtains "independent confirmation." That is why also the above difference between hypotheses on the existence of Neptune and Pluto on the one hand, and of the Vulcan on the other, cannot be described in purely logical terms, since it requires a reference to adducing time, to further progress in studies and the like circumstances of a historical and pragmatic character. On the other hand, the attempt to remove, based on the model of stationary Universe, the gravitational paradox (the Neumann-Seeliger paradox) that was suggested by C. Neumann and H. Seeliger (see Neumann 1896) was undoubtedly an *ad hoc* procedure. Their modification of the law of gravitation consisted in introducing into an equation of a potential – apart from the gravitation constant – an additional constant called "the cosmological constant" W. However, no empirical data justify its introduction.

Analogical attempt was made later by Einstein who introduced *ad hoc* a cosmological constant (the so-called cosmological term) to the equations of the general theory of relativity. Even the aim was the same: to assume the stationary character of the Universe. It was in 1917 that Einstein applied the general theory of relativity to the description of the Universe (see Einstein 1917). Seeking to avoid the gravitational paradox – like Neumann and Seeliger – he introduced a cosmological term into his field equations: Ω_{gik} (where $\Omega > 0$), which was equivalent to introducing a hypothesis of the existence of the cosmic repulsing force, acting on great distances.[7] It is interesting that Einstein's equations containing an expression with the cosmological constant are more general than Einstein's classical equation (i.e. without the cosmological constant). Therefore, one might ask why the introduction to the equations of the cosmological term was an *ad hoc* correction if it did not lead to any complication but a logical (although not mathematical) simplification of the theory.

On the other hand, a similar problem arises with reference to a number of other modifications that have been proposed earlier. For instance, with reference to the theory of gravitation, proposed by E. Dicke in 1962, and in which the principle of relativity of inertia (Mach principle) was to be fulfilled thanks to the introduction of a certain scalar field which also led to the generalization of the initial theory (Newtonian, respectively Einsteinian).

A correction is in both (and the similar) cases an *ad hoc* one; it is because there are no empirical data which confirm – in a way that would

---

[7] See Skarżyński (1969), pp. 171-177. One of the reasons of rejection of modified equations of Einstein's field with a cosmological constant was the fact that they do not lead – in Newtonian approximation – to the Neumann-Seeliger equation but to another one. This indicates that not fulfilling the principle of correspondence may be a certain indicator (diagnostic feature) that a given modification has a character of an *ad hoc* hypothesis.

be independent of the phenomenon which it tries to explain through the introduction of a given correction – the existence of "objects" assumed by these corrections (repulsing forces in case of Einstein's theory, scalar field in case of Dicke's theory and the like) (see Bażański, Demiański 1967, p. 158). Thus, regardless of whether a given correction means a logical complication of the initial theory or not, the fact that it does not find confirmation (as a component of the explanans) that would be independent of the explanandum, prejudges its character as an *ad hoc* hypothesis.

On the other hand, also Lambert-Charlier's model of hierarchical Universe was constructed *ad hoc* (its task was to overcome both cosmological paradoxes: a photometric and a gravitational one in a totally different manner) since the hypothesis that galaxies form greater and greater concentrations, between which there occur ever greater areas of space where there are no galaxies, has not been confirmed. The first non *ad hoc* way out of this situation which appeared to be heuristically fertile was pointed out by Friedman in 1922 when he showed the possibility of removing the gravitational paradox without modifying the field under the condition that the assumption of stationariness of the Universe is rejected. In connection with the observed "shift towards red" of the spectrum of distant galaxies (interpreted in a Dopplerian way as "an escape of the galaxies") and observed two years later (i.e. in 1924), that rejection appeared to be just heuristically fertile. The gravitational paradox does not appear in any theory of evolution of the Universe (the kinematic theory of relativity, the theory of the steady state, Jordan's theory, Dirac's theory or Eddington's theory). Hence the fertility of Friedman's hypothesis on the non-stationariness of the Universe.[8]

The discovery of the so-called cosmic background radiation additionally confirms this hypothesis, revealing its further heuristic possibilities.

II. An ad hoc *Auxiliary Hypothesis of the Second Kind*

Some of the examples considered above point to a certain significant feature of the *ad hoc* auxiliary hypothesis which we have already mentioned: they point to the fact that it is a hypothesis that generally (although not always) logically complicates the theory which it is to defend.

I think that the above feature should be considered as a defining feature of the auxiliary ad hoc hypothesis of the second kind. Considering the above allows one to understand why the hypotheses of Fitzgerald-

---

[8] However, it should be emphasized that also certain theories of the constant Euclidean Universe somehow manage the gravitational paradox (e.g. Keres and Kipper's theory who were astronomers from Tart).

Lorentz contraction and retardation of the Lorentz-Larmor time, which were to save Lorentz's initial theory of electrons were *ad hoc* hypotheses. As we know, thanks to those hypotheses, Lorentz's theory assuming the existence of ether, appeared no less compatible with experience, and even not less heuristically fertile – but only as far as the then known and only observational consequences – than the special theory of relativity.[9] However, it owes this to the considerable logical complexity of its foundations when compared with Einstein's theory. And this is due, among other things, to the above mentioned Lorentz's hypotheses of contraction and of time dilatation, which are introduced into it independently of the initial postulates of the theory while in the special theory of relativity they are an outcome of the postulates of the latter. One can conclude from the above that hypotheses are *ad hoc* not only when it is not possible to test them independently (in the sense discussed above)[10] and do not ensure a theory of the proper heuristic fertility but also when they make it immensely and needlessly complex, the proof of which may possibly be the existence of considerably logically much simpler competitive theories. These theories explain all the facts that are explained by this more complex theory (which refers to *ad hoc* hypotheses).

Therefore, what comes into play is the concept of the logical simplicity of a theory which is used in its methodological considerations by Einstein, among others. He noticed that the criterion of "conformity with experience" must be supplemented with the "simplicity" criterion, since through introduction of additional *ad hoc* hypotheses which complicate the theory one can always save it from experimental facts which refute it.

According to Einstein, the logically simpler a theory, the greater its informational content, on the one hand, and the smaller the number of initial statements (postulates), on the other. Thus, the following formula expresses the degree of logical simplicity of a theory:

$$Ls = \frac{Inf}{N \cdot Post}$$

---

[9] The latter appeared undoubtedly much more fertile as far as the development of theoretical physics is concerned. It owes its fertility in the theoretical aspect first of all to the postulate of relativity, and also, which is its expression of the principle of the invariability of statements with respect to Lorentz's class of transformations. Only on its basis – and in no instance on the basis of Lorentz's theory – could Einstein's general relativity theory (theory of gravitation) arise.

[10] According to A. Grünbaum, the Fitzgerald-Lorentz and Lorentz-Larmor hypotheses are not *ad hoc* in the sense of this concept "impossibility of independent testing" which was introduced – following Hempel – at the beginning of the previous paragraph. However, in this case a somewhat different concept of the "impossibility of testing" may come into play.

It can be put differently as follows: the logically simpler a theory, the lesser number postulates (initial statements) it is based on, and the more logical consequences (derivative statements) can be educed from it:

$$Ls = \frac{N \cdot Kons}{N \cdot Post}$$

An auxiliary *ad hoc* hypothesis of the second kind is a hypothesis which leads to the logical complication of a theory: increasing the number of its initial postulates, it does not lead to a proper increase in the number of its logical consequences, including observational consequences (e.g. predictions). Following some authors, we might introduce a concept of dynamic simplicity derived from the concept of logical simplicity. This concept would concern a series of modifications of a (given) theory following in succession,[11] and not a theory at a given moment (a given stage of its development). An *ad hoc* hypothesis of the second kind is a hypothesis the introduction of which leads to the shaking of the criterion of dynamic simplicity.

So what does the *ad hoc* nature of hypotheses modifying Lorentz's theory consist in if, as far as the experimental scope is concerned, the theory explains and predicts essentially the same phenomena as the relativity theory? First of all, the *ad hoc*-ness in due to the fact that, from the logical point of view, the hypotheses considerably complicate the existing theory and in sum lead to explanations which are immeasurably complicated as compared with those provided by the relativity theory. Also, the theory modified by them is characterized by a considerably lesser heuristic fertility in the theoretical scope than the latter one and obviously does not "fit" the remaining physical knowledge which it actually falsifies.[12]

In the process of assessing the degree of *ad hoc*-ness of a given hypothesis one should take into consideration not only the heuristic power of the hypothesis as such and the heuristic power which it "provides" to the modified theory but to compare them with the heuristic power (both as far as experiment and theory are concerned) of the competing theories, and also take into account the degree of logical complication which it introduces to the "mother" theory and to compare the degree of logical simplicity of that theory with the degree of logical simplicity of competing

---

[11] Such a concept is introduced, for example, by Mierkulov (1971).

[12] What we mean here is the indirect falsification of a theoretical nature, i.e. incompatibility with the recognized theories and, in addition, to an extent to which those theories appear to be well confirmed and thus true. In the case of a weaker confirmation of the latter, instead of falsification, we can speak of the disconfirmation of the theory being tested (in this case modified Lorentz's theory).

theories. Such a comparative analysis of the logical simplicity of competing theories and of their heuristic power (before and after the introduction of a given hypothesis) may play a no lesser role in the process of establishing whether a given hypothesis is *ad hoc* than the decision on whether it is "independently confirmed" or not. One should bear in mind that in the first place just from this side of the hypothesis of Lorentz's shortening and Larmor's dilatation of time are not faultless. However, giving up these hypotheses (to put it more precisely, reinterpreting them on the basis of the special relativity theory, which after all indicates that these were *ad hoc* hypotheses only to a certain, limited degree and in a certain specific sense, and not the typical *ad hoc* hypotheses) scientists were also guided by the fact that they are basically unavailable to critical argumentation on the basis of experiment. Any measuring instrument itself used an analogous shortening for the measurement of Lorentz's shortening experiences; the same applies to a clock which is submitted to the same retardation as processes whose decrease in the tempo of occurrence one would want to establish with it. Anyway, in the case of Lorentz's theory we have encountered an exceptional and strange situation whereby if not for the special relativity theory (a victorious competitor) and further development of physics which it permitted then perhaps no one would accuse Lorentz's theory of its lack of simplicity and heuristics (if only theoretical one), while its modifications would not be qualified as *ad hoc*.

One should recognize that a theory which explains in a better and simpler way a well-known set of facts must also explain future facts better. As it is, a logically simpler theory is at the same time more fertile heuristically; among other things, it provides better prognoses and as a result allows a better explanation of the future facts (i.e. predicted facts) as well, it fits better in the practice of tomorrow, and not only to that of yesterday.

A hypothesis which (1) cannot be "independently tested" and which additionally (2) complicates the initial system logically (by introducing additional assumptions or additional beings that are redundant from the point of view of other theories) could be called a doubly *ad hoc* hypothesis, or a squared *ad hoc* hypothesis, as it were. Only "hypotheses" that are completely devoid of test implications, hence hypotheses which, within empirical sciences, are actually either apparent statements, devoid of the cognitive meaning or in any case statements which are empirically untestable, may be considered to be "worse" than the type of hypotheses which are being doubly *ad hoc*. However, hypotheses that are devoid of empirical content do not deserve, within empirical sciences, the name of hypotheses at all (some people also refuse to use the term *ad hoc* hypotheses and treat both the above kinds of hypotheses rather as speculative statements than as candidates for proper statements of special sciences).

Obviously, in the case of stating unexpected facts or those that are not in accordance with the prediction, it is worth trying to modify the theory before it is rejected, so that it would assimilate the new facts. But the price which is paid for that modification cannot be too high. It can neither lead to excessive complication of the theory nor to the increase in its "flexibility." Especially that the price which is to be paid for the extension of the initial limits of applicability of the theory is often such an increase in the "plasticity" of the theory that it becomes incapable of making unequivocal predictions. Scientists know very well that a theory is worth anything only if it is sufficiently unequivocal and informative. Hence, some of them even believe that the principle (criterion) of heuristicity does not have any value without being supplemented with the principle (criterion) of falsifiability.[13]

It is sometimes possible to provide a reconstruction of a theory whereby the number of its assumptions not only does not grow but, conversely, decreases (an example may be Kepler's reconstruction of Copernicus's theory. Instead of many epicycles "enforced" by the assumption that planets move with a constant velocity on circular paths, Kepler introduced elliptical paths of planets moving with variable velocity; or a modification of the general relativity theory consisting in deriving equations of motion from equations of the field). However, reconstruction of this kind is performed usually only at transition to a new theory by means of correspondence relation: thanks to the decrease in the number of fundamental assumptions a new theory becomes more general and more precise than the previous one. Then we have a process of "correspondence generalization" of a theory through decrease in the number of initial assumptions of the theoretical system.

One should bear in mind that sometimes a modification of a theory only apparently does not lead to its logical complication or in any case to the complication of knowledge accumulated in a given discipline. For instance, according to conventionalists, in the case of the incompatibility of a theory with experience it can be established through changing certain definitions or through adoption of additional hypotheses. The second way obviously leads to the complication of a theory (unless it considerably increases its informational content). But also the former one is not as

---

[13] Cf. e.g. Zotov (1969, p. 74). "Heuristics is not a theory – Zotov writes – which opens a very large number of possibilities but the one which more or less precisely gives the location and characteristics of a so far unknown phenomenon, distinguishing it from those which should not have occurred. In other words, a theory should not foresee too much. Paraphrasing a well-known logical rule, we can say: a theory that foresees too much, does not foresee anything. Its fertility is transformed into sterility, its expansion into an attempt at self-preservation. If a theory is so liberal that *an optional* experimental result may be agreed with it, if it is not able to point to its own limits – it is *not a good* theory." In this case it is an empty, sterilized construction (p. 71).

innocuous as one might think if taking a purely formal approach to the matter, considering only the fact that the change of a definition in such a case mostly means merely a substitution of one expression with another, which is either logically non-equivalent or does not have the same meaning.[14]

Let us suppose, after Poincaré that the sentence "Phosphorus melts at the temperature of 440°C" is a part of the definition of phosphorus. Having established experimentally that a body called that thus far melts at a somewhat different temperature, we could save the theory which includes that sentence through modifying the definition of phosphorus. In scientific practice this kind of modification, however, may lead (and usually does) to complicating the existing knowledge. In this case, it would lead to such a complication if it appeared that the scope of the theory is narrowed down or that, on the strength of functional relations (quantitative laws) assumed by the theory that is being defended, the other definitional features of phosphorus are connected with the melting point of phosphorus which comply with its initial definition, and are incompatible with the modified definition. As it is in this case the change of definition of phosphorus would involve further modifications of the theory concerning its component laws, and these would be modifications which would make it more complicated.

Poincaré's conventionalism is accused of allowing to save theories which are not compatible with experience, also at the price of modifying a theory by means of introducing *ad hoc* hypotheses and changes of definitions also in cases when these procedures lead to a logical complication of a system. This objection is the all the more serious that it points to some incoherence contained in the assumptions of a doctrine. On the one hand, an attempt at saving theories endangered by experience even at the price of their obvious complication through introducing *ad hoc* hypotheses and, on the other hand, an emphasis on simplicity and elegance of a system as the basic (or some of the basic) criteria of selection of a "correct" theory.

III. An Independent ad hoc *Hypothesis*

Apart from the already discussed concepts of the *ad hoc* hypotheses as auxiliary hypotheses that are not "independently confirmed" and/or do not strengthen the heuristic value and informational content of a given theory but rather impair it, which leads to the logical complication of a theory, it is worth introducing the third concept of an *ad hoc* hypothesis, as the one

---

[14] The problem of defending a theory against refutation through a redefinition of concepts (change in the scope applicability of a theory) and numerous examples of this kind of defence are described in (Lakatos 1976).

that itself provides merely apparent explanations and is unable to predict new facts, besides the facts for the explanation of which it has been constructed. An *ad hoc* hypothesis of the last (third) kind is a hypothesis optionally adopted for the explanation of a phenomenon of a given type. However, the hypothesis is not capable of providing any explanations or predictions concerning phenomena of a type other than those taken into consideration in the process of its creation. In principle, descriptive statements (e.g. reporting generalizations) may be of such reporting nature but explanatory statements (laws or let's say historical generalization) cannot. Hence, the problem of *ad hoc* statements appears only in connection with explaining, and not with description as it is connected with *ad hoc* explaining, i.e. explaining where the explanans does not obtain its confirmation independently of the explanandum.[15] Assumptions and hypotheses which postulate "something" and find their confirmation exclusively in facts for the explanation of which they have been formulated are sometimes called methodologically vicious (cf. e.g. Braithwaite 1954, p. 327). Sometimes this "something" is assumed beforehand as the one which has no other features than those which are to explain a given phenomenon for the explanation of which the hypothesis has been formulated. Such a postulate of "something" which has no other features than those which serve to explain facts with respect to which the postulation takes place is a "methodological vicious circle." And if, in addition, such a "something" is given features that are not empirically testable, then such a postulation deserves the name of a metaphysical (speculative) procedure.

How to avoid formulating this kind of independent *ad hoc* statements, or at least how to establish at least that we have to do with them in a given case? The simplest way to detect them consists in an attempt to derive from them various empirical consequences which can be confronted with experience. Jevons claims that we should not trust any universal statement (law) before we treat it deductively and show that of the conditions assumed, predictions result, which have been foreseen by the law (Jevons 1960a, p. 413). Even the largest number of individual facts as such does not allow us to draw a conclusion concerning new facts (pp. 414-415). Facts may suggest an idea to us but the actual establishing of a new general law always occurs through pure deduction (p. 417).

---

[15] Empirical hypotheses concerning phenomena that can be observed directly and thus unable to perform explanatory functions are – according to Jevons – almost totally devoid of heuristic value. Jevons calls such hypotheses descriptive ones and says that they serve almost exclusively to provide us with appropriate names (Jevons 1960b, p. 200).

An example of a theory which provides only apparent explanations (*ad hoc* explanations) may be, as shown by Jerzy Kmita, the model theory defining the "structure" as constructed by Lévi-Strauss for the needs of ethnology (see Kmita 1971, p. 134).

The scornful attitude of Jevons to general statements, which have not been justified through deductive derivation of empirical consequences from them and their confrontation with facts, results from the fact that he treats these statements as to some extent *ad hoc* ones: they have been offered on the basis of examining (and thus also for the purpose of explaining) some (usually) narrow and rather homogeneous class of phenomena. Only deriving from them deductively ever newer and ever more varied empirical consequences and confronting them with new experimental facts "deprives" the initial statements of the character of *ad hoc* hypotheses (to be more precise, it appears that they were wrongly treated as such). As it is, each of the cases of testing the law "deductively" (in the sense indicated) represents a whole group of cases which are in some respect (but not completely) homogeneous and is their "sample" so it has – as we might say today – a greater "power of confirmation" than any of the "inductive" cases which "suggested the idea" of a law, or served to formulate it, and which do not represent any other cases. It concerns in particular those cases the occurrence of which had previously been theoretically predicted and then confirmed by observation. Compatibility with facts is a test of a true theory but when a conclusion from a theory has been predicted beforehand, then there is no doubt that a theorist interprets the results of his/her own theory impartially (Jevons 1960b, p. 221). According to Jevons, it is just that the superiority in the testing process of the law of experimental facts which have first been predicted on the basis of the law and only then observed, over facts which were known previously and to which the law may be partially and *ad hoc* "fitted in." As an example of such a hypothesis that is "fitted in" to the fact that Encke's Comet each time returned a bit earlier than had been predicted, Jevons provides a hypothesis that there is an environment (ether) which fills in the space through which the comet runs and which resists it. According to him, the hypothesis is a *deus ex machina* serving to explain that single phenomenon and its probability must be slight until it is possible to show that other phenomena may be concluded from it (Jevons 1960b, p. 274). It is characteristic of the independent *ad hoc* hypotheses and theories that in order to make them agree with new experiments they in turn require the introduction of further *ad hoc* corrections, i.e. auxiliary *ad hoc* hypotheses. That is why the *ad hoc* hypotheses, both independent and auxiliary, often accompany each other. Introducing into science one *ad hoc* hypothesis (theory) – whether as an auxiliary or an explanatory independent hypothesis – involves the necessity to introduce ever newer *ad hoc* hypotheses in the process of confronting such a hypothesis with the new facts. There arises a chain reaction, as it were, a whole avalanche or torrent of *ad hoc* hypotheses, the next of which is to save its predecessor (or predecessors) from the fall in view of "the tribunal of experience." Here, a typical example is Aristotle's doctrine that nature is afraid of vacuum. According to Jevons, advocates of the old doctrine, according to which nature abhors vacuum could not predict a significant fact that water in an ordinary suction pump does not rise higher than 33 feet. When attention was drawn to this fact, they could not explain it in any other way than to introduce to their theory of a special correction that nature's abhorrence of vacuum does not reach 33 feet (Jevons 1960b, p. 186). When, in turn, Pascal and others stated that the higher above sea level the pump (or the barometer), the lower the altitude of the water in the pump (mercury in the barometer), another *ad hoc* hypothesis was needed in order to save the theory of nature's abhorrence of vacuum; it claimed that the fright of vacuum decreases with altitude. In this manner, a chain of *ad hoc* hypotheses was created, the next of which was to be a supplement to the previous one, and actually was a more and more obvious evidence of the bankruptcy of this whole complicated and increasingly artificial explanation. On the other hand, Torricelli's hypothesis on air pressure was further confirmed independently (e.g. by the fact that a partly pumped balloon swells the more the higher it is).

It seems quite probable that auxiliary *ad hoc* hypotheses of both kinds lead to a logical complication of the system modified by their means, or to the impairment by the system of the criterion of dynamic simplicity. Accordingly, one may think that the definition of the auxiliary *ad hoc* hypothesis as the one that complicates the mother theory is applicable to both kinds of auxiliary *ad hoc* hypotheses. In turn, an independent *ad hoc* hypothesis may be defined as a hypothesis which for its defence in view of experimental facts of new types requires introduction of ever newer hypotheses which complicate it (and which actually prove to be also *ad hoc* hypotheses themselves; in this way, some *ad hoc* hypotheses are often saved by other of *ad hoc* hypotheses). Thus, the introduction of all three kinds of *ad hoc* hypotheses is connected – as one may think – either directly or indirectly with the violation of the postulate of the dynamic simplicity and thus, due to the connection of this postulate with heuresis, with violating the postulate of the heuristic nature of knowledge. That is why one can hope that in future it will be possible to explicate all the three kinds of *ad hoc* hypotheses in terms of logical and dynamic simplicity of a theory. It is closely connected with the problem of the efficiency of criteria for judging whether a given hypothesis is an *ad hoc* one or not. Solving this at the stage of formulating a hypothesis presents great difficulties: usually, it is only in retrospect and after attempts at applying a hypothesis to new problems, or even only after conducting appropriate experiments, i.e. *ex post*, that are we capable of determining its nature in a well-grounded way.

IV. Diagnostic Features and Cognitive Criteria of ad hoc *Hypotheses*

Initial diagnostic features that allow one to distinguish *ad hoc* hypotheses are well-known (even though they do not constitute any strict criterion of differentiation). Nevertheless, stating them requires further tedious studies, both theoretical and experimental. And in practice there is possibly no better "criterion" of the *ad hoc*-ness of a hypothesis (either an independent or an auxiliary modifying one) than the oft-repeated "patching" a given system by means of newer and newer hypotheses. In each case such patching is an important component of a procedure indicating the *ad hoc* nature of the hypotheses introduced. However, the very fact that patching actually occurs may be usually established only when a new, simpler, more general and more precise alternative system arises. Therefore, only in the light of that new system do "hypotheses-patches" reveal their true nature of *ad hoc* hypotheses. When there is a need to introduce more and more qualitative hypotheses in order to make a certain basic conception compatible with new data from the experiment then the system as a whole becomes finally so complicated that it must make room for a simpler alternative conception when it is proposed (Hempel 1968, p. 49). As a matter of fact only then is it clearly visible that we have been involved in a dangerous and hopeless procedure of defending an outdated system by means of introducing subsequent *ad hoc* hypotheses. No wonder that the periods when new fundamental theories arise, which solve the existing problem situations, and thus convince the scientists that the theoretical possibilities of the hitherto existing theories have been largely exhausted, are preceded by periods of *ad hoc* explanations which create whole "chains" of modifications of existing theories. Therefore, if the formulation of new theories and the modifications of the hitherto existing ones that are compatible with the criterion of dynamic simplicity lead to increasing the informational content (and the degree of testability) of knowledge, according to the tendencies in the development of science, then the introduction into the system of *ad hoc* hypotheses "gravitates towards" the opposite direction: it leads to the loss by a theory of that degree of informational content, logical simplicity, testability and heuristic power which it initially possessed.

Obviously, a rigid division into *ad hoc* and other hypotheses is a simplification of some kind; hypotheses may be ordered with respect to the degree of *ad hoc*-ness which is vested in them. The lowest cognitive status seems to be assigned to those hypotheses from which no test implications result, i.e. conditional statements of the following kind: "if test conditions C are fulfilled then the result E will occur which is empirically perceptible" and, in addition, they do not result just because of technical problems in supplying (already now) test conditions C but because of the fact that those hypotheses are totally devoid of any empirical content, so they are empty and cognitively useless. A somewhat higher position is occupied by non-testable hypotheses, not just because of fundamental reasons but technical ones, and so they are untestable at present. The third could be *ad hoc* hypotheses in their ordinary sense, i.e. hypotheses that are testable in a certain context but that do not, for example, allow "the possibility of independent testing" (in the sense discussed here). These are hypotheses "fitted" into one type of facts and not to various types of phenomena.[16] The more a hypothesis is devoid of features of an *ad hoc* hypothesis, the greater the variety of empirical facts which the hypothesis concerns. What is concerned here is not so much the variety of facts taken into considerations in the process of formulating a hypothesis but first of all the variety of facts which the hypothesis is capable of explaining or (still better) of predicting.

Thus, how much a given hypothesis is free of *ad hoc*-ness depends in great measure on a degree to which a new phenomenon, which it predicts or explains, differs from the phenomena known thus far. Leverrier discovered a new (eighth) planet, so his prediction concerned a class of objects already known to people (planets). On the other hand, Dirac predicted the existence of positive electron (positron) and in general the antiparticles of matter, thus he discovered a new class of objects, and not simply a new object of a well-known class. The knowledge that Dirac used in this case appeared to be more heuristically fertile than the knowledge which led Leverrier to his discovery. This is connected with such a circumstance that knowledge providing information on objects of a new, so far unknown kind, allows in this way to predict new regularities or relations (necessary relationships) concerning these objects: the prediction of a new phenomenon if it is at the same time properly characterized is at the same time a prediction of new laws. What can be a better sign of a great heuristic power of a theory than the possibility of formulating new laws of science on its basis? A theory reveals its heuristic power to a greater extent when it leads to laws which otherwise, especially in a directly empirical manner, would have never been discovered. A. Pap even goes as far as believing that the significance of a theory for the further development of science consists just in the fact that one may deduce from it laws that have not only been unknown so far but also such "which without this kind of theorizing would have never – or at least at a given time – been materialized" (Pap 1955, pp. 163-165).

---

[16] We are inclined to formulate hypotheses of this kind in atomistic empiricism. According to holistic empiricism any hypothesis (and theoretical interpretation) should be compatible with a number of various empirical facts which – when some adventitious assumptions are adopted – is able to describe, explain or even foresee.

An example of a theory on the basis of which new laws were discovered is Marx's theory. After all, Marx's prediction of a new phenomenon, which socialism was in the 19th century, thanks to the proper characteristics, was at the same time the prediction of a new set of features and laws governing it and specific for that political and social system. The fact that the discovery of the laws of socialism was actually their prediction within the theory is visible because at that time they were not "operational" (i.e. were empty-fulfilled).

Thereby, they could not be discovered at that time purely empirically; they could only be deduced from a theory.

The problem of the discovery of laws which were "active" (operational) already in the past (or are "active" at present) and will be "active" in future is different. Such laws may certainly be also discovered by means of empirical studies, i.e. through the analysis of empirical facts which are submitted to those laws or their consequences (e.g. concretizations). Discovering laws of this kind by means of theory, which is also possible, i.e. through their deductive derivation from the existing theories is no longer a prognosis of future "phenomena" (relations, regularities) but as if and at the same time a postgnosis of repeatable "phenomena" (dependencies) occurring both in the past and in future. The deduction of all features and relations of a general temporal nature has this character, as well as deductive discovering of new types of phenomena occurring in the past and in the future (every type of phenomena is characterized by a certain constant, repeatable set of features and relations), in contradistinction to the prognosis of particular phenomena which nevertheless occur once at a given time.

The heuristic role of theory is connected not only with directly manifest heuristic role of predicting (prognosis, postgnosis and indirect diagnosis) but also, which is much more difficult to perceive, the heuristic function of explaining: by answering the "why" question with reference to (experimental) laws, we construct a theory. Theory lets us explain well-known experimental laws and formulate new ones. What is more, theory allows in general to discover more precise laws than the appropriate experimental laws, earlier formulated through experience, i.e. derived directly on the basis of an experience. Thus, the heuristically fertile theory allows both to discover new laws which have so far not been known as well as to enhance the precision of some of the hitherto known ones. This kind of theory is certainly "a full contradiction" of an *ad hoc* hypothesis. One can accept that a theory is the more fruitful heuristically when: (1) it predicts and explains more phenomena of a known kind, (2) the more varied these phenomena are, (3) the more phenomena of a new kind it predicts and explains, (4) the more new types of phenomena it predicts and explains, (5) the more new properties and laws (relationships) it allows to establish (or correct), (6) the more all those predictions and explanations are precise, and finally (7) the more difficult it would be to come to discover all these new objects, types of objects, features and relations, and especially laws concerning them, without using this theory. Obviously, what comes into play in each of the above points is not only the number of cases predicted and explained but also their quality, first of all, how significant are the discovered phenomena from the theoretical and practical point of view. Also the "quality" of predictions and explanations themselves: the more precise, and more in agreement with reality the prognoses and explanations are, the more exact the numerical measures they give of magnitudes which characterize the studied objects, the more heuristically powerful the knowledge which makes possible such prognosis and explanation.

On the other hand, it is worth noting that the heuristic value of a theory may be revealed from the quite unexpected side: it may also consist in eliminating unnecessary and apparent problems, which were presented as real ones by the previous theory. This negative "heuresis" is significant at least as far as it allows scientists to concentrate on solving other, real and vital problems. Here new great revolutionary theories, such as those of Copernicus, Newton, Einstein, Darwin, Marx and quantum mechanics are in particular of great heuristic importance. Thus, for example, classical mechanics allowed to eliminate the problem of dynamic causes (force) of the inert motion, special relativity theory, the problem of dynamic causes of the "relativist effects" (the growth of mass along with velocity, time dilatation, Lorentz contraction and the like).

V. Conclusion

In the article, three kinds of *ad hoc* hypotheses are described. Because of certain similarities that occur between them (especially between auxiliary *ad hoc* hypotheses of the first kind and independent *ad hoc* hypotheses) we can cherish hopes that in the future it will be possible to give a general definition of *ad hoc* hypothesis, which would encompass all kinds of provisional (*ad hoc*) hypotheses.

So far, the lack of high heuristic fertility and, perhaps, the impairment of the criterion of dynamic simplicity in connection with the redundant logical complication (either directly or indirectly) of the recognized system of knowledge seem to be the only features that all the *ad hoc* hypotheses considered here have in common. In this connection, short-lived hypotheses have been set in this paper against heuristically fertile hypotheses.

Another problem, which is significant from the point of view of practice, is that of evaluation of *ad hoc* hypotheses and *ad hoc* procedures. In this article, I generally assumed their negative evaluation. However, it may be feared that excessive rigour in this respect may inhibit the development of creative thinking. The problem is that either the provisional (*ad hoc*) or non-provisional (that is heuristic) nature of a hypothesis is not, at least in more complicated cases, revealed at once but only in the course of further investigations, i.e. *ex post*. Thus, it is difficult to postulate that *ad hoc* hypotheses be avoided as early as the process of the initial formulation of new hypotheses. It is rather at the stage of further criticism and selection of hypotheses allowed for testing that the postulates of this kind may play a positive role.

It seems that *ad hoc* hypotheses never become scientific theories. This inability to become a fully valuable theory may be the best evidence of the provisional (temporary) nature of a hypothesis. However, it is evidence which – like Minerva's – appears only at dusk. It does not testify to its infertility but only to the fact that it means much more at the stage of justification than at the stage of heuresis.

Part V
Philosophy of Physics and Cosmology

ON THE PECULIARITY OF PHYSICS
AND ITS DIVISIONS*

I. The Peculiarity of Physics

The most important distinguishing factor in physics which makes it the foundation of the natural sciences is undoubtedly the fact that it concerns – as the only science – all currently known levels of the structure of matter, starting from virtual particles and ending with the cosmological Universe. All the other empirical (i.e., natural, social and technical) sciences can only be used to a limited degree, which generally does not extend beyond one level of the structure of matter: microlevel, macrolevel or cosmic level. The above is closely connected with the other distinguishing factor of physics: its universal space-time scope that is the same – as one might think (at least potentially) – as the scope of the whole Universum. Physics studies the whole Universe in two senses. First, it investigates local processes but processes which probably constitute either a link of the whole hitherto evolution of the Universe (starting with the Big Bang), or elements of the current structure of matter that repeat on the scale of the whole Universe. Second, it studies global (or in any case, large-scale) properties of the Universe which characterize either the Universe as a whole or the part of the Universe which is accessible to our observation (sometimes called the Astronomical Universe or Metagalaxy).

Even chemistry is superseded by physics in this respect, in spite of the fact that chemical processes and phenomena currently seem to occur also on the scale of the whole Universe. As for biology, even if it appeared that life processes are scattered all over the whole Universe at the present level of its development, still thermodynamics imposes rigid restrictions on the density of occurrence of the islets of life or, even more so, on the islets of civilization in the Universe. It can be seen from calculations, based on the second principle of thermodynamics that islets of life may occupy at most one thirty-thousandth part of the whole area of the Universe which constitutes for them an indispensable source of low entropy. This means

---

* Translation of "O swoistości fizyki i jej działów" by J. Such, K. Matraszek. In: J. Such, M. Szcześniak (eds.) (1996), *Osobliwości przedmiotowo-metodologiczne w nauce* [Peculiarities in the Subject-Matter and Methodology of Science]. Poznań: Wydawnictwo Fundacji Humaniora, pp. 45-54.

that every possible islet of life must be surrounded by an area that is several tens of thousands larger, at the expense of which it is alive.

However, neither chemistry nor biology, the latter one even more so, can pretend to be time universal as the chemical and biological evolution appeared at the definite stages of the development of the Universe, preceded by a stage of purely physical evolution. Neither of them studies global properties of the Universe, in spite of the fact that the phenomenon of the beginning of life and its long development (the rise of civilization on the Earth) imposes certain significant constraints on the possible cosmological models of the Universe. This is expressed by the anthropic principle which, in its almost commonly accepted weak form, assumes that life (or civilization, respectively) requires for its rise and development, apart from local conditions, also certain global conditions which characterize the Universe as a whole or, in any case, on a large scale.

The distinguishing factors of physics outlined above characterize its subject matter.

In turn, the main methodological distinguishing factor of physics seems to be a close connection between experimental-measuring methods (or, as in the case of astronomy and cosmology, of the measuring ones) with mathematical methods. It brought about an extremely extensive use for research purposes both of mathematics (in theoretical physics) and of experimental and laboratory procedures (in experimental physics). As a result, a powerful method of mathematical hypothesis was created, which, through idealization and isolating abstraction (separation), led to the strict mathematical description of simplified models of various aspects of physical phenomena: the mechanical, electromagnetic, gravitational, thermodynamic, and other.

Physics, like no other science, succeeded in taking advantage of two powerful mathematical tools of penetrating natural phenomena: geometry (after all it was Galileo who called his research method a geometric one) and differential and integral calculus (developed by Newton just for the precise description of the course of physical phenomena). Harnessing mathematics, experiment and measurement for research work and making them aspects of a uniform scientific method was the greatest achievement of modern science since it meant the rise of exact natural sciences.

One of the important functions which it started performing became the homogenization of natural sciences, the reduction of qualitative differences between them to quantitative differences. In this process of levelling differences, physics has gone much further than any other science. Theories of unification of contemporary physics aim at total obliteration of differences between physical phenomena occurring under certain extreme conditions, which occurred in the first moments of the Big Bang. However, as the evolution of the Universe progressed, the subsequent breaking of the symmetry and other processes led to differentiation of phenomena, i.e. objects and interactions which occurred between them, which led to the present state of great variety. The reflection of this state is the rich development of ever greater number of specific sciences, including the ever greater number of physical sciences.

II. The Diversity of Physics

In physics, theories that are constructed have a varying degree of applicability. The most general, and at the same time the most fundamental, seems to be quantum mechanics, which from the theoretical point of view combines physics with chemistry, and it may further appear to be important also for understanding of the nature of life phenomena. Other most general and most fundamental theories of contemporary physics (quantum electrodynamics, the theory of small unification, the general theory of relativity, thermodynamics of open systems) each seem to unite several disciplines considered to be separate in classical physics. However, the vast majority of theories of contemporary physics concerns only relatively narrow class of physical phenomena. At the same time, each of those classes constitutes usually a subject matter of a different physical discipline and thus, of a separate specialization. Obviously, this does not preclude the possibility of combining them into larger divisions. However, those divisions are often not compatible with classical divisions of physics into disciplines such as mechanics, the theory of gravitation, electrodynamics, thermodynamics, the kinetic theory of matter or the physics of solid bodies.

On the other hand, the division into experimental and theoretical physics has remained almost unchanged and is certainly determined not by the subject matter but by the method. Actually, physics (just as e.g. chemistry and medicine) is said to be an experimental science. However, if we interpret physics extensively – as a whole of disciplines using methods of studies that are typical for physics, ranging from microphysics through macrophysics to astrophysics (astronomy and cosmology), only some of the disciplines of physics may be considered to be experimental in the full sense of the word. The others, such as astrophysics and especially cosmology, are experimental sciences only in the (indirect) sense that they use physical knowledge obtained by means of experimental research methods. The problem is additionally compounded just by the fact that one of the significant internal divisions in physics runs between experimental physics and theoretical (mathematical) physics, whereby the latter one, as is well known, applies mathematical rather than experimental research methods. Thus, apart from experimental physics, we have mathematical physics – a division which occurs in physics so distinctly as in no other science, not even in chemistry.[1]

The experimental-laboratory research in many fields of physics does not essentially differ from laboratory studies in other sciences, such as chemistry, biology or medicine. Still, in some areas of fundamental studies demarcating new frontiers of scientific cognition such as physics of elementary particles and physics of high energies, which is connected with it, laboratories, and even particular experimental devices, assume the size of large industrial plants, which does not normally happen in the case of other sciences. It is connected with the fact, mentioned above that physics as the only one of the sciences reaches the deeper and deeper levels of the structure of matter, and in this respect, it is universal. We know well that also the penetration of the cosmic level requires ever more powerful observation instruments, and that consequently science which is involved also in the study of levels of the structure of matter other than the macrolevel is an extremely expensive enterprise. It requires both international co-operation (in sharing the cost of research) and usually painful selection of those scientific problems which should be solved first and, therefore, for which financial means must be found now. No wonder that just in these fields of physics competition of national teams gradually gives way to cooperation of research institutions of the supranational character (such as CERN in Geneva or the research centre at Dubna).

The construction of ever more powerful laboratory equipment which penetrates (from the experimental point of view) ever lower and ever higher levels of the structure of the matter is not able to prevent the considerable outpacing of experimental physics by theoretical physics as far as the scope of the studied phenomena is concerned. The lowest level of the structure of the matter accessible to contemporary experimental physics is within 10^{-26}sec as far as time is concerned and 10^{-16}cm as far as space is concerned. By contrast, theoretical physics operates in a reasonable way with magnitudes of 10^{-44}sec for time and 10^{-33}cm for space. Magnitudes of this range appear in cosmology (where they determine the so-called Planck threshold, closing the quantum cosmology era) as well as in the theory of elementary particles (e.g. in the string

---

[1] There are, however, fields of physics in which the division is partly some obliterated. This is the case e.g. in the field of elementary particles where analytical methods are more and more often replaced with numerical methods whose significance grows considerably in connection with the growing potential of computers, which are already capable of performing tens, and even hundreds of billions operations per second. No wonder that, just in the field of elementary particles, an apparently paradoxical expression "experimental theoretical physics" has appeared for the first time (see Weingarten 1996, p. 76). Because of the complexity of physical theories, enormous expenditure on experiments and the growing potential of computers – American physicist, D. Weingarten writes – experimental theoretical physics may become the most practical research tool in certain disciplines of physics (p. 76).

theory). The same concerns the largest objects. While astronomical observations cannot exceed the so-called horizon of events determined by galaxies and quasars moving away with the speed of light and actually encompass the area of the Metagalaxy described by definition just as the observable part of the Universe and sometimes called the Astronomical Universe then some cosmological conceptions (e.g. A. Linde's hypothesis of the so-called chaotic inflation) interpret our Universe, started by the Big Bang as one of the infinite number of bubbles initiated by quantum processes (like, for example, fluctuation of the vacuum) in the boundless areas of the primary scalar field of the zero spin.

To conclude, we shall outline some possible divisions of contemporary physics.

III. Possible Divisions of Physics

Some of the possible divisions of physics have already initially been outlined. One of them (the methodological one) consists in distinguishing two of its sections: experimental physics and theoretical physics.

Another (subject matter) is connected with the penetration by physics of three levels of the structure of matter which leads to segmentation of physics into microphysics, macrophysics and astrophysics (with astronomy and cosmology). Sometimes, physics interpreted from that point of view is reduced to picophysics and megaphysics.

Another division, which in principle also concerns the subject matter, is obtained when we distinguish particular aspects of physical sciences. The mechanical aspect then serves as a basis for distinguishing (classical, relativist and quantum) mechanics, the gravitational aspect for the gravitation theory (the older gravitation theory authored by Newton and the contemporary one, i.e. the general theory of relativity), the electromagnetic aspect for electrodynamics (classical electrodynamics founded by Faraday-Maxwell and quantum electrodynamics), the thermodynamic aspect for thermodynamics (classical one, dealing mainly with closed systems approximate to equilibrium, and modern one whose main subject are open systems which are far from being in equilibrium). This division is not exhaustive since a number of important disciplines of physics (e.g. atomic physics, nuclear physics, or elementary particles physics) simply study specific kinds of physical objects (or structures), taking into consideration all the important aspects which characterize them.

It was no coincidence that enumerating the basic theories which investigate specific physical aspects of phenomena, we have listed (in pairs or in threes) older theories which penetrated a given aspect with contemporary theories. It appears that older theories (classical mechanics, Newton's gravitation theory, Faraday-Maxwell's electrodynamics, thermodynamics of closed systems) have by no means been abandoned or forgotten. They are still being studied, made more precise and more extensive, both because of certain practical applications and the heuristic possibilities that are still inherent in them. For instance, newer studies of classical mechanics revealed certain important significant "indeterministic" feature of classical dynamic systems (their basic instability), which became one of the important theoretical premises of the theory of chaos. Cosmological applications of Newton's theory of gravitation still appear to be fertile. Those are 19th-century theories which showed the irreducibility of physical phenomena to mechanical phenomena, i.e. classical electrodynamics and classical thermodynamics. Neither has totally exhausted its heuristic power. The distinctive nature of particular disciplines of physics, and particularly of the theories which represent those fields, comes in a great measure from the peculiarity of the mathematical apparatus applied in them. The mathematical apparatus which is the language of one fundamental physical theory usually appears to be useless for the expression of contents of other fundamental theories. The above fact undoubtedly reflects the peculiarity and uniqueness of the subject matter of particular physical aspects of phenomena and of particular levels of the structure of matter.

In this way, the internal differentiation of physics, especially of theoretical physics, has its equivalent in the internal differentiation of mathematics. One could even express it in a stronger way, when claiming that the abundance of mathematical structures (used in physical studies) shows or even exaggerates the natural differentiation of physical phenomena. However, the integrative trend in physics, which consists, among other things, in a corrective (correspondential) joining of the hitherto existing theories into more and more capacious theoretical systems would not have created proper conceptual means of mathematics, points to a great integrating role of mathematics in physical studies.

One more division of physics that is important from the methodological point of view is connected with a notion of an exact science. Although physics is a typical example of an exact science, this still does not mean that the whole of physics deserves that name. Contemporary cosmology, being undoubtedly a part of physics (thus a scientific discipline) is not by this very fact an exact science, and consequently physics is not a homogeneous science from this point of view. The failure to fulfil the criteria of an exact science by cosmology is connected with the fact that, as emphasized by M. Heller, the central problem of the discipline is the methodological question, i.e. the problem of extrapolating local physics onto non-local areas (see Heller 1978, p. 3). As a result, one of the essential components of this discipline is the so-called internal methodology of cosmology. Although it is constructed by cosmologists themselves, it undoubtedly brings cosmology very close to philosophy, not only from the point of view of the subject matter (which has always been the case) but also from the methodological point of view (Heller 1978; Such 1994).

We do not have to add that the impossibility of including contemporary cosmology within exact sciences does not result from the scarcity of mathematical methods used in it but is caused by the scarcity of sufficiently precise and sufficiently extensive empirical basis. The scantiness of observational data which are the basis of cosmological models cannot be compensated either by a powerful mathematical apparatus, used in this science, or by the power of the theoretical background which is provided to it by modern physics. This scantiness is all the more visible that the theoretical interpretation of empirical data, concerning levels of reality other than macrolevel, plays a considerably greater role than the interpretation of such data concerning the macrolevel.

MODELS OF RATIONALITY IN PHYSICS*

I. Models of Rationality in Human Knowledge

Beyond doubt, models of rationality have often changed in the history of human knowledge. But have they changed in the history of sciences? The latter is much shorter than the former, and science, in its every stage, seemed to be a sufficiently homogeneous system of thought to make us doubt whether the changes in science, even the most profound of them concerning e.g. its methods of explanation, are radical enough to result in the change of the model of rationality.

Such a change undoubtedly took place when science was emerging, when mythological and common-sense thinking gave way, in some areas, to scientific thinking (also called rational thinking), which might suggest that the first model of rationality was formed when science came into being. Such a suggestion, however, would be pervaded by ahistoricism and would be merely a result of the fact that everything that does not meet the requirements of the model of rationality in force at a given time and in a given field, seems to be, at least at the first glance, simply irrational and unreasonable.

The predecessor of scientific explanation, the mythological explanation, was not contradictory to reason as such but to some rules which underlay scientific knowledge. For instance, it did not follow the principle that all the phenomena of nature and especially of society were determined by causes of natural character, or the principle that everything was subject to the laws of nature. Since the above principles have been confirmed with respect to an increasing range of phenomena, one may claim that the model of rationality created by science at its outset is "more rational" than the mythological model of rationality of thinking (knowledge). "More rational" means here "more rational epistemically," i.e. helping to discover the truth, to implement epistemic progress more than the previous one.[1]

---

* The paper appeared originally in *Dialectics and Humanism*, **2**, 1984, 329-337.

[1] Apart from cognitive rationality and its various types (models), one may distinguish several other kinds of rationality (each having various models of its own) connected with other kinds of activities following other systems of values. One of them is no doubt the utilitarian rationality connected with effective practical activities, while the other – the

On the other hand, one may ask what the essence of the mythological cognitive rationality of thinking (cognition) is. I believe that it is the (actual, not necessarily conscious) following of the principles which made it possible to gain mythological knowledge, and thus made mythological knowledge a form of the cognition of reality.[2] Their list could include e.g. the principle that the external world exists objectively and can be known (at least partially), or the principle of supranaturalism, which admitted the influence of supernatural phenomena on the phenomena of nature and society. Thus, some of the principles of the model of mythological rationality are observed also in the model of scientific rationality while others are not. This proves both the existence of mythological rationality and its (partial) dissimilarity to scientific rationality.

Various models of cognitive rationality differ not only in ontological and epistemological principles but also in methodological and logical ones. Scientific rationality follows, above all, the methodological principle according to which all scientific reasoning (i.e. scientific justification) has to refer to empirical or logical facts. That principle was not observed in mythological cognition.

The principle of identity and the principle of contradiction (inconsistency) which states that if there are two contradictory statements, both of them cannot be true, are the examples of the principles of logic which are satisfied by scientific cognition. The above principles were notoriously violated in mythology which allowed to identify different things, such as e.g. human beings with animals or specimens of plants.

The fact that mythological thinking (cognition) neglected both the philosophical (ontological and epistemological) principles and the principles of logic, some of which have been mentioned above, proves that the model of mythological rationality was less developed than the model of scientific rationality. This explains why mythological thinking may seem irrational, i.e. why it may seem to fail to implement any model of rationality. Particularly, the violation of some rules of logic, traditionally called the principles of logic (the principle of identity and the principle of contradiction) by mythological thinking may be understood as its failure to implement any model (any principles) of rationality. However, mythological thinking did follow some other principles (also the principles of logic) which make it possible to establish whether this system of thinking is cognitively rational or not; without them, it could not have become a form of human cognition.

---

axiological system of rationality (or a few of its variants) connected with values such as the ethical value or the aesthetic value.

[2] One must agree that all the forms of human cognition have their cognitive rationality ; thus, if one accepts the existence of mythological cognition, one must concede that it has some kind (model) of cognitive rationality.

Obviously, the models of rationality in mythology and in science vary not only in that they follow partially different logical, methodological, ontological, and epistemological principles. The model of scientific rationality is superior, among other things, because it follows the principles which are: (1) more numerous; (2) more precise; (3) more rigidly observed; and 4) applied with greater consciousness than the principles of mythology. It also shows that the model of scientific rationality is more complete and more defined than the model of mythological rationality.

II. Models of Rationality in Physics

Let us now discuss a more controversial issue, i.e. whether the models of rationality in science undergo changes. If they do, what is the essence of these changes, and also, whether the changes advance the progress of cognition similar to that which was due to the origin of scientific cognition (scientific rationality) as such.

Let us consider the above, taking physics as an example.[3] At first, there occurs a preliminary hypothesis, i.e. that if the models of rationality did change, then it must have happened at least twice: when ancient physics became modern physics and when modern physics became contemporary physics.

1. Models of Rationality in Ancient Physics and in Modern Physics

When we compare ancient physics to classical physics, a great difficulty arises. Namely, in ancient physics there were four different systems of theorems, symbolized by the names of Aristotle, Ptolemy, Archimedes, and Eratosthenes. Thus, we have Aristotle's physics (4th century B.C.), Ptolemy's astronomy (the geocentric system; 3rd century B.C.), Archimedes' statics (3rd century B.C.), and Eratosthenes' discoveries concerning the shape and size of the Earth (3rd century B.C.). The two former systems, Aristotle's physics and Ptolemy's astronomy, which closely correspond to each other, are typical of ancient science. But the two latter ones, Archimedes' statics and Eratosthenes' calculations concerning the Earth, are – due to their accuracy and high degree of reliability as well as the research methods applied, e.g. the methods of modelling and idealization – the first examples of modern physics, and are, in a way, well ahead of their epoch. Thus, to grasp the model of the rationality of ancient physics and to compare it with the model of rationality of modern physics, it is necessary to concentrate on the two former systems, especially on Aristotle's physics.

---

[3] I understand physics as a system that includes astronomy.

In contrast to the mythological view, Aristotle's physics does not introduce, as we know, any supernatural factors which would directly influence the course of the phenomena of inanimate nature.[4] However, physical processes are anthropomorphized, especially that many of them are given a teleological character which is manifested, among other things, in the notion of entelechy and of the final cause. From the above point of view, the difference between animate and inanimate nature appears to be of secondary importance in Aristotle's theories.

Another important difference between the models of rationality of Aristotle's physics and of classical physics (the physics of Galileo, Newton, and their followers) is that they apply different principles of methodology. Aristotle's physics is a qualitative physics,[5] it is based on common experience and on the methods of induction by enumeration. Modern physics (and also Archimedes' statics, and Eratosthenes' "geophysics") is concerned with mathematical physics of quantity, based on experiment and observation with instruments (including astronomical observations), and it uses modelling, idealizing abstraction, and thought experiment on a large scale.

There is no essential difference, however, between the principles of logic applied by ancient physics and modern physics (Aristotle is believed to be among the first to formulate such principles of logic as the principle of contradiction and the principle of the excluded middle which are the foundations of classical logic).

Thus, one may draw a conclusion that the models of rationality of ancient and classical physics apply different philosophical and methodological principles but similar principles of logic. Since we admit that mainly the differences in the principles of logic adopted prove the difference of the models of cognitive rationality, we come to the conclusion that the models of rationality of ancient and modern physics, though not identical, do not differ so much as the models of scientific rationality (even in its ancient form) and mythological rationality.

Of course, one may doubt whether the differences of extralogical principles are sufficiently great for us to speak about different models of rationality in ancient and modern physics. If we follow our intuition, it seems that the above mentioned differences are sufficient since the notions of Aristotle's physics of natural place, natural motions, enforced motions, and four kinds of causes as well as his division into the physics of the sky and the physics of the Earth seem to be irrational or, at least, not very rational in the light of the discoveries of modern physics. There are also

---

[4] Such supernatural phenomena are, e.g. the Prime Cause of Motion of extramaterial character but its "activity" took place far in the past and is not felt at present.

[5] Some quantitative definitions and descriptions, e.g. concerning certain proportions, are also present in it but do not play any important role.

some substantial reasons why a given model of cognitive rationality in empirical sciences cannot be established only on the basis of the principles of logic (such a possibility exists only for the models of rationality in logic and, perhaps, in mathematics.) The reason is that in empirical sciences it is not possible to acquire knowledge without referring to empirical facts, and thus, to empirical methods (extra-logical principles), which are vital for establishing those empirical facts.

On the other hand, one must agree that the principles of logic are the indispensable components of the models of cognitive rationality. One must also admit that models of rationality have different capacities in various sciences, and in various historical periods, i.e. the number of principles which constitute them varies.

2. The Models of Rationality of Modern (Classical) Physics and Contemporary (Non-classical) Physics

Let us now consider the differences between the rationality of modern physics and contemporary physics. Some problems arise here, for it is difficult to define the notion of contemporary physics due to the lack of agreement as to its content and scope. From the formal point of view, contemporary physics is the 20th-century physics. However, such a strictly chronological division is unsatisfactory and that is why the terms "classical physics" and "non-classical physics" were coined in physics and in methodology of physics. Although classical physics is sometimes understood to be the physics before our century and non-classical physics is defined as all the physical theories developed after 1900, including the relativity theories (general and special), the majority of physicists and specialists in methodology believe that non-classical physics includes only microphysics with quantum mechanics, thus qualifying the relativist physics as part of classical physics.

The latter attitude seems to be the right one, for it is quantum mechanics which turned out to be not only a new physical theory but also an entirely different theory which did not fit into any of the methodological models of scientific theories existing up till then. Thus, we shall not follow, as we did in the case of ancient physics, the purely chronological criteria but shall establish the divisions based on the substantial features (e.g. the methodological ones).

Quantum mechanics (non-classical physics) breaks away, at least in two points, from the classical ideal of full and purely objective description, which is part of the model of the rationality of classical physics. Firstly, the description in terms of quantum mechanics by means of a wave function is not a purely objective (and thus fully objective) description of the state of the microobject but is also a metadescription of our knowledge of the given state. Secondly, it is also a probabilistic description which (1) either directly violates the principle of strict determination (which was put forward by the earlier determinism, the so-called dynamical or Laplace's determinism), according to which "the same happens under the same conditions," or (2) from the point of view of this principle, it is not a complete description. As we know, the former attitude (1) is maintained by the Copenhagen School and the latter (2), by A. Einstein and other opponents of the Copenhagen School. Although this problem can be finally solved only after the future theories of microworld (or submicroworld) are introduced, one may presume that quantum mechanics violates two important principles: an epistemological one and an ontological one, undoubtedly two crucial elements of the model of rationality of classical physics: the postulate of objective description and the principle of the strict determination of the course of phenomena. Some experts on the subject, such as Heisenberg, Weizsäcker, J. von Neuman and Reichenbach, believe that quantum mechanics is even more radical, since it exceeds the limits of empirical science and enters into the domain of logic. It is there, as a result of the introduction of Heisenberg relation, the principle of classical logic that two values are to be used is violated, and a form of multi-valued logic (which introduces a new value, i.e. an unspecified one), usually called quantum logic, is created.

It seems that research on the logic of microworld carried up till now does not confirm the above hypothesis; it has not been proved that quantum mechanics cannot be properly interpreted on the basis of the principles of binary classical logic, either. However, due to the great difficulties there, no conclusive solution in this respect has been forwarded so far.

Even if quantum mechanics "follows" the same logic as classical physics, the differences between them, both as far as their philosophy and methodology are concerned, are so great that one may say, without risking a mistake that they follow two different models of rationality. It is obvious if we look at the problem following our intuition, since the adherents of purely objective description of classical physics (including relativity theory), such as Einstein, found quantum mechanics and the description in terms of quantum mechanics unacceptable, as it violated the postulates of the "rational description" of classical mechanics: the postulate of objectivity, completeness, and strict determination. One may even presume that the models of rationality of classical and contemporary physics differ more than the models of rationality of ancient and modern (classical) physics.

Did the followers of the new probabilistic description in terms of quantum mechanics, i.e. the adherents of the interpretation of the Copenhagen School (with Bohr), think that the classical ideal of fully objective description was irrational, or, at least, not fully rational? Perhaps not, but it is only a question of time: the description in terms of quantum mechanics seems to be strange not only from the point of view of classical physics but also of common sense until we get used to it. The classical system would seem to us more natural and more rational. That the dispute about quantum mechanics is concerned with two models of rationality is, in my opinion, also proved by the character of the dispute itself and the arguments used. It is, as Einstein stressed, not a dispute about physics but about philosophy, and the arguments are intuitive in nature, as they refer to such subjective notions as the scientist's intuition, his/her sense of physical reality, etc. The physical facts and arguments of experimental and theoretical nature influence the dispute only indirectly, and that is why it cannot be settled by means of physics but philosophy, at the present stage. Thus, the philosophical and methodological principles are the most important, and it is they that, along with the principles of logic, form the models of cognitive rationality.

It is worth noting that in the present essay modern (classical and relativistic) physics is assumed to represent a uniform model of rationality. This assumption is probably controversial, as in my approach the term "modern (classical) physics" embraces three centuries of physics and five fundamental theories: classical mechanics, including Newton's gravitation theory, classical thermodynamics, classical electrodynamics, the special relativity theory, and the general relativity theory, which differ considerably. In my opinion, the question of whether the formation of field theory (Faraday-Maxwell electrodynamics) which was, according to Einstein, the most radical change in the history of physics (even more radical than the one caused by the relativity theory or by quantum mechanics) was the reason of the origin of a new model of rationality in physics is an open question. Such a view can be justified by the fact that field theory started the decline of mechanicism in physics (until then, classical mechanics was the fundamental theory of physics or even of the whole natural history). It formed a new "field" image of the world in physics and a new way of explaining the phenomena. In particular, due to that theory the mechanistic concept of central forces, linking the centres of masses in straight lines, and the concept of "acting at a distance," *actio in distans,* were abandoned. Besides, field theory refuted the famous Newton's concept of absolute time and space as substantive self-existent entities and reversed the course of physics to an attributive concept of time and space, which was finally established in relativistic physics. Field theory also refuted the notion of force, the role of which became gradually less important as field theory concepts were introduced to various branches of physics.[6] It was gradually replaced by the notion of the action which is one of the most fundamental notions of modern physics.

---

[6] The first concept of physics in which the notion of force was entirely eliminated was put forward by H. Hertz at the close of the 19th century. The concept, however, was premature.

The adherents of the mechanistic approach in natural sciences believed that the field-theory way of explanation which broke away from mechanicism was so irrational that they tended to consider only the mechanistic explanation to be the actual and not apparent explanation in physics. The above attitude gave rise to the sustained attempts, which however turned out to be useless, on the part of Maxwell and others to build mechanical models of phenomena of electromagnetics and on this basis to explain those phenomena. The concepts of ether and other "imponderables" (agravic liquids) were, as we know, introduced into physics (and chemistry) in order that the phenomena of non-mechanical character (light, electricity, magnetism, gravitation, heat) could be explained within the mechanistic theory.

On the other hand, the differences of the mechanistic physics and of the field theory do not seem to be marked enough to allow us to speak about a field model of rationality in physics. It is all the more so since there are no less pronounced differences in this respect between other branches of classical physics, e.g. between the dynamic physics and statistical physics (the former is concerned with the physics of temporarily non-directed processes, whereby the rules of statistics do not play any role there, whereas the latter is a physics of directed processes, and is governed by a new type of laws, i.e. the laws of statistics). It would thus be necessary to multiply models of rationality in physics instead of adopting a more proper attitude, i.e. instead of treating the differences in question as belonging to the same (one) model.

There is one more question, namely of whether the transition from the model of rationality developed within ancient physics to the model of rationality espoused by modern physics and henceforth to the model of contemporary physics is a progressive sequence. The problem is beyond dispute in the case of the first transition; modern physics and the technology based on it became such a powerful means of reshaping and also explaining the world that it is impossible to deny the progressive nature of the change, including the purely cognitive aspect. The question of whether the second transition is progressive in nature is more complicated and, for some scientists, controversial. Such adherents of the classical ideal of purely objective description as Einstein, von Laue, or Planck, are ambivalent in their estimation of the cognitive value of the transition. On the one hand, they are aware that quantum mechanics is much more adequate for experimentation (characterized by a hitherto unknown degree of accuracy and subtlety) than all the previous theories of physics. On the other hand, they are of the opinion that quantum mechanics departs, without good reason, from the ideal of purely objective description realized successfully by classical physics. From the point of view of this idea, the intuitions and premises of philosophy which underlie it, the second transition is a step backwards rather than forwards on the path to the complete and fully objective description of physical reality. A scientist who believes that it is necessary to give up the above ideal (and the principle of strict determination) in the development of more precise description of the quantum phenomena expresses a different opinion. He will be justified to argue that even though the description in terms of quantum mechanics violates the postulates of classical physics of purely objective description, it is much more complete than the rigid spatio-temporal description of the course of phenomena applied by classical physics. It takes into consideration e.g. the so-called phase dependence between the microobject and macroinstrument, which constitutes certain probabilities, which classical description in spatial categories must omit (cf. von Weizsäcker 1971, pp. 485-487). It is thus more precise and complete and in this way close to the truth in its classical (correspondence) sense. The fact that it has to "pay the price" for it in that it has to depart from the ideal of fully (purely) objective description only indicates that the cognitive progress accomplished here, as it often happens in the case of a progressive line of development, is not progress in every respect (absolute progress). In order to come closer to the truth in quantum, mechanics one has to give up the ideal of purely objective description.

In my opinion, the direction in which microphysics has been developing proves that the latter standpoint is the right one since microphysics and especially quantum mechanics, as compared to the classical macroscopic physics, undoubtedly foster cognitive progress.

III. Conclusions

The conclusions and inferences reached in this essay can be summed up as follows:

1. There are various types of rationality connected with various types of activity and various kinds of axiological system related to them, e.g. epistemic, pragmatic (utilitarian), ethical, or aesthetic rationality.

2. Epistemic rationality, like all the other types of rationality, has been changing in history; different models of epistemic rationality are applied not only at different stages of human knowledge in general but also of scientific knowledge.

3. Models of cognitive rationality are established by means of four types of principles: logical, ontological, epistemological and methodological principles.

4. The greatest differences occur between the models of epistemic rationality which follow different principles of logic (such a great difference in the principles of logic occurs, in particular, between the

model of rationality of mythological knowledge, which preceded scientific knowledge, and the models of scientific rationality).

5. At least three models of rationality occurred in physics: the model of rationality of ancient physics (Aristotle's physics), the model of rationality of modern (classical) physics, and the model of rationality of contemporary (non-classical, quantum) physics. This list could, perhaps, be expanded, since the origin of classical electrodynamics may have given rise to a new model of rationality, and classical (modern) physics may turn out to be linked with two successive models of rationality: the mechanistic world-view philosophy and the field philosophy.

6. The models of rationality in physics differ mainly in their philosophical (ontological and epistemological) and methodological principles and, in the case of the model of rationality of quantum physics, perhaps in logical principles as well.

7. The successive models of epistemic rationality in physics are certain to foster cognitive progress, which occurred also when the prescientific model of rationality (mythological cognition) was abandoned and the scientific one was adopted.

Translated by Sylwia Twardo

TRANSCENDENTAL PHILOSOPHY AND PHYSICS
OF THE MICROWORLD*

I. Introduction

Physicists and philosophers have been impressed by certain analogies between modern physical theories, especially the most fundamental one i.e. quantum mechanics, and old philosophical conceptions.

In this essay, I would like to outline, following a rather free convention, an analogy between transcendental philosophy and the physics of the microworld. Certain parallels on the subject were conducted by two well known German physicists (who were also philosophers), Heisenberg and his disciple Weizsäcker, assuming that modern physics, just as Kantianism, looks for certain general statements concerning the "conditions of all the possible experience." In undertaking this topic, I shall first try to present certain basic ideas of transcendental philosophy in so far as necessary for further considerations.

II. Idea of Transcendentalism

I understand transcendental philosophy as a creation whose main rudiments had been developed by Kant and Fichte. Further on, certain important aspects of that philosophy were developed in their own ways in the classical epoch by Schelling and Hegel, and later by Marx and some neo-Kantianists, and in a more modern epoch, mostly Husserl's and Heidegger's transcendental phenomenology as well as that of their disciples, and also those currents which were under this influence, especially hermeneutics (the so-called post-phenomenological hermeneutics) and existentialism, and within Marxism such researchers as Gramsci and Lukács.

Adherents of transcendental philosophy claim that this philosophy started in principle the new theoretical perspective of philosophizing, to be more concrete, interpretations of the proper problem of philosophy which

* Translation of "Filozofia transcendentalna a dialektyka mikroświata." In: T. Maruszewski (ed.) (1986), *Filozofia-Poznanie-Psychologia* [Philosophy, Knowledge, Psychology]. (*Poznańskie Studia z Filozofii Nauki*, **10**), Poznań-Warszawa: PWN, pp. 219-239.

is – according to them – the problem of knowledge (or as Kant said, the problem of reason). Kant calls transcendental any cognition which deals in general not with objects but rather the manner in which objects are recognized by us, so far as this manner is to be *a priori* possible. Kant contrasts his "transcendental" perspective with the "empirical" perspective, which is typical of the exact sciences including mathematics, and with the "metaphysical" perspective, which is characteristic of the hitherto existing (i.e. pre-Kantian) philosophy. One of the central, if not fundamental, ideas of transcendental philosophy, which may be termed, following M.J. Siemek, the idea of transcendentalism, is a view that the separation (division) into the objective and subjective sphere of reality that of object and that of subject, into the sphere of being and thinking, of things and consciousness, as distinct essentials, so to say, it is an initial fact of only empirical consciousness or metaphysical consciousness (which corresponds to it). However, this fact should be overcome by a more fundamental consciousness, which may be called transcendental consciousness (cf. Siemek 1977). According to this view, a separate interpretation of thinking and being, assuming the duality of the one being which "recognizes" and the one which "is recognized," which is typical of the pretranscendental philosophy, especially the modern one from Descartes to Hume, leads to dualism in philosophy, to breaking it into the theory of cognition and ontology. The former considered "purely" cognitive problems; it asked about cognition of "things," i.e. its subject matter was located on the "purely" epistemic level. By contrast, the latter considered "purely" ontological problems (it asked about "things themselves" whereby the subject matter concerned the "purely" ontic level, in the sense given to this term by transcendental phenomenology).

The idea of transcendentalism removes this division as an initial division, introducing into philosophy a new and, as it were, higher level of considerations which Siemek terms epistemological and which he contrasts with the epistemic and ontic level of the pre-Kantian philosophy. According to this idea, thinking about reality, at this deeper epistemological level, cannot be separated from thinking about thinking itself. On the other hand, all considerations on cognition, conducted at the transcendental level, must lead to the analysis of conditions necessary for the possibility of cognition, which entangles them inevitably in the ontological problems. The subject matter of "transcendental" reflection is not so much knowledge itself, as in the traditional theory of knowledge, but the conditions of the possibility of all subjective-knowledge-about-an object. Thus, a transcendental question is a question not about cognition but about the knowability, about the conditions and the necessary structure of the relation itself (i.e. the relation between a "subject" and an "object," and between "cognition" and "reality") (Siemek 1977, p. 64). Between such a question and its new "object" a specific relation occurs which constitutes another peculiarity of epistemological thinking which is worth attention. Namely, the question is not external to its object, nor is the object independent of the question. On the contrary, the question changes the object, and actually it only creates it as such. . . . Such an "object" arises only together with such a question and thanks to it. It is constituted by that question. And conversely, constituting just such an object gives the question itself the status of an epistemological question (pp. 64-65).

As a result, the theory of cognition and ontology are not obtained separately as two metaphysical fields of philosophy but as a kind of the ontology of cognition dealing with the understanding study of knowledge and the overall foundations of its "existence" (p. 44).

Epistemology as the ontology of knowledge must "carve out" its subject matter differently in the beginning. It is "no longer 'cognition' (as a part of the epistemological field, separated from things) but just the relation of cognition and things, thus the field itself in its overall structure" (p. 44). In (Kant's) transcendental philosophy

"knowledge-about-knowledge" or the "cognition of cognition" cannot be differentiated from "cognition" itself, from "knowledge-about-the-subject-matter." . . . Philosophical reflection on cognition is at the same time, and directly, a reflection on the manner of existence and fundamental forms of objectivity, thus of this "being" which is presented to such a cognition (p. 48).

Such a philosophy

asks about conditions of our cognition of the subject matter but as long as they are at the same time conditions of being of the same subject in their fundamental constitution and structure. Its main question, therefore, is neither epistemic nor meta-epistemic: it concerns neither Being itself in its ontic reality (like questions of traditional metaphysics or modern positive science) nor only knowledge itself as a logical instrument or medium, making accessible to us this ontic reality of Being (as questions of the traditional "theory of cognition") (pp. 48-49).

The question of such a philosophy

always concerns Knowledge and Being simultaneously: their unity, their common foundation, their mutual relation within the larger whole, in the bosom of which they are only distinguished and comparable; . . . what is meant here is the understanding of the epistemic forms and structures of knowledge as such, that is as already always opening an access to certain forms of ontic reality, to Being of a definite type of objectivity. But also an epistemological question is at the same time, and in the same theoretical act, a truly ontological question: understanding knowledge is always created here, and only, together with the understanding of being (p. 49).

According to Kant, the empirical-human world is the world whose "objectiveness is always given together with its sensual-conceptual schematization" (p. 50).

Kant's criticism "is a philosophical theory of empiricity as the overall form of existence" (p. 50). It equally removes and at the same time equally absorbs in a new form *both* the "fields" of traditional philosophy which complement each other: the classical "metaphysics" and the classical "theory of knowledge."

This is so at least in the sphere of theoretical philosophy since if practical philosophy, i.e. ethics and the philosophy of action, is taken into consideration, then an even larger whole is received. Gramsci and Lukács, thus Marxists who were under a substantial influence of transcendental philosophy and who emphasized the processes of socialization, gave this larger whole the name of "the philosophy of practice" (Gramsci) and "the ontology of social being" (Lukács). These philosophers see in the fact of unification of being and consciousness or, to stylize it metalinguistically, in the fact of combining ontology with the theory of cognition, the deep dialectics whose adherents, in their opinion, included Marx. It abandons the assumptions of "metaphysical materialism" whose subject matter is "pure being," being in itself, abstracted from the recognizing consciousness and in opposition to it absolutely (cf. Siemek 1980, pp. 4-18).

It is easy to notice that transcendental philosophy in some sense undermines the assumptions of realism and the classical definition of truth and concurrently the ideal of objective knowledge outlined clearly particularly in classical physics. It is not accidental that scientific knowledge is included, on a par with pre-Kantian philosophy, in the epistemic and ontic level of knowledge which is to be overcome by certain procedures of transcendental philosophy which also lead in this way to the criticism of scientific knowledge. Among other things, the procedure which Husserl calls the phenomenological reduction aims in this direction, and in it is usually distinguished an eidetic reduction, which is, so to say, set on resolving everything to experience, and a transcendental reduction which, in turn, submits to criticism the experience itself.

III. The Relativity of the Division into a Subject and an Object in the Studies of the Microworld

Overcoming the subject-object dualism assumed in the idea of transcendentalism has its equivalent in quantum mechanics, showing the relativity of the division into the subject and the object in the research of the microphenomena. While summing up his considerations of what has been established by quantum mechanics on this question, W. Heisenberg writes that the hitherto commonly used divisions of the world into an object and a subject, into the external and internal world, into body and soul are no longer adequate and lead to insurmountable difficulties. Also in natural sciences the object of studies is no longer nature "in itself" but nature given over to human questions, and also here man encounters himself again. Elsewhere, Heisenberg writes:

in classical physics, science started from the belief, or should one say from the illusion, that we could describe the world or at least parts of the world without any reference to ourselves. This is actually possible to a large extent. . . . Its success has led to the general ideal of an objective description of the world. Objectivity has become the first criterion for the value of any scientific result. Does the Copenhagen interpretation of quantum theory still comply with this ideal? One may perhaps say that quantum theory corresponds to this ideal as far as possible" (Heisenberg 1959, pp. 54-55).

However, one cannot argue here simply that quantum mechanics concretizes on a certain material the content of the idea of transcendentalism but only speak of a certain analogy in the approach and results of both these fields. The fact of the matter is that the unity of being and thinking, according to the principles of the transcendental philosophy – is realized not at a common empirical level of knowledge and cognition but at a deeper level, that of, so to say, transcendental consciousness. And the approximation of the object and the subject in quantum mechanics occurs already at the level of empirical knowledge and with respect to dual relationships between them: theoretical and experimental.

Specifically, a dual "influence" of the object on microphenomena occurs, i.e. both theoretical and practical (experimental) influence. Theoretical influence, i.e. influence through knowledge, consists in the wave description of the microstate containing not purely objective characteristics of the state of an microobject but also the characteristics of our knowledge concerning that state is at the same time a metadescription to some extent. By contrast, practical (experimental) influence, exerted through experimental equipment, consists in the influence of those instruments on the state of the microobject. That is why the wave description of a microsystem is not a pure description of the state of the system itself but a description of a larger whole from which a microsystem cannot be isolated either in thoughts or in reality. The lesser the studied material system, the greater influence is exerted, the greater the disturbance on the part of the technical instruments of research, so that it is more and more difficult to conduct a distinct and precise dividing boundary between the behaviour of the material systems studied "in themselves" and their reaction to the experimental apparatus.

In this connection, we have already had a long discussion on the subject matter of quantum physics, started by Niels Bohr. According to the founder of the Copenhagen School, as distinct from the classical physics, the subject of which are macroobjects as such, the subject of quantum physics is, generally speaking, an interaction of a microobject with a macroscopic experimental instrument. Bohr states that while in classical physics the interaction between an object and an instrument might be neglected or, if needed, compensated for, then in quantum physics the interaction constitutes a part of the phenomenon which cannot be separated. According to this, a unequivocal description of the quantum phenomenon proper should in principle include a description of all the existing parts of the experimental instrument. The above understanding of the subject of the tasks of quantum physics, requiring an insertion into the description of microphenomena of the characteristics of instruments and means of observation prompted a well-known Russian physicist, W. A.

Fock, to formulate the so-called principle of relativization of quantum description to the means of observation.

Opponents of the Copenhagen School, on the other hand, evaluated the above understanding of the subject matter and the tasks of microphysics as an expression of subjectivism and submitted it to criticism. A. Einstein, in particular, who was the main proponent of that trend, considered such an understanding of the subject matter of quantum physics as "basically unsatisfactory" from the point of view of what can be considered as the chief aim of the whole physics: full description of the real state of an optional system (which, as is assumed, exists independently of an act of observation or existence of an observer).[1]

In my opinion, it is now out of question that Einstein was wrong and that the quantum description is a full (complete) description of the phenomena of the microworld, in spite of the fact that it is essentially statistical in nature.[2] Microobjects "by themselves," their "behaviour as such" cannot be the subject of quantum physics due to the inevitable "transgression" of the recognizing object into the structure of physical knowledge. This transgression does not completely preclude the value of objectivity of quantum description. Its subject matter is not actually the microsystem "in itself" but phenomena which arise as a result of mutual interaction of the microsystem and the experimental apparatus. However, on the one hand, it is just the subject that chooses the definite object of study from the whole reality which surrounds her, and the definite research instruments on the other. In addition, on the basis of the knowledge that s/he has acquired, develops schemes of experiment and measuring instruments and experimental equipment still the very interaction of a

---

[1] It is worth noting that while many physicists and philosophers, following Einstein, view the quantum-mechanical description as an impairment of the principles of objective description, formulated within classical physics, Eddington, a well-known British astrophysicist and the founder of the doctrine of "selective idealism," claims that it is only quantum mechanics that shows that the scientific description of a structure and regularities of the world have a value of objectivity. Due to the extraordinary strangeness of properties and the unexpected behaviour of microobjects it is difficult to admit the very idea that the regularities of the microworld were simply a creation of the cognitive mind in spite of the fact that just such an idea seems natural in reference to the laws of macroworld which are with great obviousness imposed on us.

[2] It can be shown that, just on the contrary, the quantum description in spite of the fact that as a statistic description it does not fulfil the ideal of full description in the classical sense, is still a fuller description than the spatial description of classical physics. It is even if only because it takes into account the so-called dependencies of phases between a microobject and the macroinstrument, being certain probabilities from which the classical description in spatial terms must abstract. For this issue, see Weizsäcker (1978), pp. 503-504.

microobject and macroinstrument remains the physical interaction just between physical objects, independent of their origin.[3]

In connection with the increased role of the subject in microphysical research, four attitudes to this cognitive situation may be distinguished among researchers (physicists and philosophers). Representatives of the first orientation believe that quantum description contains results of observations understood as interactions of the recognizing subject with microobjects realized through the mediation of some or the other instruments. The group also includes supporters the interpretation supplied by the Copenhagen School. Those in the second group are of the opinion that quantum description expresses mutual objective physical interaction between an instrument and a microobject, independently of any observer. Those from the third group assume that quantum description expresses information on microobjects obtained by a observer. Finally, the fourth group suppose that quantum description reflects the behaviour and properties of microobjects "themselves" (as such).

When analyzing this problem and evaluating the corresponding standpoints one should remember that, as shown by M. Bunge (1977, pp. 1-13), none of the basic equations of quantum mechanics contains any variables or magnitudes which would concern an observer or if only an instrument. On the other hand, even some of those researchers who think that the subject of quantum mechanics is an interaction of a microobject with an instrument think that on the basis of results of this interaction the properties of quantum objects themselves can be directly established. Thus, for instance, W. A. Fock states as follows:

> Such a formulation of the problem fully allows to introduce magnitudes which are characteristic of the object itself independently of the instrument (charge, mass, spin of a particle and also other properties described by quantum operators) but at the same time allows a varied approach to the object: an object may be characterized from its side (e.g. corpuscular or wave) the appearance of which is conditioned by the structure of an instrument and by external conditions which it produces (Fock 1969, p. 194).

Thus, it appears that numerous characteristics of a microobject, e.g. of an electron, are not connected with one or another type of macroscopic instrument: mass, charge, spin, being submitted to Fermi statistics belong to the electron as such. And the others, including the position and momentum of a microobject and the other space-time features, are actually devoid, in this sense, of a univocal macroscopic determinacy (see Markov 1976, p. 47). In spite of Heisenberg's and some other physicists' view, it

---

[3] This kind of "subjectivity," connected with the role of an observer, may be called, following B. Russell, physical subjectivity. Russell, speaking about the physical subjectivity, sees the role of an observer and the significance of choice of a reference system in the special relativity theory. In it different results coming from the difference between one "point of view" and another one are connected with the notion of the "point of view" in such a sense that they are right in the same way in reference to physical instruments and to people and their observations. See Russell (1949, p. 123).

follows that microobjects as such exist not only in the potential sense but in the fully real sense, having the characteristics which are totally independent of man, i.e. mass, charge, spin and others. However, the number of their space-time characteristics as well as dynamic-energetic ones depends on the interaction with instruments and indirectly with the recognizing subject. It is a situation which has its correspondents in relativity theory, and even in everyday experience. For instance, the so-called secondary qualities (sound, colour, taste, scent) of objects depend not only on the characteristics of the object itself but also on the condition of perception organs of the recognizing subject. The unusualness of the situation in quantum mechanics consists, however, in the fact that what is meant is not "secondary" but "primary qualities" of objects, since such are its dynamic-energetic and space-time characteristics.

One way or the other, we encounter the fact that contemporary physics, starting with relativity theory, and ending with quantum mechanics, is an evidence of the increasing activity of the subject in the process of the cognition of the more and more subtle phenomena and the necessity of taking into account its activity both at the level of experiment and that of theory. In this sense, the description of the studied phenomena – as physics develops – contains more and more subjective elements, dependent on the subject, in spite of what is usually thought about it. To come back to quantum mechanics, we can agree that the quantum description does not fulfil the ideal of the purely objective description in two respects, as propounded by classical physics. It contains a subjective moment connected with the state of our knowledge (or better still: the lack of knowledge) about a microobject and the second moment as it seems also of a subjective kind, connected with the experimental influence of an observer through a macroinstrument on the microobject. The latter fact may be expressed by stating that in the area of microphenomena a usual "passive" observation is impossible because of the inevitable influence on a microobject. All contact with a microobject is by necessity active experimenting. According to the prevailing Copenhagen interpretation, abandoning the ideal of a fully objective description, forced by cognitive situation in microphysics, is connected with the fact that we investigate objects from another level of reality than the one we are in. While we are macroscopic beings employing macroinstruments and the language which is well adapted to the description of the macroworld, the subject matter of microphysics are the phenomena of the microworld.

IV. The Problem of Agnosticism

Trying to find affiliations between results of transcendental philosophy and results of quantum mechanics one cannot ignore that both these fields of knowledge – to put it mildly – border on agnosticism. Let us take

Kant's approach on the one hand, and the Copenhagen interpretation of quantum mechanics on the other. Indeed it is not much justified to ascribe to the Kantian conception of "things within themselves" of such an interpretation, according to which Kant simply recognized the existence of things unrecognizable for us (to our minds). The Kantian category of "things within themselves" is rather an expression of a conviction that the research results in the fields of necessary conditions of our cognitive possibilities demarcate boundaries within which speaking of cognition has any sense. However, delimiting as if *a priori* boundaries of that kind is undoubtedly to some extent agnostic.

On the other hand, the fact that the quantum mechanics description by means of the wave function is a statistical description, raised suspicions that were expressed on many occasions, particularly by Einstein, that it is not a complete, full description, that it leaves out of consideration certain individual characteristics of microobjects, and that is based, like a probabilistic description in classical statistical physics, on the procedure of averaging of values of certain magnitudes (e.g. energy or velocity), which belong to individual microobjects. Many times, this conviction also led some opponents of the Copenhagen School to search the so-called hidden parameters, i.e. magnitudes that are "responsible" for this or that "individual" behaviour of microobjects, which do not find their expression in the "averaged" statistical description.

Abstracting from the above, the fact interpreted by Heisenberg's principle of indeterminacy, consisting in the impossibility of simultaneous precise determination of the value of two canonically connected magnitudes, i.e. represented by the non-commutative operators, e.g. the position and momentum or time and energy of a microsystem, may be interpreted either in an indeterministic way (thus ontologically) or agnostically (thus in a theory-cognitive way). The former interpretation assumes that a microobject actually (objectively) does not have any of the two characteristics which come into play (e.g. position and momentum). That is why the impossibility of simultaneously capturing (establishing) them precisely in the act of measuring (cognitive one) is a reflection of the ontological actual state consisting in a certain "indeterminacy" of the microobjects themselves and involving a fundamental qualitative change, the peculiarity of the phenomena of the microworld as compared with the phenomena of the macroworld.

The latter interpretation assumes that although objects of the microworld have, just as like macroobjects, both characteristics (canonically connected magnitudes), still – because of the inevitable and uncontrollable influence of a macroinstrument on the microobject in the process of its empirical exploration – the procedure of empirically determining the value of one magnitude of a given pair inevitably leads to breaking the state of a microobject characterized by the other (second) magnitude of a given pair. Therefore, the impossibility of simultaneously establishing the value of both magnitudes is – under this interpretation – a cognitive fact (cognitive limitation), and not an ontological one (existential indeterminacy), and simply indicates the limitation of our cognitive possibilities.

Initially, adherents of the Copenhagen interpretation, especially its most eminent representatives, i.e. Bohr and Heisenberg, placed more emphasis on the second of the indicated interpretations (the agnostic one). This was reflected in the well-known thought experiments of Heisenberg adduced in order to "justify" the principle of indeterminacy which he formulated and in Bohr's idea about the uncontrolled interaction of the microobject and the macroinstrument. Later on, more emphasis was placed on the ontological (indeterminist) interpretation of Heisenberg's indeterminacy principle and Bohr's principle of complementarity. It was assumed that a complete description of a microsystem is possible neither exclusively in the wave picture (language) nor exclusively in the corpuscular picture (language) but should constitute a complementary combination of both those pictures (languages) which from the point of view of classical physics exclude each other. The indeterministic (ontological) interpretation cannot be left out of consideration completely even if only due to the so-called causal anomalies appearing in the physics of the microworld both at the level of theoretical considerations and in the experimental studies conducted.

One of such anomalies is as follows: if we pass the elementary particles through screens with openings then in the case of closing some of the openings the number of flowing particles increases instead of decreasing or remaining the same. Facts of this kind do not occur in experiments with macrobodies and seem inevitably to lead to a conclusion that the principle of univocality (also called the principle of necessity) of univocal determinism, stating that "the same happens under the same conditions" (or, to use a different wording, that "the same causes under the same conditions evoke the same results") cannot be maintained when applied to microphenomena.

Nevertheless, it is almost commonly accepted – not only by the advocates of the Copenhagen interpretation – that the quantum theory description of the state of a microsystem by means of the wave function is not a "purely objective" description of the state of a microobject, but at the same time it is a metadescription of the state of our knowledge concerning the microsystem and that both these descriptions cannot be separated from each other, indicates that both those interpretations, ontological (indeterministic) and theory-cognitive (agnostic), come into play to some extent "at the same time."

V. Transcendental Cognition and Scientific Knowledge

Thus, in quantum mechanics we have to do with the approximation of the object and the subject, and with, as it were, mixing in the quantum-mechanics description of the state of microobject with the state of our knowledge on the microobject, that is mixing of the objective with the subjective. In this connection, a question arises whether quantum mechanics realizes, at the level of empirical study and at the level of a particular study, what transcendental philosophy wants to realize at the transcendental level, which, according to it, is deeper. Do both of them go in the same direction? I think it is so, in spite of the fact that transcendental philosophy went that way much further, assuming the far-reaching unity of the objective and subjective "world" and aiming at the study of a relation which combines them. This relation constitutes both those "worlds" as a certain larger whole in which they "lose" their own autonomy, which is assumed in metaphysics. However, the fact that in quantum theory and, in general, in the physics of the microworld that unity is manifested already on the level of empirical consciousness of the ("ordinary") scientific cognition, and that not just – as is assumed by transcendental philosophy – at the level of "pure consciousness" of transcendental cognition, is greatly thought-provoking and makes one reflect on the justifiability if not of the idea of transcendentalism itself, then in any case of the transcendental method. Furthermore, since it is characteristic of transcendental philosophy that both these "things" (idea and method) cannot be separated from each other, one can also doubt the reliability of the idea of transcendentalism. Let us take into consideration that Kant, in *The Critique of Pure Reason,* distinctly opposes considerations at the transcendental level both to considerations at metaphysical level as well as those at the empirical level. Both of them are treated as cognitive perspectives of the epistemic field of theory. However, according to Kant, the empirical nature ("the empiricity") of non-transcendental knowledge does not simply consist in its aposterioric, i.e. directly empirical, character since also "pure mathematics" and "pure natural sciences," being *a priori,* constitute empirical cognition. Similarly, the "metaphysicality" of non-transcendental knowledge does not have to mean its obvious transcendental transgression in the direction of apparent (speculated) beings which, in principle, is imperceptible to any experience (such as, for instance, God or soul). But it means knowledge which is a response to any question concerning any objects (even if "empirical" ones) which asks about their "being in itself," abstracting from the manner and conditions of their encounter and "being given" (see Siemek 1977, pp. 57-58). Both the empirical and metaphysical point of view are directly a certain "experience" and just because of this none of them can "see" "experience" itself as such whereas the transcendental point of view – and only it – is just "seeing the experience itself" (p. 63). The very fact that Kant's transcendental philosophy, while invalidating the metaphysics of the hitherto existing philosophy, at the same time constituted a theoretical justification of modern science, and quantum mechanics abandons the ideal of the purely objective description formulated on the basis of that science makes one think. As it is, jointly with the criticism of modern science, its aims and ideals, quantum mechanics seems to have undermined – in spite of the marked parallelisms – also transcendental philosophy providing a theoretical justification to modern science. In my opinion, we are dealing with a situation that is rather often encountered in various cognitive situations, whereby principally the new fact, apparently confirming some hitherto existing view, actually leads to its rejection and, to put it more precisely, to its basic modification.[4] It is also worth noting that in order to overcome the difficulty discussed above some researchers attempt to give quantum mechanics and other "basic theories of physics already existing now or expected" (Weizsäcker) also the status of transcendental knowledge. In his work *Die Einheit der Natur*, C.F. Weizsäcker, for instance, formulates a hypothesis that the future "uniform physics results from conditions of the possibility of carrying out experiments,"[5] and that its final findings will be conclusions concerning those conditions.

According to this hypothesis, quantum mechanics, the theory of elementary particles and cosmology, would have to create from together something like such a "uniform physics" of a transcendental dimension. In this programme of "transcendental physics," quantum mechanics would play a (distinguished) role of "the general theory of movement of any objects," the theory of elementary particles the role of "the theory of actually existing types of objects" and, finally, cosmology would play the role of "the theory of all objects actually existing" (pp. 255-256). Here it

---

[4] In the light of quantum mechanics, classical mechanics with its ideal of a purely objective description of real phenomena – even though it constitutes an indispensable element also of modern physical knowledge without which this knowledge could not "contact" reality – is only an approximate boundary case of the former which is realizable (applicable) in areas in which the quantum character of physical phenomena may be neglected. This means that macrobodies – but only they – may be submitted to a purely objective (macro)physical description.

[5] Cf. Weizsäcker (1978, p. 254). Elsewhere in that volume, Weizsacker formulates the hypothesis in the following way: "Who would be able to analyze under which conditions experience is possible this person would have to be able to show that all general laws of physics result already from these conditions. The physics derived in such manner would be the expected uniform physics" (p. 250). Further on, Weizsäcker formulates in a similar way a "supposition" that the "basic postulates of the last closed theory of physics will not formulate anything else but only conditions under which the experience is at all possible" (p. 253). "Deduction of the quantum theory from the conditions of an experience, as it seems, cannot be treated beforehand as a hopeless undertaking" (p. 292).

is hypothetically assumed that quantum mechanics is a fundamental physical theory ("the general framework of physics") from which both the theory of elementary particles[6] and cosmology may be derived (as logical consequences). The former, as a fundamental physical theory of a transcendental dimension (as "in some sense uniform physics"), would reply to Kant's question about "the possibility of experience itself," i.e. about "the *a priori* conditions of the possibility of getting to know any objects."[7] Reaching in its scope "the area of recognizing all possible objects," i.e. "the area of possible experience" – quantum mechanics, interpreted in such a way, would replace transcendental philosophy and through this would lead to joining –at the transcendental level – philosophical knowledge with scientific knowledge.

Further development of microphysics, penetrating ever deeper levels of the structure of matter leaves no doubt however – as it seems to me – that giving quantum mechanics the status of a transcendental theory is unjustified. On the other hand, there is no doubt that what has been established by quantum mechanics and in general by physics of the microworld, in spite of the fact that it has been obtained at another level of considerations than thoughts contained in the transcendental idea will have a significant meaning for the further fate of the latter. Just as in transcendental philosophy, epistemological studies are not limited to the analysis of the cognitive process itself. They encompass a larger whole, including conditions of cognitive possibilities so the analysis of cognitive processes realized in microphysics includes the whole which consists of: a physical object, the conditions of cognition and an observer (researcher); or – to use a somewhat different terminology – the cognitive relationship in microphysics occurs between three segments: "object – instrument – subject." The cognitive process, taken as a whole, forms a unity of those

---

[6] Elsewhere in that volume, Weizsäcker "treats seriously the possibility" that "quantum theory and the theory of elementary particles will finally appear equal as to the range and then probably – identical" (Weizasäcker 1978, p. 275).

[7] The following reasoning is characteristic of Weizsäcker's approach to this matter: "For general laws I can imagine the only one justification in view of an experience, which is not an *a priori* dogma and which is not *petitio principii* either. This is contained in Kant's idea that general laws formulate general conditions which make experience possible. Recognized by us, they will still not seem necessary by themselves, as we do not know whether it is necessary that the experience be possible. However, they will be valid in such a scope in which experience is possible and that is why everybody who is ready to recognize the evidence of an experience will have to consent to it. Thus, a question arises whether the basic assumptions of physics are not just indispensable assumptions so that an experience may exist" (Weizsäcker 1978, p. 281). Weizsäcker includes among such necessary conditions of any possible experience, outlined by quantum mechanics "the structure of (historical) time" described by quantum logic ("logic of times") and the "concept of objective probability" concerning exclusively the future. According to him, both those conditions are "non-classical," i.e. quantum in nature (p. 291).

three segments so that, for instance, considerations on the active role of a subject have any sense only in the context of study of the process of cognition as a whole.

VI. The Problem of the "Infinity" of Being and the "Openness" of the Future

To conclude, I would like to draw attention to one more analogy between transcendental philosophy and microworld physics, an analogy which is after all connected with the two previously discussed ones. Thus, it concerns not so much transcendental philosophy itself in its early classical version but rather contemporary "extensions" of this philosophy which can be observed at least in some trends of hermeneutics, Marxism or existentialism.

In those trends (and also, as it is known, in Bergsonism) – in each of them in its own specific manner – emphasis is put on the idea of the "infinity" of existence, its "disposability," an ability always to bring out new and basically new definitions that were unavailable in the earlier stages. Under such an interpretation, the history of nature, and in particular that of the society and human being, is the rise of ever newer alternatives which make the world ever richer as far as possibilities are concerned, and causing the future not only to be uncertain from the point of view of cognition but also o be ontologically "open." What comes to the fore here is a deeply dialectic category of the "potentiality of being," which is not unfamiliar also to the very old philosophical conceptions, such as Aristotelian philosophy and Thomism.

This category has been very successful just in quantum mechanics and has been applied with invariable predilection by all microphysicists, no matter what their philosophical and theoretical-methodological orientation, but in particular by followers of the Copenhagen interpretation. This is connected with the fact that the quantum theory description by means of the wave function is not so much a description of the present state of a microobject in the classical sense but rather a probabilistic description of all the possibilities of future states of a microobject. The wave function simply supplies a probability of all the future states into which a given microobject may pass, in which it can find itself in the future. Therefore, it describes literally not the present state of a microobject but its possible (statistically determined) transformations resulting from this state or, in other words, its possible states in the nearest future. It describes in certain sense the tendencies that are contained in it, its potential chances that it describes as if its future (thus it is not simply the present).

Some physicists, Heisenberg for one, simply claim that microobjects as such, taken in isolation from an experimental system – as distinct from macroobjects – do not exist currently, factually, but only in potency. Only a microobject that is in an appropriate experimental situation which is inevitably – like our technical research means and we ourselves with our senses – macroscopic in nature, is something that is fully real (and not potential). In contradistinction to e.g. a table which, according to Heisenberg, really exists independently of the observer and the experimental situation, this or another electron really exists in combination (connection) with a macroinstrument which only – thanks to the irreversible macroscopic process which we call measurement – as if "codefines it," transforming it from the potential being to the actual being.

A philosopher being permeated with the idea of transcendentalism – as distinct from a physicist – will not accept such a rigid separation of (the only potentially existing) objects of the microworld and (fully really existing) objects of the macroworld from the point of view of status and character of the kind of existence (being) vested in it.

Again, along with the similarity between transendental philosophy and physics of the macroworld we can see – as if at the same point – a fundamental difference. This time I also think that it is not a difference resulting exclusively from the different perspective or plane of formulating the problem or, to put it differently, from another level of considerations conducted, but that different standpoints underlie it, which – disregarding the analogy outlined – are in disagreement with each other.

VII. Conclusion

Quantum mechanics is – when compared with the earlier physical theories – not only a new theory but a theory of a completely new type both with respect to its internal structure and the content of its statements, and consequently the functions that it fulfils. I think that it does not only go beyond the framework of mechanics and that of physics (at the same time being, for instance a chemical theory constituting the foundation of quantum chemistry) but also beyond the limits of science in general. It becomes the first natural theory which to some extent deserves the status of a definite philosophical conception (or theory).[8] As such, quantum mechanics seems to indicate that the problems undertaken by transcendental philosophy and its continuations in the from of phenomenology and postphenomenological philosophy on the plane that is different from the plane of science may and should be considered (and solved) on the philosophical and scientific basis at the same time. It does not mean that one can agree with Weizsäcker that quantum mechanics should take the place of transcendental philosophy that its fundamental concepts characterize conditions of all the possible experience. It is

---

[8] Similarly to Marx's theory which is a scientific theory (in this case in the field of the humanities) and at the same time a certain philosophical conception.

impossible if only for the fact that quantum mechanics is in no sense an "ultimate" or "closed" theory,[9] but constitutes only the next (although a very significant) link in the development of physics. Also, and no less importantly, it constitutes a fundamental step in the direction of bringing together of science and philosophy. However, there is still a long way to go before these two fields of cognition are ultimately combined.[10]

It can be seen at least from the fact that, even within science, physical experience is not the only kind of experience that comes into play. For instance, Piaget shows in a convincing manner that apart from the physical experience, consisting in "manipulating objects and obtaining knowledge through abstracting just from these objects" (Piaget 1977, p. 80), no lesser role is played in the scientific (and not only scientific) cognition by logical-mathematical experience, also consisting "in manipulating" objects. However, "cognition is here derived from the activity itself, and not from the objects themselves. In this case, activity begins from ascribing features to objects which they did not possess by themselves (and which, after all, retain the previous properties of objects) and experience concerns relations between features which an object is endowed with by an action (and does not concern previous properties of an object). In this sense, cognition is really abstracted from activity as such, and not from the physical properties of objects."[11]

----

[9] According to Weizsäcker, if not the quantum mechanics itself then the unified physics that is based on it, and extended in a proper way, will see it end before the close of century, in the sense that its fundamental laws will be formulated. These will be laws formulating conditions of any possible experience: "... physics can be finished off if or as far it will be possible to enumerate conditions of the possibility of experience" (Weizsäcker 1978, p. 254). In spite of the admission that scientific knowledge is not complete in three fields: in the theory of elementary particles, cosmology and biology, Weizsäcker speaks in favour of a hypothesis that "in all the three areas, quantum theory remains valid exactly in the form which we know now" (p. 218). If it were really so, then further development of physics would not be already a development of theoretical physics but of "concrete" physics and would consist in conducting perhaps an unlimited number of structurally different experiments, exploring "surely also an unlimited sequence of the ever higher physically possible structures," leading, e.g. to biology and cybernetics (p. 254).

[10] Weizsäcker is more optimistic on this issue. He states that although quantum mechanics has so far been referred, by means of its physical semantics of an empirical character, "to particular experience," yet "now we try to find for it a reference not to specific experience but to common features of all experiences, and maybe even to every possible experience" (Weizsäcker 1978, pp. 293-294).

[11] Piaget (1977, p. 80). It is known from other Piaget's works that he recognizes yet other kinds of experience playing an essential role in some sciences, e.g. social experience or psychological one (introspection). The latter one has to do with the properties of the subject him/herself, becoming conscious through introspection, or the analysis of one's own mental states. Cf. Piaget (1972). Cf. also Zamiara (1977, pp. 186-188).

The fact that there are two kinds of experiment is admitted, in a sense, by Weizsäcker himself, when he writes that "it seems that mathematics has to do with problems which are analogous to the problems of physics considered here (I personally think that they are identical as to their essence)" (Weizsäcker 1978, p. 233).

Anyway, coupling – at present already – "any possible experience" with one or another physical theory or at least with physics as such seems to be to narrowing down the philosophical problem to the problem of a particular science. And, as such, it is not in a position to give contemporary physics – in spite of its fundamental and distinguished role in the system of empirical sciences and its partly philosophical nature – a full status and rank of a philosophical discipline.

THE UNIVERSALITY OF SCIENTIFIC LAWS AND THE EVOLUTION OF THE UNIVERSE*

Let us pose the following question: can the evolution of the Universe lead to the rejection of the widely accepted claim that scientific laws are universal (strictly general)? Let us then consider first what the thesis about the universality (strict generality) of scientific laws states, and then try to decide whether the evolution of the Universe does not undermine this thesis.

I. The Thesis About the Universality (Strict Generality) of the Laws of Science

That thesis, at least in my view, means something else within formal sciences (logic and mathematics) than within empirical sciences (e.g. physics or biology).

Within the framework of formal (mathematical) sciences the strict generality of a statement is understood as non-equivalence of the finite conjunction of individual statements. If we submit this requirement to the so-called laws of logic or mathematics then it will mean that none of them may be replaced by a finite class of individual statements (I do not consider here the problem of whether the laws of formal sciences are really statements or perhaps expressions of another kind, e.g. sentence functions, i.e. certain schemes of statements).

In empirical sciences, the strict generality of a statement is understood as a spatiotemporal universality, i.e. it encompasses within its range of applicability all the phenomena of a given class, independently of where or when they occur. The strict generality of a statement thus understood does not prejudge of an area in which the law is non-emptily satisfied (fulfilled) – whether it is a finite or an infinite area.[1] That is why a strictly general statement may appear to be an individual one in the sense that it is non-emptily fulfilled only once, it "rules" over only one (unrepeatable) phenomenon. Strict generality claims only that the law is potentially

*The translation of the "Uniwersalność praw nauki a ewolucja Wszechświata" *Zagadnienia Naukoznawstwa*, **3-4**, 1987, 312-328.

[1] I disregard the problem of the idealizational nature of most laws formulated within developed empirical sciences. As is well known, idealizational laws are strictly speaking emptily fulfilled in the world of empirical phenomena which can be directly observed.

fulfilled (either emptily or not) in the whole Universe, and it "acts" (is non-emptily fulfilled) wherever and whenever there exist conditions that are conducive to its occurrence (usually formulated in the antecedent of the law). The conception of the strict generality of the laws of science as of their spatiotemporal universality fits well into classical physics in which there hold Newton's views on space and time that are devoid of structure (everywhere the same) backgrounds of phenomena. It is expressed by the so-called Maxwell principle which guarantees just spatio-temporal universality to the strictly general statements.

II. Maxwell's Principle

According to that principle, the course of physical phenomena (and, as we can presume, all the others) does not depend on the place and time of their occurrence. To put it differently, Maxwell's principle excludes time and space from a set of "conditions" which influence the course of physical phenomena (and others). The time and place in which a phenomenon occurs do not matter for its course, and also for the laws to which it is submitted. Therefore, Maxwell's principle may obviously be applied not only for phenomena but also for the regularities of nature and – in a metaformulation – to the laws of science. In the nomological version, Maxwell's principle states the independence of any regularities from the place and time of their occurrence, otherwise, their spatio-temporal universality. This means that regularities are unchanging with respect to space and time or that they occur always then when appropriate conditions appear while the laws which describe them are given unchangeability with respect to transformations consisting in either a space or time shift of the beginning of the system of reference. This makes it possible also to speak of the spatiotemporal universality of laws (under the condition of their strict generality).

Maxwell's principle in the nomological version in which we are interested here may also be formulated as a principle which excludes from a set of conditions such "conditions" as space and time, thus a principle that makes it possible to apply laws for all the areas of space and time in which "proper" conditions are satisfied. At that, conditions which appeared sufficient for the occurrence of a regularity in one area of space and time are sufficient for its occurrence in all the other areas (obviously, the same concerns conditions which are indispensable for the occurrence of regularities). It follows from the above that Maxwell's principle excludes all the spatiotemporal evolution of laws or their replacement by other laws, assuming that the conditions remain unchanged, i.e. that the possible evolution or replacement of laws is not caused by the change (evolution) of the conditions themselves, among which space and time, obviously, do not play any role. To put it differently, according to

Maxwell's principle, evolution (or an exchange) of regularities cannot be caused by space and time by themselves, independently of the other factors. The principle discussed in its nomological version may therefore be expressed briefly in the following statement: "under the same conditions always and everywhere there are the same regularities."

A characteristic feature of the spatiotemporal universality of laws that are strictly general, implied by Maxwell's principle, is that it belongs to all the strictly general laws in the same degree, so it does not distinguish any category of laws formulated in strictly general terms: each of them is applicable in all the areas of space and time in which conditions proper for it are met. In the areas where those conditions are absent every strictly general law appears to be empty-fulfilled. Spatiotemporal universality, assuming non-empty fulfilling of the law always when proper conditions are realized, thus assumes – as in the case of regularities – both: (1) the unchangeability of laws in space and time, i.e. the non-validity of the laws of evolution caused by space and time and (2) the irreplaceability of laws – when the assumption of unchangeability of conditions is assumed – by other non-equivalent laws. Thus, the laws cannot "leave the stage" or "start working" if conditions have not undergone a change.

This leads us to the conclusion that the spatiotemporal universality of laws might appear theoretically shattered – in case Maxwell's principle were refuted – in two ways. (1) It can occur in a "continuous" manner by a slow evolution of laws when going to new areas of space and time in which conditions do not differ from those in an initial area. Such a case might have taken place if, for instance, it appeared that the physical universal constants are functions of either space or time. Then, one could attempt to formulate more general laws (than those which appeared functionally dependent on space and time, thus fulfilled only in the local area), taking into account this functional dependence on either space or time of the universal constants. However, if this dependence appeared to be very complicated, it would perhaps not be possible to formulate such more general laws. (2) The second way of undermining the spatiotemporal universality would occur in a case of a sudden (leap) replacement of some laws with others, in a transition to other spatiotemporal areas in which identical conditions occurred.

In both cases, the reality would reveal a certain indeterministic "flaw." However, in case (2), this "fissure," shaking the deterministic order of the world would appear to be more severe: the possibility of formulating more general laws would probably be excluded in this case. In physics or, to be more exact, in cosmology, only the first possibility is considered (the evolution of laws) and almost exclusively within the temporal aspect (the evolution of laws in time), surely rightly thinking that this case of shaking the spatiotemporal universality of strictly general laws is the most probable one (which does not mean, after all, that the probability of its occurrence is considerable).

After all, a possibility cannot be excluded that the spatiotemporal universality of strictly general laws only apparently constitutes the type of universality which belongs in the same measure to all the strictly general laws. Maxwell's principle may appear strictly fulfilled only when applied to a certain category of strictly general laws: if not all of them are universal from the spatiotemporal point of view, then not all of them must evolve or undergo replacement caused by space or time. Finally, it may appear that it is fulfilled in reference to all strictly general laws but with a varying degree of approximation.

Taking into consideration both possible cases of shaking the space and time universality of regularities (or laws, respectively) let us adopt the following formulation of Maxwell's principle in the nomological version: under the same conditions there are always and everywhere regularities (respectively, there "act" the same laws) – regularities (laws) which neither undergo evolution in space or time nor are replaced – without any change of conditions – by other regularities (laws).

The following question arises: does Maxwell's principle while guaranteeing the spatiotemporal universality of regularities (and strictly general laws) exclude thus any their evolution or only a spatiotemporal evolution, determined by the change of place or time? There appear in literature the conceptions of changeability of regularities under the influence of a change of objects which these regularities concern; they assume a negative answer to this question: not all the evolution of regularities leads to undermining their spatiotemporal universality (cf. e.g. Kedrov 1965, pp. 137-143; Rutkiewicz 1965, pp. 25-34).

This problem is closely connected with the following question: does showing the possible changeability of one or another universal constant lead unavoidably, as is often thought, to the refutation of Maxwell's principle, and thus to undermining the spatiotemporal universality of at least some regularities of nature, or does it not?

There is no doubt that the positive answer to that question would not be right. Let us assume, along with some physicists and cosmologists (e.g. with Dirac), that at least one of the universal physical constants (such as constant speed of light c, gravitation constant k, electric load constant e, Planck constant h), playing a fundamental role in physics and occurring in formulations of many laws of main importance, is not a constant in the exact meaning of the word, since it is submitted to a slow uniform time evolution. Let us further assume that we have succeeded in expressing this change in a simple way as a function of time ("the age of the Universe"). Therefore, for the formulation of law in which this "constant" appears, we explicitly introduce a time coefficient t as an independent variable. As a consequence – as it seems – we should accept that this law, while evolving in time, leads to undermining Maxwell's principle. Indeed, this will become obvious when we take into consideration the following lucid metaformulation of Maxwell's principle in the nomological version: "A formal expression or a metaformulation of Maxwell's principle is considered to be a statement concerning laws of physics which states that these laws do not contain, or cannot contain, independent variables, thus: time *t* and spatial coordinates *x*, *y*, *z* *explicitly* but exclusively *implicitly*" (Augustynek 1964, p. 173).

It is obvious, based on the formulation quoted above that one should consider Maxwell's principle in the case discussed here as an undermined one. Let us assume that this law evolving in time (thus non-invariant with respect to the shift of the zero point of time counting) – e.g. with respect to the changeability of the gravitation constant *k* – is the law of universal gravitation. On the strength of the metaformulation of Maxwell's principle quoted above and an assumption that the unchangeability of laws with respect to the shifts of zero point of time counting is equivalent to the spatiotemporal universality of regularities which they describe, we come to a conclusion that also the gravitational regularity is not a spatiotemporal universal since it is submitted to evolution in time.

In this way, to establish the changeability some universal constant (and to show that it can be expressed as function of time) seems to lead unavoidably to the conclusion that Maxwell's principle is undermined. In any case, such a conclusion is sometimes drawn when changeability of universal constants is assumed and when they are opposed to such "idiographic constants" which are of purely local significance, as the acceleration of the gravity constant *g*, occurring in Galileo's law on the free fall of bodies. The change of those constants does not evidently lead to the evolution of laws.

However, the problem is not so simple. As can be seen from Maxwell's original formulation ("The difference between phenomena does not depend on their space and time location but only on the nature, configuration or motion of proper bodies"), the fate of the principle called by his name depends – in the case of showing the changeability of one or another universal constant – on the answer to the question: what causes the change in the value of this universal constant (and – as a consequence – the "evolution" of a given law) "character, configuration or motion of proper bodies" or simply the change of "the space or time location"? If the gravitation constant *k* appears to be dependent in some way on the structure of the world, e.g. distribution of matter in the Universe or its part, and changes its value, let us say, together with the change in the medium density of matter in some sufficiently large area (or in the whole Universe), then this change is caused by "motion and the change in the configuration of relevant bodies," and not simply by the "lapse of time" *t*. Maxwell's principle appears in this case unshaken while the introduction to the law of gravitation of time *t* *explicitly* is unjustified. Therefore, Maxwell's principle would appear to be unshaken if, for example, the change in the gravitation constant could be explained by means of the principle of the relativity of inertia (Mach principle). According to the latter, the magnitude of the inertia of bodies is completely defined by the masses which occur in the Universe and their distribution. Thus it is determined exclusively by mutual interaction of material objects, and not by the independent properties of space and time. Therefore, it depends on the structure of the Universe and may change when the world transfers from one state to another. At present, attention is drawn to the fact that by means of Mössbauer effect which made it possible to check, for the first time, the general theory of relativity under laboratory conditions, it will also be possible to check the validity of this principle. That principle has so far been controversial: if Mach principle is right, then the forces of inertia should appear to depend on the direction of the motion with respect to the Galaxy (some for the motion of the body in the direction parallel to the Galaxy plane, and others in the direction perpendicular to this plane). The change in the acceleration of particles, caused by the change in the direction of motion should be manifested in the spectral lines sent by them, which can be checked by means of the Mössbauer effect. The positive result of experiments of this type would therefore lead to saving Maxwell's principle in the case when the gravitation constant would appear to be "variable in time."

And in the case when the gravitation constant would really appear to be the function of time and would change together with "the age of the Universe," independent of any courses of physical processes and on changes occurring in the structure of the Universe so that nothing would "stand" behind this change except for the time variable *t*, then we would have to consider Maxwell's principle as a shaken one. The fate of Maxwell's principle depends thus on whether the possible change in the universal constant is caused by the change in the structure of the world which may be treated as a change in the conditions of holding of laws (and not as a change of laws themselves) or it depends simply on "the age of the Universe" or "the spatial location." In the first case, the time variable *t* and the spatial variables *x*, *y*, *z* should occur in the formulation of laws exclusively *implicitly* as dependent variables (on "the structure of the world") since Maxwell's principle retains its power, and in the second case the above variables should occur *explicitly* as independent variables, which leads to the principle being undermined.

However, showing the variability of any universal constant may easily be interpreted, especially in the initial period of studies, as undermining Maxwell's principle also in the first case, namely, if we were not able to "capture" the influence of the structure of the world on the change of this constant and, which follows, on the "action" of laws, introducing on this basis the conclusion that Maxwell's principle does not hold, one can always suspect that the apparent undermining of Maxwell's principle has been considered to be real. This suspicion would be even more justified if it appeared that to magnitude which has been accepted as the universal constant does not actually include not only an attribute of constancy (in time) but also the attribute of (spatial) universality, i.e. if it appeared that that "constant" is only of local significance (occurs only in a given limited area). On the other hand, however, even in this case we would not have full certainty that the shaking of Maxwell's principle is purely specious. It might appear that the possible inconstancy (in time) and non-universality (local nature) of this or the other "universal constant" are the result that this "constant" is at the same time both the function of time and the function of space. The suspicion that it is not so would practically be transformed into a certainty only when it would be possible to prove that the constant that is taken into consideration is non-invariant with respect to the conditions that prevail in the world. On the other hand, in order to make sure that the time or space evolution of the universal constants leads to the shaking of Maxwell's principle we would have to be certain that these magnitudes actually have the rank of universal ones as well as invariant ones with respect to the conditions, i.e. that they will not in any case appear to be variables dependent on conditions. In this case we would also have to be sure that these constants are really either functions of time or of space.

It cannot be precluded that no universal constants, characteristic of the Universe as a whole, exist. In this case, undermining Maxwell's principle might consist in the dependence of local constants on time or space. Neither can we totally preclude the possibility that universal constants are functions of both time and space as well as of changing conditions so that the existing regularities may change both in the sense of space and time and under the influence of changes occurring in the structure of the physical Universe.

In each of the two cases of the changeability of universal constants at least certain laws would appear non-universal, spatiotemporally in the first one, and with respect to conditions in the second (because of non-occurrence in the whole Universe of relevant conditions), whereby the first one leads to undermining Maxwell's principle. The fate of Maxwell's principle appears to be closely connected with the fate (status) of universal constants. If these constants actually appeared to be constants only under certain definite conditions, thus constants of a local significance and, in this way became similar e.g. to the constant of gravitation g, then showing their variability would not at all mean the refutation of Maxwell's principle. However, even if in spite of showing their changeability (in time) preserved the status of universal magnitudes and independent of the structure of the Universe but dependent only on the age of the world or the spatial location, then to show their changeability would equal the refutation of Maxwell's principle.

It seems that only in the last case would there be any point in speaking of an evolution of regularities and interpreting the hitherto existing laws in which universal constants appear as inadequate to these regularities and as non-invariant with respect to spatiotemporal shifts of the start of the system of reference. And in the first case, all the existing laws – including those which in the earlier interpretation, assuming unchangeability of physical constants, were considered to be universal – should be considered to be local laws ("conditionally" non-universal), i.e. non-applicable under these conditions in which universal constants assume other values than the values that belonged to them before. The laws that have been known so far might also be submitted to a freer interpretation, assuming that appropriate relationships also occur under conditions in which universal constants assume other values, if only while applying laws one substituted those "constants" with their proper values on each occasion (by analogy to Galileo's law of the free fall of bodies, in which in the consequent occurs a formula $S = gt^2/2$, instead of g we substitute appropriate values dependent on the location and altitude of the falling body with respect to the centre of the Earth). However, in both cases the existing laws would lose their importance and one could attempt to formulate more fundamental laws as well as more general ones, even universal laws, in the first case from the point of view of time and space, and in the second case "conditionally" (with respect to conditions).

Therefore, one should clearly differentiate between the case in which a restricted universality of laws with respect to non-repeatability of appropriate conditions (their non-occurrence in other areas of space and time), or the so-called conditional ("conditional") universality of laws and the case in which the spatiotemporal universality of laws is undermined. Certainly, only in the last case could we interpret regularities as those which evolve or are submitted to exchange either in time or in space, while the laws which describe them may be interpreted as non-invariant with respect to spatiotemporal shifts.

III. Maxwell's Principle and the Problem of the Homogeneity of Space and Time

It should be noted that the homogeneity of space and the homogeneity of time are the only features which cause the understanding of the strict generality of lawlike statements as their spatiotemporal universality to be natural and unproblematic. What follows is that without ascribing these features to the spatiotemporal continuum, Maxwell's principle could not hold.

How in this light should one assess the fact that space and time – contrary to what Newton thought – actually have a complex and variable structure (i.e. are in a sense "non-homogeneous")? This fact, considered by the general theory of relativity which establishes that the spatiotemporal metrics depends on the intensity of the gravitational field and consequently on the distribution and density of the masses results in the fact that space and time can no longer be considered to be an "indifferent background" of events that take place. However, this does not mean that the conception of strict generality as a spatiotemporal universality is defeated together with Maxwell's principle. They may be saved and this is because the local modification of metric properties of space and time, evoked every time by an appropriate modification of physical conditions (change of distribution and density of gravitating masses), does not deprive time and space of the fundamental property which we call the homogeneity of space and time.[2] This property, which allows any shifting of the zero point of counting space and time, is – as is well known – strictly connected with the law of the conservation of energy and the momentum conservation law. It also guarantees the fulfilment of Maxwell's principle and, along with it, the spatiotemporal universality of laws of science and other strictly general statements. The difference consists only in the fact that at present (i.e. after the general theory of relativity has been formulated) one can no longer interpret space and time as an "indifferent background" of the course of phenomena. What is "indifferent" is only a certain *residuum* which is created by these properties of space and time which always and everywhere remain the same. On the other hand, those of their properties which are changeable and connected with the distribution of masses, at present enter into physical conditions which have an influence on the course of phenomena and that is why they should be taken into consideration in the process of formulating laws (taken into account in their antecedent). It appears that this suffices to salvage the conception of the spatiotemporal universality of laws.

A more complicated problem arises when one accounts for the evolution of the Universe, theoretically justified by the relativistic cosmology and solidified by many empirical data.

---

[2] Although some physicists, e.g. W.A. Fock, claim that the general relativity theory showed that time and space are not – contrary to what was thought before it was created – homogeneous, what they mean is not the non-homogeneity that is meant here (which is connected with the laws of energy conservation and momentum) but simply one whereby space-time ceases to be considered as an absolute (indifferent, devoid of dynamic properties) background of the physical events.

IV. The Evolution of the Universe

The Evolution of the Universe itself may be considered today as something indisputable. What is greatly disputable and sensitive depending on one's wider philosophical and religious outlook are the various philosophical interpretations of that fact, and even its cosmological interpretations which go beyond factual data, on the basis of which attempts are made to establish what this fact exactly consists in.

Let us first consider its interpretation which seems the least favourable for the conception of the spatiotemporal universality of laws. According to this interpretation, the evolution of the Universe should be understood as a directional transformation of the whole physical Universe (the whole matter) which not only causes the unrepeatability of the conditions (both on the scale of the whole Universe and on the scale of local conditions) but is also connected with the change in the universal constants (c, k, h, e and the like): their values would be different not only in various studies of evolution ("growing old") of the Universe but also in various areas (taken at the same time); and the universal constants would be functions both of space and time.

Maxwell's principle would in this case be obviously undermined, which would shatter the very conception of the strict generality of laws, understood as their spatiotemporal universality.

Maxwell's principle and the above-mentioned conception of the strict generality would not fare better if the evolution of the Universe showed that universal constants are functions of time although they are not functions of space (location). In this case, the course of events would indeed not depend on the place of their occurrence, and at the same time it would depend on the time of their occurrence (on the "age" of the Universe).

Theoretically, one might also imagine an opposite situation: the Universe is submitted "not to a temporal but to a spatial evolution"; the universal constants are functions not of time but of space (location, place). However, there are no data which would allow one seriously to consider this (intuitively) not very real possibility.

After all, all the three possibilities considered do not seem very probable. Accepting that the value of universal constants does not depend on the actual conditions which are realized in the Universe, and on the other hand depends on space and time, introduces a strange kind of indeterminism and is an assumption bordering on the absurd.

On the other hand, a possibility that is considerably more probable is that the evolution of the Universe modifies the general conditions which are found in the fact that the universal constants change their value just under their influence. The first and the least probable variant of this possibility would be realized if the universal constants would change their value both at the transition to new time areas as well as to space ones. It would mean that in none of the two (various) areas similar (the same) conditions would be realized. In this case, assuming that any law may be non-emptily fulfilled (may "act") only under strictly defined conditions, one should accept that every (true) law is non-emptily fulfilled "only one time." Obviously, the above fact would not by itself deprive any law of its strict generality. However, a question arises of what would become in this case of their spatiotemporal universality. Indeed, one might still claim that they are fulfilled in the whole Universe but it would be an empty fulfilment (except for one case for every law).

Formally speaking, even in this case, the conception of spatiotemporal universality might be saved. After all, one can still claim that any law is fulfilled wherever and whenever propitious conditions occur but the evolution of the Universe precludes the rise of "the same" conditions (and thus a non-empty fulfilment of one or the other law) if only in two different spatiotemporal areas. However, it is easy to see that the conception of the strict generality of laws, interpreted as their spatiotemporal universality, would in this case be devoid of any cognitive value.

It seems, however (and the earlier experience fully confirms this) that the evolution of the Universe which actually takes place, does not lead to such drastic results as the unrepeatability of all the phenomena (and conditions) not only from the point of view of time but also from the point of view of space.

Let us then consider another, more probable possibility. The Universe evolves, and along with its evolution conditions under which laws act change as well as do the values of these magnitudes which are considered to be universal constants. However, this does not shatter the possibility of the occurrence of "the same conditions" in various space areas of the Universe which are at a given stage of evolution, neither does it cause that the value of universal constants depends on place. This signifies a change in the conditions of the acting of laws as time elapses (in the course of the "ageing" of the Universe) but it does not mean the unrepeatability of conditions of a non-empty fulfilment of laws in various areas of space. In this case, the conception of the temporal universality of laws would lose its cognitive value but the conception of space universality of laws would still show its usefulness. One has to state, however, that in spite of numerous and convincing data – both observational and theoretical – indicating the evolution of the Universe, it has not been possible so far to discover facts that would point to the changeability of any (at least one) universal constant. This may mean that the evolution of the Universe does not preclude the repeatability of conditions of non-empty fulfilment of laws also as time elapses (in the course of time) (not only in the space aspect as has been assumed in the recently considered variant of the evolution of the Universe). Thus, the evolution of the Universe itself, in spite of some opinions expressed by various authors, does not have to lead to undermining the conception of strict generality, understood as their spatiotemporal universality, of laws of empirical sciences.

On the other hand, one should not disregard the fact that the evolution of the Universe may lead to the necessity of an essential modification of this conception, and in an extreme case even to its complete abandonment.

However, the probability of the last case occurring is very slight. The fact that we successfully use the most general of the known laws for the phenomena which occur in the whole space and time area of the Universe that is accessible to astronomical observation, indicates that even if the evolution of the Universe leads to the modification of conditions under which laws act in space and time, this modification is still so slow that we can practically assume at least a "local" (in the spatiotemporal sense) repeatability of these conditions. One should conclude from this that having no certainty whether the conception of strict generality of laws as their spatiotemporal universality is based on true premises, we can justly assume that none of the future discoveries in the field of astronomy and cosmology will impair its practical and heuristic values. At most, it may appear that it has an only limited field of application and that its extrapolation onto the whole Universe shares the fate of those extrapolations which have not passed the test of time. The hitherto observed repeatability of conditions in various areas of space and time which are accessible to us sufficiently justifies the fruitfulness of the conception discussed, its heuristic fertility (and the fertility of Maxwell's principle) independent of its further fate.

V. Final Remarks

Two conclusions may be drawn from this discussion. These are as follows:

1. It is not very probable that the evolution of the Universe would lead to undermining Maxwell's principle and, along with it, the conception of the strict generality of the laws of science understood as their spatiotemporal universality.

2. On the other hand, what is more probable is the possibility that the evolution of the Universe leads to a variability of universal constants, which is caused, however, not by "space" and "time" as such but by the change (evolution) of the structural conditions of the world (as e.g. the density of the mass of the Universe). However, carrying out this possibility, without undermining Maxwell's principle allowing the interpretation of exact generality of laws as their spatiotemporal universality would in practice lead to the conclusion that the conception of the spatiotemporal universality of laws is not as useful as it has so far been considered to be.

ASPECTS OF THE PROBLEM
OF THE SPATIOTEMPORAL INFINITY OF THE WORLD

The problem of the spatiotemporal infinity of the Universe has a separate status among different problems of the infinity of the world addressed by physics, cosmology and philosophy. The problem itself can be subdivided into eight distinct, independently formulable problems.

The first is the problem of the temporal infinity of the Universe, which can be stated as the question: is time infinite? Assuming a definite unit of time (for instance, a second or a year), we can formulate the problem as follows: is time composed of a finite (i.e. expressable with the aid of some natural number) or infinite (i.e. non-expressable in this way) number of the units of time (e.g. seconds)? The second problem is one of the genetic infinity of the Universe, which can be expressed as the question: does the world last a finite or an infinite number of time units? The problem of the spatial infinity of the Universe (is space infinite) is the third one. Having assumed a definite unit of space (e.g. cm^3 or km^3), one can ask a question: does space include a finite or an infinite number of such units (e.g. km^3)? The fourth problem is the one of the material infinity of the Universe. It can be expressed in a twofold way: (*a*) does matter occupy a finite or an infinite space? Having assumed a definite unit for matter (any unit of mass such as gram or kilogram), one can formulate the problem as follows: (*b*) is matter composed of a finite or an infinite amount of such units (e.g. grams)? The question arises whether the two above-mentioned formulae of the fourth problem (*a* and *b*) are equivalent. It turns out that they are equivalent if once accept a natural assumption that is usual among the physicists of a density of matter being finite in any area. Should this assumption prove wrong, then, for instance, an infinite amount of matter would fit into a finite area, which would mean that the Universe is infinite in the sense (*b*) but finite in the sense (*a*).

The problems identified above are the problems of the "extensional" infinity of the Universe. They appear when we pass from the macroscopic objects (or areas) to increasingly large ones, and ask if there is an "upper" limit for the existence of the objects. These problems deserve the label of the problems of cosmological infinity of the Universe. Directing our attention in the "opposite" direction, towards increasingly small objects, we arrive at four analogous problems of the infinity of the Universe "inwards": the problems of the infinite divisibility of time, space and matter (in temporal and spatial aspects). We shall call them the problems of the microphysical (or microcosmical) infinity of the Universe.

The first of them (and the fifth in general series) is the problem of the infinite divisibility of time. Applying mathematical terminology, we can ask: is time a continuous set, i.e. a continuum (which usually has been assumed so far), or at least a dense set, or, on the contrary, a discreet set? Physicists usually call this problem the problem of the quantization of time, and formulate it as follows: is there an elementary indivisible unit of time, elementary persistence (time quantum) or, in other words, is it meaningful to introduce the smallest possible units of time? Then, there is the second (sixth) problem of the infinite divisibility of matter in temporal aspect: are there absolutely elementary (the shortest possible) processes, or can any process be divided ad infinitum into increasingly short processes? Then comes the third (seventh) problem of the infinite divisibility of space. In mathematical terms, it can be formulated as follows: is space a continuous set (as previously has usually been assumed), or at least a dense set, or, to the contrary, a discreet one? Physicists call this problem the problem of the quantization of space, which is closely connected with the problem of the quantization of time, and ask the following question: is there an elementary, indivisible unit of length (and, respectively, the unit of volume), elementary length (the quantum of space), or in other words can one introduce any units of length (or volume) which are as small as possible? The fourth (eighth) problem is the one of infinite divisibility of matter in spatial aspect: are there absolutely elementary particles, i.e. particles with no internal degrees of freedom that are indivisible, have no internal structure and no parts they could be dissected into (absolutely simple)?

All these problems are clearly linked in pairs (equivalent in pairs) according to some natural and almost universally accepted physical assumption, namely that there is no empty time and no empty space (absolute vacuum). In this case, the temporal infinity of the Universe would imply a genetic infinity (for all of the infinite time would have to be "filled" with matter) and vice versa (for infinite temporal persistence of the Universe would not fit into finite time). The same holds true for spatial and material infinity, and also – or so it seems – respective pairs of the problems of "inward" infinity. If so, then only formulae are independent, at least in principle, while the solutions are closely connected in respective pairs. Theoretically, we have eight problems of the spatiotemporal infinity of the Universe but because of the equivalence of the problems in pairs we are actually facing the necessity to solve four independent problems, which, even if linked, do not appear to be related. It seems that none of these problems have been solved by physics, cosmology or philosophy until the present date. Research conducted A. Friedman, A. Einstein and others on cosmological applications of the general theory of relativity suggests that an average density of matter in the Universe is a decisive parameter, which determines whether the Universe is spatially and materially open or closed, in the Universe. There is no such distinguished parameter for the other problems of infinity.

On the other hand, it should be remembered that – because of the necessity to distinguish (as pointed out by Riemann) the concepts of infinity and boundlessness with respect to time, space and matter (in temporal and spatial aspect), there will appear, along with the above-mentioned problems of cosmological infinity of the Universe, four respective problems of the cosmological boundlessness of the Universe (the problems of microcosmic infinity do not have such counterparts in problems of boundlessness). These are:

(1) The problem of the temporal boundlessness of the Universe, which can be stated as the question of whether time is unlimited, i.e. whether it has no moment which would be topologically distinguished (each has a normal neighbourhood), and whether it possesses border moments that are limiting ones (particularly with respect to the first moment, i.e. "the beginning of time" and the last one, i.e. "the end of time"). Let us note that, in topology, the above-mentioned approximate definition of the boundlessness of time (whereby the same holds true for the definitions of the boundlessness of space and matter in temporal and spatial aspect) can be expressed in a precise manner, due to the explanation of the concept of "normal neighbourhood" and other concepts in question.

(2) The problem of the genetic boundlessness of the Universe, included in the question: is the world unlimited in time, i.e. is no moment in the history of the world topologically distinguished in that each has normal neighbourhood and no border, limiting moments (especially the first moment called "the temporal beginning of the world" and the last moment called "the temporal end of the world")?

(3) The problem of the spatial boundlessness of the Universe: is space unlimited, i.e. is no point in it distinguished topologically (each has a normal neighbourhood)?

(4) The problem of the material boundlessness of the Universe: are there no material points (or material bodies) which would be on the border and might be called the (spatial) "beginning" or "end" of the world?

The above problems can also be linked in pairs (with the same assumptions of time and space being non-empty), since the boundlessness of time implies the boundlessness of the persistence of the world in time, and the boundlessness of space implies the material boundlessness of the world (the converse cases are also true).

Discussing different aspects of the problem of the spatiotemporal infinity of the Universe, we have thus formulated twelve problems (including eight problems of infinity and four problems of boundlessness), but, due to their "lucky" pairing, we are practically dealing with six independent problems (i.e. those that are independent at least with respect to our present knowledge): four problems of infinity and two problems of the boundlessness of the Universe. All of the above-mentioned problems of infinity of the Universe are usually considered (rightly, I think), to be open (unsolved), while some of the above-mentioned problems of the boundlessness of the Universe are usually considered – also rightly, I think – to be closed (solved).

The above-mentioned problems and their mutual links (which from pairs of them) can be presented in the following table.

Problems of spatiotemporal infinity and boundlessness of the Universe:

Cosmological problems of infinity:
   1. Temporal infinity of the Universe
   2. Genetic infinity of the Universe   } I.
   3. Spatial infinity of the Universe
   4. Material infinity of the Universe   } II.

Microcosmic problems of infinity:
   5. Infinite divisibility of time
   6. Infinite divisibility of matter in temporal aspect   } III.
   7. Infinite divisibility of space
   8. Infinite divisibility of matter in spatial aspect   } IV.

Cosmological problems of boundlessness:
   9. Temporal boundlessness of the Universe
   10. Genetic boundlessness of the Universe   } V.
   11. Spatial boundlessness of the Universe
   12. Material boundlessness of the Universe   } VI.

It is worthwhile to think about the relationship between these problems of the infinity and the boundlessness of the Universe and historically the most significant problem of the eternity of the Universe, which is still in the centre of discussions about the spatiotemporal infinity of the Universe. Since the eternal world is opposed to the world which has a temporal beginning, the problem of eternity of the world can be identified, following Einstein's suggestions, with the problem of the genetic boundlessness of the Universe (a complete identification of these problems is not possible, for, till the second half of the 19[th]century, the concepts of the genetic and temporal boundlessness were indistinguishable from the concepts of the genetic and temporal infinity of the world, and the controversy about the eternity of the world, although essentially the argument about the genetic boundlessness of the world, was also involved in the controversy about genetic infinity of the Universe).

THE PECULIAR STATUS OF COSMOLOGY
AS A SCIENCE*

I. Introduction

There are scientific disciplines whose road to achieving the status of scientific knowledge is long and arduous. They include numerous areas in the exploration of humanist phenomena, and also some areas in the penetration of natural phenomena.

As far as the latter are concerned, it should be said that the process of making more scientific the so-called Baconian sciences, such as chemistry and some branches of physics such as the study of heat, electricity and magnetism, took a relatively long time.[1] However, it seems that the most strenuous and thorny road to being a science was that of cosmology. It may sound strange if we take into consideration that its close relative, astronomy, had already achieved the status of science in ancient times, becoming at that time one of the first and very few but very prestigious scientific disciplines.

II. Peculiarities of the Subject Matter and the Method of Cosmology

1. Looking at the peculiarities of cosmology as science, one should begin with the peculiarity of the subject matter that the science deals with. Generally speaking, it is the Universe understood (1) as everything that exists or (2) as the world of nature (material one) which is inhabited by human beings (when we assume that there also exists another separate supernatural world, inhabited by personal beings of a kind of deities and other spirits) or finally (3) as a considerable fragment of the (material) Universe, no smaller than its part that can be observed (called the observed Universe or the Metagalaxy).

As compared with the subject matter of an "average" science which in general studies recurring phenomena that take place on a "medium" scale

---

*First published in J. Such, M. Szcześniak (eds) (1998), *Z epistemologii wiedzy naukowej* [On Epistemology of Scientific Knowledge]. Poznań: Wydawnictwo Naukowe IF UAM, pp. 161-171.

[1] The division of physical sciences into classical physical sciences, and "Baconian sciences" was introduced by T.S. Kuhn (cf. Kuhn 1977, part I, chapter 3).

comparable with human being, the subject matter of cosmology is distinguished by its enormity (both in space and time) and uniqueness and, therefore, non-recurrence.

The uniqueness of the subject matter of cosmology ("there is one world") at once raises a question whether cosmology may be a science.[2] The affirmative answer may seem to be very dubious especially to the advocates of the inductionist conception of science. This also applies to the other peculiarity of the subject matter of cosmology that has been mentioned before: an enormous, perhaps even infinite, extension of the Universe, both in the spatial and temporal aspect, and only "fragmentary" accessibility to it of the experimental procedures (i.e. observation, measurement, experiment) poses a serious methodological problem, particularly but not exclusively to an inductionist.

2. Furthermore, there is no doubt that both those peculiarities of the subject matter of cosmology considerably impeded the application of scientific methods in this field, and at the same time distinctly prolonged the process of transforming cosmological knowledge into scientific knowledge.

What was decisive here were undoubtedly both some theoretical achievements and the definite experimental (observational) discoveries. One of the most significant theoretical achievements was the formulation of the general theory of relativity (Einstein 1915) and the creation on its foundations of relativistic cosmology (Einstein 1917). The general relativity theory created theoretical foundations of cosmology as a science: cosmological variants of solving its equations led to the formulation of the Standard Model of Evolution of the Universe with the Theory of the Big Bang (Friedman's, Lemaitre's and Gamov's works).

The chain of decisive experimental discoveries was initiated by a discovery of a shift to the red spectrum of distant galaxies, interpreted (according to Doppler) as the "expansion of the Universe" (Hubble 1929), which confirmed the relation of the relativist cosmology of a dynamic (rather than stationary) nature of the Universe (Friedman 1922 and 1924).

Another link of the experimental chain was the discovery of the microwave (relic) radiation (Penzias, Wilson 1965), which was what had remained after the Big Bang and which strengthened the Standard Model

---

[2] It is obvious that there are a number of physical conceptions which usually refer to certain interpretations of quantum mechanics and cosmological ones which refer, for instance, to the so-called strong anthropic principle that assume the multitude of worlds. However, even if it appeared that at least one of them is right, i.e. there are more than one world (our Universe) then it would still not mean that cosmology is a science of the multitude of worlds. The question of the accessibility of those other worlds to our research methods always remains as a separate, and in particular that of their accessibility for our means of observation.

of Evolution of Cosmology (Gamov et al. 1948). Finally, the third significant result of experiments was the discovery by the COBE (Cosmic Background Explorer) of subtle heterogeneities (temperature differences) of this radiation (1992) which are an excellent proof of the inflational version of the Big Bang Theory developed in the recent decade (Guth, Linde, Steinhardt, Hawking).

The achievements presented here enhanced the status of cosmology as an experimental science that is superimposed on physics, astronomy and astrophysics. Cosmology takes from physics the results on which it is based, and research methods which it applies.

3. However, it does not mean that the attitude of contemporary cosmology to contemporary physics is simply a relation of a subordinated (secondary) science to a basic science.[3] Two circumstances caused cosmology to stop being an ordinary addition to modern physics and begin to play a more fundamental role in it. The first one is that the Universe in the initial stages of its development (in tiny fractions of the first second after the Big Bang) was at the same time a cosmic object (with respect to the amount of mass and energy) and a microobject (with respect to spatial dimensions and perhaps time dimension). As such, it therefore constitutes in this period the object of both physics (microphysics) and of cosmology (megaphysics). The other circumstance is that unifying theories of modern physics such as the theory of unification of weak and electromagnetic interactions, the Theory of Great Unification which investigates conditions under which weak-electric interactions become identical with strong interactions, or the Theory of Superunification, establishing conditions under which the interactions of the three kinds mentioned before became indistinguishable from gravitational interaction are directly applied in principle only to those early stages of the evolution of the Universe. This is because it is just there that these extreme conditions of enormous densities, pressures and temperatures are realized under which the fundamental interactions become indistinguishable.[4]

Therefore, checking those theories is (at least at the present stage of development of physics and technology) impossible without reference to the cosmological-physical theories which describe early stages of the evolution of the Universe and to observations which confirm these theories.

---

[3] The problem of the relationship of basic sciences to secondary (ancillary) ones is considered by A. Szczuciński on the example of the relation of physics to astrophysics (cf. Szczuciński 1986, pp. 61-68).

[4] Indeed similar extreme conditions occur also inside black holes (if the latter actually exist), however, this inside (interior) is inaccessible to an observer from the outside.

4. However, the peculiarity of the relationships of physics and cosmology does not end here. Becoming dependent on cosmology and its results, physics also becomes indirectly dependent on certain extracosmological (and non-physical) principles which play quite an important role in modern cosmology. Here I mean first of all the anthropic principle in its various formulations. This principle results from the answer to the question: what must be the Universe (as a whole or on a large scale) so that life (man, civilization) could appear in it? Cosmologists agree rather commonly that at least in its weaker formulations this principle should be taken into account in cosmological studies.

The role of the anthropic principle in cosmology flows from a rather obvious fact that some of the conditions necessary for the appearance of life and civilization on the Earth are cosmological in nature, i.e. such which are connected with global characteristics concerning either the Universe as a whole or at least the Metagalaxy (understood as that part of the Universe that is accessible to observation).

These conditions are said to include the flatness of the Universe, i.e. an almost ideal convergence of the actual density of matter in the Universe with the critical density (the ratio of these densities marked with a symbol Ω is close to one: $\Omega \approx 1$),[5] since in the case of significant difference between these densities the phase of expansion of the Universe would last either too short for life to begin, especially in the developed form (a case when $\Omega \gg 1$) or expansion of the Universe would occur at such a great speed that concentration of matter in the form of stars, planets and galaxies would not be able to occur (a case when $\Omega \ll 1$).[6]

According to Hawking, an explanation of the cosmological phenomena referring to the anthropic principle (in its weak version) is not really a full explanation, i.e. such with which one could abandon any further research but still it is a form of explanation which should not be given up by a cosmologist, at least until he has no better explanations.[7]

---

[5] The consequence of this co-incidence (proximity) of the values of both densities, actual and critical, is the fact that the rate of expansion of the Universe is very close to the rate of critical expansion (the critical rate separates the ever expanding models from those which will undergo shrinking) (cf. Hawking 1988).

[6] This perfect proximity of the expansion of the Universe with critical expansion is, as is well known, responsible for the fact that so far it has not been possible to establish which geometry "rules" the world as a whole ("in infinity"): Lobaczevski's, Euclidean or Riemann's, without which it will not be possible to resolve the question of whether the Universe is spatially infinite (open) or spatially finite (closed).

[7] In his work A Brief History of Time, Hawking, when speaking of the application of a weak anthropic principle in order to explain why the Big Bang occurred about 10 billion years ago ("simply more or less so much time is necessary for the intelligent beings to develop through evolution") put the word "explanation" in inverted commas (cf. Hawking 1988).

5. The further methodological peculiarity of cosmology which is connected with the specificity of its subject matter concerns the relation of cosmology to methodology. Generally, neither methodological consciousness of scholars nor, to even greater extent, the methodology (or philosophy of science) developed by methodologists (philosophers) and being (to some degree at least) the reconstruction of this methodological consciousness of scholars are considered to be science.[8] However, Michał Heller thinks, and not without justification that along with *the general methodology* (the philosophy of science) dealing with problems of explaining and studying methodological concepts which are bound with every empirical science (resulting, truthfulness, regularity, the relation of theory to experience) there is also *a detailed methodology* (which is sometimes called a philosophy of a given science), studying methodological concepts that refer to the concrete empirical science (axiomatization of a given science or of its part; defining of some terms that occur in the language of a given science by means of other terms; analysis of measuring techniques applied in a given science and the like) (cf. Heller 1978, p. 66). The latter, in its concrete development concerning a given science, becomes a constituent part of that science and thus "an internal methodology" (p. 66).

The exceptional position of cosmology in this respect consists in the fact that the peculiarities of its subject matter require application of very special research methods and it gives internal methodology of this science some remarkable significance for its practising. If we interpret cosmology

---

[8] It is obvious that not everybody excludes methodology (and in particular methodological consciousness of scientists) from the scope of science. For instance, this is not done by proponents of historical epistemology, as according to this branch of studies consciousness which accompanies a given type of practice becomes part this practice as its so-called socio-subjective context which controls its activity. With such an interpretation methodological consciousness of scientists is a socio-subjective context of scientific practice and, as such, becomes a part of it and thus of science (cf. Banaszak and Kmita 1991, Chapter IV). Sometimes representatives of historical epistemology, when speaking of science as one of the fields of culture pay attention not to the scientific practice (as a whole) but just to the methodological consciousness of the scientist understood as "a whole of principles which institutionally regulate and make uniform the process of cognition..." (cf. Pałubicka 1992, p. 65).

A. Zinovev does not exclude methodology from the area of science either. He writes: "*special methodology* of this or that concrete science is a part of this science. However, this part has a number of special features as compared with other parts of science. First of all methodology of a concrete science is not a localized part of this science in a sense in which organic chemistry is part of chemistry, and optics part of physics. It can be dispersed all over this science, even to such an extent that taking it together in one place is impossible ... A methodology of a given science is lost in this science itself, and is not its separate branch. When along the science still another separate science develops which is its methodology then it is more of a sign of a poor condition of this science or methodology of science (and possibly of both) than of its progress" (Zinovev 1976, pp. 307-308).

as an empirical science that serves the reconstruction of the structure-evolution of the currently observed Universe then at once appears as a central methodological problem of this science the question of extrapolation: how to test models of global structure-evolution of a system called Universe by means of local observations.[9] In practice, this means that methodology (and to be more precise, "internal methodology of cosmology"), becomes a part of cosmological studies; it becomes one of the branches characteristic of this science, inherently connected with its other branches. Thus, a cosmologist is at the same time a methodologist of his/her own science since being a methodologist (of cosmology) becomes an important aspect of his/her professional work as a cosmologist. It points to exceptionally close relations of cosmology and philosophy which result not only from the subject matter of their studies, and this is obvious but also because of the strong methodological involvement of cosmology.

What is it that differentiates "the internal methodology of cosmology" from "the general methodology of sciences"? According to Heller it is the task of the former to improve the quality of the research work on cosmology and to make the results of this research more precise. Among other things, this goal is achieved because in the process of reconstructing cosmological theories it does not use idealization (or stylization) of the theory which it is to help (as it is done by the general methodology) but it "stands up to methodological problems with their whole complexity" as much as it is only possible (see Heller 1978, pp. 66-68).

Heller also includes among the tasks of internal methodology its "critical function." It results from the fact that "not only the central cosmological problem extrapolation of the local physics to non-local areas

---

[9] Cf. Heller (1978, p. 66). The complexity of the problem of extrapolation is due to the fact that "there is no method of unequivocally achieving a global structure only on the basis of knowledge of local properties" (or, to be more precise, there is no way to unequivocally reconstruct the global structure-evolution of the Universe on the basis of knowledge of local physics and observations that are made locally) (p. 68). The problem of extrapolation would not exist if it were possible to prove that there is one and only one set of global assumptions of which local physics results (together with the results of locally made observations). The very fact of the multitude of various cosmologies of a deductive type is an evidence against such a possibility: the variety of initial postulates for cosmological deduction does not seem to be smaller than the abundance of different extrapolations coming from the same local properties (p. 69).

According to Heller, the deceptiveness of extrapolations of local properties to global properties (or only to large-scale ones) of the Universe; consequently all kinds of cosmological principles should be restricted to extrapolation of our knowledge to the spatiotemporal regions which are in principle observable, i.e. to the areas of our effective horizon of events. However, we have to bear in mind that "even our knowledge concerning the observed part of the Universe does not only result from observations but from observations that are interpreted by means of our local, laboratory physics extrapolated far beyond the laboratory area of time-space" (p. 72).

is based on tacitly adopted assumptions" but also "specific cosmological theories or conceptions are as a rule based also on assumptions that have different power. Very often, the scientific value of such conceptions, especially in view of the shortage of current testing data, is reduced to the values of initial assumptions. ... It happens occasionally that what decides about recognizing a scientific opinion for certain hypotheses is, for instance, the authority of its creators" (p. 73).

Thus, the critical function of the methodology of cosmology would consists in "unmasking hypotheses of this kind by showing their assumptions and discussing in detail the role which they play in their logical structure" (p. 73).

Internal methodology of cosmology, though it has to take advantage of the achievements of general methodology of sciences to some extent, still in principle arises "not from an *ex post* reflection on science which has already gone through a certain stage of development but in confrontation with the problems to be solved which currently face cosmology" (p. 66). It is obvious that as such it is a product of cosmologists and not of professional philosophers.

The very central rather than peripheral position of methodology of cosmology in the science of the structure-evolution of the Universe results, according to Heller, from the fact that "the central issue of cosmology" is just the methodological problem: the problem of extrapolating local physics to non-local areas (p. 73). A similar "central" place of "internal methodology" is not observed in principle in other sciences and this is another subject-methodological peculiarity of cosmology.

III. Conclusion

The peculiarities of cosmology concerning its subject matter and its methodology that were presented above make that discipline extremely closely connected with other fields of knowledge, and first of all with sciences such as physics, astrophysics and astronomy and also biology and philosophy.

Being an important component of modern physics, it encompasses elements of philosophy and methodology in the form of the "internal methodology of cosmology." As such, it constitutes an important link that ties physics with philosophy.

However, its ties with philosophy do not end there. When taking up the problem of the "origin" of the Universe, modern quantum-relativist cosmology becomes more and more involved in the "threshold" studies which have so far been reserved, as it were, for philosophy. In this manner, a "competitor" of ontology is developing, partly taking over its problems in order to give it a strictly scientific form. Since the general theory of relativity and quantum mechanics, which are both involved in the problems concerning the origin of the Universe, are the two most fundamental theories of modern physics, the whole modern physics (along with astronomy and cosmology) may be considered to be a rival of philosophy (as far as certain important problems are concerned). It looks as though such a "rivalry" has very good chances of becoming a "sound" competition which assumes a form of co-operation between philosophy which is traditionally more speculatively disposed and more and more empirically oriented cosmology.

THE ORIGIN OF THE UNIVERSE
AND CONTEMPORARY COSMOLOGY
AND PHILOSOPHY

I. Introduction

One of the determinants of scientific rationality is the condition that science undertakes only those problems whose solution is within the range of possibilities of research methods that are currently available. Simply speaking, scientists are attracted by solvable problems. If this is really so then the discussions widespread since the 1970s among physicists and cosmologists on the subject of "the beginnings of the Universe" seem to be a sign that also this unusually intriguing problem has matured for its scientific solution.

The purpose of my paper is to attempt to answer the question of whether the problem of the origin of the world currently evades philosophers (and theologians) and passes completely to the realm of science (i.e. physics, astronomy and cosmology), or whether science by itself is not able to solve this problem. In the latter case, one would have to acknowledge that metaphysics, the philosophy of nature and epistemology, provide important premises, assumptions and methods that are indispensable for this solution.

II. Two Approaches in Cosmology

The task that cosmology has to perform is to explain the structure of the Universe as it is observed. Contemporary cosmologists carry out this task by means of two, to some extent contrary, approaches. One seeks to explain the structure of the Universe observed through what the Universe was at the very beginning; the other, on the other hand, tries to show its present structure as an inevitable consequence of past physical and chemical processes, irrespectively of what was at the beginning. A permissible possibility here is that the Universe is so constructed that no observable traces of its quantum origin remain. The Theory of the Big Bang in its classical, purely relativistic interpretation, is an example of the former approach. However, its inflational development which refers to the physics of high energies and to quantum mechanics is the fullest expression of the latter approach. Irrespectively of how the Universe began (and whether it began at all) that which we observe at present on a cosmological scale is the result of an extremely short period of inflation in which, due to the unparalleled intensity of extension at an exponential rate, it was transformed from a microobject (with respect to its spatial dimensions) into a macroobject.[1]

Obviously, to solve the problem of the origin of the world, of fundamental importance is the former approach which assumes that further evolution of the Universe did not preclude the significance of its initial states for what we now observe on a large scale.

It does not mean that the prevalence of inflational models in modern relativist and quantum cosmology renders cosmology incapable of dealing with questions of the origin of the Universe. It is only in recent years that scientists have focused on the former approach. There has been a search for fundamental laws which would determine the initial conditions of the Universe.

III. The Dichotomy of Laws and Initial Conditions

In classical physics (and, in general, in the modern approach to explanations of phenomena) a principle of the dichotomy of laws and of initial conditions is obligatory.[2] According to this principle, initial conditions (understood as conditions in which laws may be applied) are totally independent of the laws of nature, in the sense that from the point of view of laws their distribution is completely random.

Quantum mechanics seems to be the first theory which started threatening this dichotomy. According to Heisenberg's principle of indeterminacy, initial conditions of a mechanical system (i.e. position and velocity) are canonically joined and as such cannot at the same time be determined with optional precision. This imposes certain restrictions on the distribution of these conditions and leads, as is well known, to the undermining of the so-called Laplace determinism which respected the principle of dichotomy.

In modern cosmological considerations concerning the earliest stages of the Universe, the problem of the dichotomy of laws and conditions is given first-rate significance. As it is, the problem of the beginnings of the

---

[1] It can be seen from calculations that at present the Universe is 10^{60} times larger (10^{27}cm) than it was in Planck's era (10^{-33}cm). During inflation it expanded by about 10^{30} – 10^{50} times. This means that in a very short period of inflation (10^{-35} – 10^{-32}sec) the Universe has expanded in at least the same proportion as it did during the rest of the time (which is about 15 billion years) that has elapsed since Planck's era (10^{-43}sec).

[2] I discuss the principle of dichotomy in detail in Such (1972, pp.145-163).

Universe may not be solved in any way without undermining this dichotomy.

IV. In the Search of a New Type of Laws

In reference to the initial state of the Universe one should consider critically the view that initial conditions are independent of the laws of nature. One might assume that the Universe is exceptional in the sense that it is the only nomologically coherent possibility. This means that the initial conditions are also exceptional and through this become a law of Nature themselves ("a superlaw"). Another possibility is that the Universe is genetically unlimited, i.e. is a Universe "without boundaries."

An approach of this kind leads to the search for a new type of laws. These would not be the laws which determine a permissible change of the state of the world between one moment and another but laws which rule very initial conditions.

The traditional view, based on the dichotomy of laws and conditions as applied to the Universe, was often associated with the view that theologians deal with initial conditions and physicists deal with the laws of evolution. However, contemporary cosmologists try to establish whether there are any laws of initial conditions which would overcome the randomness of their distribution. Studies divide into two directions.

V. World Without Borders

The first one, which is more radical, is the Hartle-Hawking conception of the world without borders (see Hartle, Hawking 1983, pp. 2960-2975; Hawking, Penrose 1996, Chapter 5). The suspicion that time and space are not universal characteristics of the physical reality, sometimes formulated in microphysics, has been strengthened in this conception but only as far as time is concerned, and transformed into an assumption that in Planck's era, given extreme densities under which quantum effect dominates time loses its properties which differentiate it from space. First of all, it stops being anisotropic, stops running in a given direction. It means that the initial quantum state is actually a timeless state and that there had been nothing "before" the beginning of the Universe since "then" there was no time. As a result, in Planck's era the Universe behaves as a four-dimensional hypersphere (surface of a ball). In this manner, "the condition of not occurring of a border" makes it possible to avoid the singularity of the initial state. On the other hand, the ordinary nature of time as qualitatively different from space begins to crystalize in the first moments following Planck's era. Therefore, we can see how the question on the beginning of the Universe leads to the question of the nature of time itself.

The conception of time which becomes space – to be more exact, still another (fourth) dimension of space – is one of the most innovative ideas of modern cosmology. The condition of the non-occurrence of the border shifts the whole "responsibility" for the origin of the Universe onto the laws ruling the Universe (and possibly onto their source, e.g. in the form of a Lawgiver). In this way the question of the dichotomy of laws and conditions is settled through the total subjection of initial conditions to initial laws, even at the cost of liquidating initial conditions as such.

The solution for this dichotomy assumes another form in the other direction of studies.

VI. Creative Conceptions of the Universe

This is a direction which contains various conceptions of the creation of the Universe (1) out of "nothing" or (2) out of a "vacuum." However, at least one of these conceptions solves the problem of dichotomy in a similar manner to the conception of the lack of borders. This is a view that our Universe is exceptional in the sense that it is the only (logically or nomologically) cohesive possibility. This leads to the conclusion that also the initial conditions could not have been different from what they were. In this interpretation, the role of conditions is as if reduced to the role of laws, so that the conditions themselves become the law (or laws) of nature (let us note that in this respect at the extreme end of cosmological views there is a conception of a multitude of universes, according to which there is an infinite number of worlds in which all possible conditions are realized; this conception does not, therefore, undermine the classical dichotomy of laws and conditions).

However, a vast majority of the conceptions of creation assume that our Universe, in which conditions arose for the existence of "a conscious observer," due to its insignificant *a priori* probability is – from the point of view of laws which we know – only one of many possible universes. Therefore, even if the initial conditions are not totally independent of laws, the latter still do not determine them unequivocally. In the conceptions discussed, the problem of the description of initial conditions is reduced to supplying the characteristics of what the Universe arose from.

What is significant from the point of view of philosophy is the division of creative conceptions into those which assume that the Universe arose from "nothingness" in the strong ontological meaning of the word and those which lead to the conclusion that it originated from a certain "poorer" physical reality, usually called "a vacuum" or space-time. Both kinds of conceptions usually refer to quantum mechanics or the future quantum theory of gravitation which is to combine quantum mechanics with the general theory of relativity.

1. Creation From "Nothing"

Supranatural conceptions of creation out of "nothing" which refer to God are well-known to us in European philosophy and Christian theology. In cosmology, the first conception of creation out of nothing appeared and was already promulgated within the relativist theory of the Big Bang, especially after 1970 when Hawking and Penrose proved their famous theorem on singularities (Hawking, Penrose 1970, pp. 529-548). It could be seen from those theorems that if there is a sufficient amount of matter in the Universe and gravitation always and everywhere attracts, then the extension of all light rays back in time to infinity is impossible. This meant that (at least some) time lines break at the moment of the Big Bang, or in the primary cosmological singularity.

It could be interpreted as the rise of matter, time and space out of "nothing," hence as "a beginning of the world and at the same time the beginning of time." In this interpretation, the world does not arise *in* time but together with time (which is reminiscent of St. Augustine's conception). However, since the general theory of relativity could not give any "mechanism" of transition from "the state of nothingness" to "the state of existence" of the Universe, further conceptions of creation out of "nothing" began to adduce quantum mechanics (see Vilenkin 1982, p. 25; Griszczuk, Zeldowicz 1982; see also Davis 1983, Chapter 16). This theory describes at least two processes which may compete for the name of "creative processes" which rouse the interest of cosmologists. These are phenomena of "quantum tunnelling" and of "quantum fluctuation" which occur due to quantum effects; consequently, they are impossible from the point of view of classical physics.

In the conceptions of creation out of "nothing" the former phenomenon is mostly used. "Tunnelling" in its common quantum interpretation is overcoming certain energetic potentials by microobjects whose energy resources are lower than the necessary minimal energy if the process is to be interpreted in the light of the unexceptional laws of classical physics.

In quantum cosmology, the "tunnelling out of nothing" is discussed. When an analogy with the effect of quantum tunnelling is made, a mathematical description of the process in which the Universe together with time and space emerges out of nothingness is performed in the language of mathematics by an empty set. One cannot speak here about tunnelling in the strict sense of the word, since such a claim would assume the existence of two different physical states. In the meantime, the process postulating the creation out of nothingness reveals that one cannot determine any earlier states of the emerging Universe since no prior external time exists, previous in relation to the time which appears together with the world.

If we apply Schrödinger's equation defining the wave function in ordinary quantum mechanics for the whole Universe so that it also takes gravitation into account (the space-time curvature) then the Wheeler-De Witt equation is obtained, called "the wave function of the Universe" U. On its basis, we obtain the transition function T, determining the probability of the occurrence of concrete changes in the state of the Universe. In this way, the function $T[x_1, t_1 \rightarrow x_2, t_2]$ gives the probability of finding the Universe in state x_2 at the t_2 moment, if it was in the x_1 state at the previous moment t_1. One of the possibilities consists in that, according to functions U and T, the Universe tunnels to the state of existence out of nothing with a definite probability. In other words, there is a certain probability of transition which has no previous initial states. According to this scenario, there is a certain probability of a spontaneous appearance of the Universe of a definite type together with a definite time and space, created out of nothing.

However, there is a fundamental question: can nothingness (total nothingness in the strong ontological sense) exist at all from the point of view of the laws of physics? This doubt does not refer to the conception of the creation out of a "vacuum."

2. Creation out of a "Vacuum"

A well-known fundamental metaphysical question is: "why is there something rather than nothing?" which led great philosophers of various times (Aristotle, Leibniz, Heidegger) to grapple with the basic philosophical problems. It assumes the meaningfulness of the concept of "nothingness."

The question of the legitimacy of this concept is conspicuous in quantum mechanics and in the quantum theory of the field, extended around quantum electrodynamics. Sometimes a conclusion is drawn from Heisenberg's principle of uncertainty that a "vacuum" (in the sense of something devoid of any substrate or any physical properties) is impossible since its existence would allow establishing exactly the value of the canonically joined values, e.g. position and velocity.

The quantum field theory (and attempts at creating the quantum theory of gravitation made on this basis) interpret vacuum not as a nothingness but as an extremely active arena which through "fluctuation of a vacuum" constantly creates virtual particles of all possible kinds. Vacuum in this interpretation has not much in common with the philosophical concept of "nothingness" but constitutes a kind of active time and space endowed with various physical properties (which are usually divided into topological, metrical and properties of symmetry). On the other hand, the dynamic nature of the vacuum (time and space) signifies that it is also apportioned with energy.

Conceptions of the creation of the Universe out of "vacuum" assume that it arises in the process of the "fluctuation of vacuum," and therefore it is created not out of nothing but out of a certain physical reality, poor in properties and called a "vacuum" (see Tryon 1973, p. 396; Brout, Englert, Gunzig 1978, pp. 78-106; Atkatz, Pagels 1982, p. 2065). Some of the conceptions of this type, e.g. J. Wheeler's conception, impoverish even more this initial reality of which the Universe was to emerge. They assume, for instance that time as such is a complex structure which consists of simpler elements. Assuming that these elements "become a part of" time at the moment of the rise of the Universe we obtain a conception according to which time is created together with the Universe but it is created not out of nothing but is preceded by certain elements of physical reality (this brings to mind the conception of "the world without borders" where time in Planck's era also loses some of its characteristics or, in other words, becomes impoverished). Following this line of thinking, one could imagine that our Universe was not formed in one act of creation but in a number of such acts, e.g. first out of nothing (in the literal sense) elements of time and space had been created (pregeometry) which then formed space-time ("vacuum"), and then out of vacuum the Universe was created (which in the course of further development may be enriched with new essential ontological characteristics and so on without end).

VII. Conceptions of the Universe that Is Infinite in Time

Not all conceptions which deal with the origin of the Universe negate its eternity; it is not negated by (1) A. Linde's conception of a chaotic inflation, (2) the conception according to which the Universe suddenly starts an expansion from the solid state in which it rested for an infinite period of time in the past, (3) the conception which assumes that the Universe was smaller and smaller in the past, however, never achieving zero dimensions, and some others.

VIII. Cosmology and Philosophy

Let us now go back to the main problem of my paper: has the beginning of the Universe become a strictly scientific question, i.e. such that science is able to solve it itself ?

A partial answer to this question can be given by considering the cognitive status of the conceptions which assume the creation of the Universe out of "nothing."

Thus, the concept of nothingness itself and the assumption that nothingness is possible, is a philosophical concept and a philosophical assumption, which goes much beyond that which can be established by scientific research as such. The latter reaches at most the concept of a vacuum which, as can be seen, is by no means identical with the concept of nothing (in the philosophical sense). That which is understood as vacuum in physics and cosmology is apportioned with essential physical properties, and deserves more of the name of space or space-time than that of nothing. The fact that in these sciences at least several concepts of vacuum are used ("absolute vacuum," "relative vacuum"; "quantum vacuum"; "false vacuum," "true vacuum," etc.), and that actually this concept is relativized to the concrete physical theories does not change anything (e.g. the absolute vacuum is not nothingness in the philosophical sense either). Similarly, the theses on the creation of the Universe out of "nothingness," on "tunnelling out of nothingness," "fluctuation of nothingness" and the like are not purely scientific theses. Theses and conceptions of this kind which they contain must assume as a basis much more than that which corresponds to the philosophical (and also the everyday) understanding of the concept of nothingness. If one omits energy, mass and even geometry (chronogeometric properties) as characteristics not of nothingness but of the active (dynamic) space-time then one should recognize that "at the very beginning" there must be laws of nature. According to them "nothingness creates the world," which also assumes the existence of something that can be called the world of logic and mathematics. In this sense, an explanation of the origin of the Universe cannot do without an assumption of some structure of rationality (Barrow 1994, Chapter 6).

Perhaps these assumptions may by themselves not lead to theology but seem to lead inevitably to philosophy. Perhaps physics is able to explain both the origin, the order and content of the physical Universe but not of the laws of physics themselves. There must exist laws of quantum mechanics if quantum processes are to lead to the rise of the Universe. From this point of view, open to critical analysis is S. Hawking's conviction in which elimination of all questions about the so-called boundary of time and space, which has its own place in his conception, signifies as if automatic exclusion of problems which as a rule led to religious and philosophical comments.[3] From the philosophical point of view, the conception of the non-existence of borders shifts the problem of the source of initial conditions to the question of the origin of physical laws, according to which the Universe does not have borders.

Another reason which indicates an inevitable involvement of the question of the beginning of the Universe in philosophy is that cosmology as an empirical science is entitled to make statements on the structure and evolution of the observable part of the Universe but not on the Universe as

---

[3] Cf. e.g. Hawking (1988, Chapter 6). It is worth noting that not all cosmologists agree with this interpretation of the Hartle-Hawking conception which is supplied by its founders themselves. For instance, according to W.B. Drees, this conception creates, just as well as other models, a possibility of specifying boundary conditions, which leads to the very necessity of posing theological and metaphysical questions. Cf. Drees (1990, p. 114).

a whole. In turn, limiting our scientific knowledge about the Universe to its observable part (called the horizontal Universe) means that we are not able to check scientifically the correctness of a rule for initial conditions (or their lack) for the whole Universe. After all, we observe the results of the evolution of only some part (indeed a very small one, which is confirmed by conceptions of inflation) of the initial state.

However, it can be seen from the above that extrapolations of local physics onto the whole observable Universe cannot do without cosmological principles, about which we know are basically philosophical in nature (cf. Rudnicki 1995). Still, extrapolations of this kind conducted in cosmology are insufficient to make their conclusion pertaining to the Universe, understood as the whole physical Universe.

One more reason for the inevitability of philosophy in cosmological considerations is that the outcome of cosmological statements, including those that are essential for the solution of the question of origin of the Universe, may be interpreted differently, and that in the process of this interpretation an important role is played by philosophical assumptions, both ontological and epistemological (as well as those which are not connected with cosmological principles). The great significance of interpretation in cosmology is emphasized by the fact that while macroscopic physical theories such as classical mechanics – due to their relative "closeness" to experiment – have the so-called natural interpretation (which generally does not raise any doubts) microphysics and cosmology theories may be interpreted in many different ways. This is proved by numerous and extremely different interpretations of quantum mechanics on the one hand, and by the multitude of interpretations of the shift to the red in galactic spectra, the multiaspectual dispute over the nature and distance of quasars or different interpretations of the initial singularity, on the other hand.

Hawking's criticism of metaphysics seems to be a little bit outdated because it combines in itself elements of positivist philosophy of cognition with the conception of God, shared by Clark in the 18th century, filling in the gaps in natural science (cf. Życiński 1993, p. 199).

In connection with the occurrence in cosmology of numerous, and at the same time significant, philosophical and methodological assumptions which cannot be verified empirically without which it could not function as a science, cosmology may be considered as "science not only of the Universe in its largest scale but also about assumptions which one should form in order to make such a science possible" (Heller 1995, p. 106).

NEWTON'S FIELDS OF STUDY
AND METHODOLOGICAL PRINCIPLES*

I. Scientific Study and Self-Reflection

A scientist need not be preoccupied with an analysis of his/her own study nor with general methodological questions. A self-reflective nature of knowledge, i.e. its ability to study its own nature and methods by which it is acquired, thereby arriving at self-knowledge, has always been a domain of the philosophical approach. Thus, since the times of Descartes, it has been customary to assess various branches of knowledge for their underlying philosophy taking into account their capability for self-reflection.

Scientists would hardly ever devote time and energy to pursuing an analytical study of the methods which they apply and of the research activities which they undertake. In normal scientific practice, any reflection of that type would be more of a hindrance than real assistance.[1]

The above holds true, first of all, of natural sciences. In the humanities and social sciences, self-reflection is far commoner. In such fields as sociology, political economy and humanistic psychology, it is almost a routine procedure, which seems to be in agreement with the general belief in the affiliation of those disciplines with philosophy.

There are times, however, when self-reflection becomes common in natural sciences as well, when it becomes an obligatory component of scientific study. It happens so in the periods of crisis and historical turns or, in other words, when scientific revolutions take place. Then, outstanding scientists, who are aware of a critical situation in their discipline and whose responsibility it is to advance a revolution, become

---

*Translation of "Obszary badań Newtona a zasady metodologiczne." In: S. Butryn (ed.) (1991), *Z zagadnień filozofii nauk przyrodniczych* [Some Problems of the Philosophy of Natural Sciences]. Warszawa: Res Publica Press, pp. 55-63.

[1] W. Heisenberg once recalled his and Weizsäcker's failed attempt to approach a herd of goats with a movie camera. Niels Bohr made the following comment with that connection: "The goats managed to escape only because they were unable to ponder and discuss how to do it" (cf. Heisenberg 1987, p. 179).

Although in the goats' case it was more of instinctive behaviour than scholarly activity, the analogy seems quite suitable; self-reflection, in particular if it is pursued while a given action is just being carried out, is very likely to reduce its effectiveness.

involved in the philosophical disputes that are metascientific in nature –
on ontology, epistemology, and even on axiology. This seems to have been
the case of the physicists who were authors of the 20th-century revolution
in physics, i.e. of Planck, Einstein, Bohr, Heisenberg, Schrödinger, Born,
Dirac and Prigogine, all of whom have also been active philosophers.

Likewise, in the 17th century – the age of the scientific revolution,
when six great authors of modern science, i.e. Copernicus, Kepler, Galileo,
Descartes, Newton and Leibniz made a considerable effort to explain their
procedures and methods of investigation. Two of them in particular, i.e.
Galileo and Descartes, could correctly identify the methods which they
were putting in practice. It seems paradoxical that the greatest of them all,
Newton, who has frankly had no equals in the world of science, failed to
understand the methods with the aid of which he laid the foundations of
modern physics, and which he partly invented by himself and partly
borrowed from his contemporaries.

That gave rise to many biting remarks on the part of other physicists
and philosophers. Einstein, for instance, claimed that Newton had been
unable to see that the methods which he had been using with great success
had nothing in common with typical inductive generalization, contrary to
the latter's belief that they were applications of the inductive
methodology. Engels calling Newton "an inductive ass" might be hinting
at the same (cf. Engels 1953, p. 211).

It is not easy to answer the question of why Newton failed to identify
adequately the methods which he was using. Einstein was of the opinion
that it was only in the 19th century that physicists could become aware of
the methods of theoretical physics and describe them adequately. He
believed that due to "the inductive illusion" that could not have occurred
earlier. This may be true but only to a certain extent. Newton's
contemporary, Galileo, did, in fact, correctly define his own method as
"geometric," thus pinpointing its most essential property.

Newton's situation was, undoubtedly, much worse than Galileo's,
owing to the flat empiricism which had been prevailing in his native
England since the Middle Ages, having been strongly supported by such
authorities as F. Bacon and R. Bacon. Therefore, it was more difficult for
him to rid himself of "the inductive illusion" than for his continental
colleagues.[2]

It seems, however that there was yet another reason why Newton was
unable to overcome the illusion and identify his methods correctly.
Namely, unlike Galileo and some others, Newton was active in several
areas of study, some of which extended far beyond mathematical physics.
Distinct methods which had to be applied in each of them by no means

---

[2] Engels, criticizing the one-sided inductive approach, writes: "All that inductive cheat
[comes] from the English" (Engels 1953, p. 236).

facilitated the process understanding their specific properties. As has been aptly remarked by Weizsäcker, science is easier to do than to understand. Furthermore, being involved in so many basically distinct activities, Newton must have been limited both in his time and energy to be as much concerned with the methodological questions as some of his contemporaries (e.g. Descartes and Galileo).

II. Three Fields of Newton's Study

Newton was actively engaged in at least three areas of human knowledge. Only two of them may be said to belong to science proper, whereas the third one was merely in some "contiguity" to it.

1. Studies in the "Classical Physical Sciences"

The first field of Newton's interest comprised "the classical physical sciences" which dated back to the Hellenic epoch. In Newton's times, these were geometry, astronomy, geometric optics and mechanics. In T.S. Kuhn's opinion, the first three were the only branches of physics whose language and methodology, founded as far back as in the antiquity, made them inaccessible to laymen and, consequently, the works written in each of those fields were comprehensible to specialists alone (see Kuhn 1985, p. 73). According to Kuhn's criteria, they were well-developed scientific disciplines which, having abandoned the questions of fundamental nature, could concentrate on puzzle solving, thus entering the "paradigmatic" (i.e. mature) phase of their development. This is always marked by the true scientific advancement consisting in presenting solutions to given problems (see Kuhn 1985, p. 73).

In Newton's times, the above mentioned group of the sciences of classical physics represented the entire (or nearly so) strictly scientific knowledge. Accordingly, we may argue that the first domain of Newton's scholarship belonged to the strict sciences or was strictly scientific. His major work *Philosophia Naturalis Principia Mathematica* (henceforth *Principia*) belongs in this area.

2. "Baconian" Studies

Next to the traditional branches of physics, i.e. classical physical sciences, the 17th century faced the growth of the so-called "Baconian sciences." In Kuhn's opinion, the scientific status of those sciences was largely due to the emphasis which the philosophers of nature placed on experimenting and on compiling various natural historics, including a natural history of crafts.[3] The Baconian "histories" were mainly concerned with thermal, electrical, magnetical and chemical phenomena.

Although in the 17th century each of the two types of study was confined to a distinct geographical territory (i.e. England was the centre of the Baconian studies while the Continent, mainly France, was home to the representatives of the traditional disciplines), Newton was exceptional in making contribution to both trends.

The scientific revolution of the 17th century, in which Newton played a major role, radically transformed the classical physical sciences and led to the acceptance of the Baconian sciences. The latter had grown not from a scholarly university tradition but from the applied arts and craftsmanship. They were much dependent on the new experimental programme as well as on new instruments whose introduction was advanced by the crafts.

In Newton's times, the major difference between the two discussed "schools" lay in that the classical physical studies applied mathematical methods and were based on measurement and instrumental observation. Thus, they were strictly scientific, naturalist in nature or, putting it differently, they comprised the whole of natural science. By contrast, the Baconian studies used experimental methods.

However, unlike the experimental methods used in ancient and medieval times, Bacon's methodology, whose most eminent advocates were Boyle, Gilbert and Hooke, was not intended to confirm and exemplify the existing theories but instead to study nature in the conditions which had not been investigated before or, even, in conditions which could not normally occur.

While experiment was highly valued in the Baconian study, theoretical considerations were intentionally disregarded. The scholars were clearly unaware of the natural relationship between the two. Hence, the Baconian "school" of the 17th century was outside the natural sciences based on mathematics.

Until the mid-19th century, the Baconian tradition remained undeveloped in the sense that its proponents could not propose a coherent theory which would have any predictive power. According to Kuhn, its evolution as well as the character of its publications reveal a striking resemblance to many social sciences of today, yet they are totally unlike the classical physics of their contemporaries (Kuhn 1985, p. 86).

At the turn of the 18th and 19th centuries, the Baconian studies acquired (at least in physics and chemistry) the scientific status thanks to works of such authors as Aepinus, Cavendish, Coulomb, Gauss, Poisson, Lavoisier and a few others. But only during the second scientific revolution, which came at the beginning of the 19th century, could Baconian physical

---

[3] Cf. Kuhn (1985, pp. 301-302). In Kuhn's view, these sciences are a new by-product of the Baconian principles of "the new philosophy."

disciplines undergo a transformation similar to the one which had affected classical physics some two hundred years before, during the first revolution. The quantitative approach became obligatory both in experiments (measurement) and in theoretical studies (quantitative theories), which is best evidenced in works of such scientists as Fourier, Clausius, Kelvin and Maxwell.

Strangely enough, throughout all that time, including the 19th century, the two "schools" of the physical sciences, i.e. the classical school and the Baconian school, were developing independently of each other. A few exceptions such as Newton and some other scholars who contributed to both traditions could not alter the overall picture. Generally speaking, classical disciplines formed "mathematics" (in fact, they were quasi-mathematical disciplines) whereas Baconian studies were treated more like "experimental philosophy."

Not only the founder of the Baconian school but also many of his followers such as Franklin, Black and Nollet had never placed much trust in mathematics nor in the quasi-deductive structure of classical physics; therefore they never even attempted to acquire any mathematical skills. That was certainly an important subjective reason for keeping them apart from the classical disciplines.

The reverse relation was far more complex. Owing to the newly introduced experimental equipment, the Baconian movement exerted some influence upon the older, classical disciplines, in particular on astronomy. The result, however, was rather a gradual increase in experimental rigour on the part of the proponents of classical physics than any substantial change in their orientation. As most adherents of the Baconian tradition did not confide in mathematics, so many classical physicists were often reluctant to appreciate the significance of empirical experiment and, accordingly, they used it rather sparingly. Needless to say, Newton was an exception, together with some continental scientists (e.g. Huyghens and Mariotte) who, like him, belonged to both traditions.

Another factor which contributed to the separation of the two schools was the fact that, unlike the Baconians, classical scientists were by no means amateurs. They were scholars with academic status, most of them university lecturers and members of the Academy (Kuhn 1985, pp. 92-93).

The long-lasting split within the physical disciplines into those of the classical and those of the Baconian traditions was additionally strengthened by the independence of the mechanical arts from the non-mechanical ones. The latter became a subject of the scientific interest much later than the former, and had far more utilitarian objectives.

In spite of the above mentioned numerous factors whose importance cannot be underestimated, the integrating tendencies were already quite strong in the 19th century. As a result, the 20th-century physics has emerged as a well-developed unified science. Two processes seem to have had a crucial role here, namely the mathematization of Baconian disciplines on the one hand, which transformed them into strictly scientific research, and the introduction of experimental methods into the classical physical sciences on the other.

3. Parascientific Studies

Newton was equally active in the third area of his studies, which not only exceeded the boundaries of natural science but in fact lay outside any science. We shall call them activities of a parascientific nature. They ranged from alchemy through the attempts to explain gravitation and the nature of absolute space, to the hypothesis of "the first push," and the role of God and of supernatural phenomena in the physical world.

Typical of this tendency was his treatise *De gravitatione et aequipondio fluidorum* (hereafter *De gravitatione*) written in the 1660s. In the treatise, he first discussed the conception of immaterial (unsubstantial) ether, which would be recurring in a number of his later works. In the treaty, Newton departed from the idea of material ether but not for good; in other numerous works and letters he would be again considering the possibility of explaining certain physical phenomena in terms of ether, either material or immaterial, i.e. spiritual. In *De gravitatione* Newton attributed the function of material ether to God Himself, thereby initializing his search for the divine medium or, in other words, for the "unsubstantial ether."

Surprisingly enough, in the same work he introduced his concept of absolute space and, concurrently, the idea that the absolute space was the physical attribute of the infinite presence of God ("immanent effect of God," "constituted by the infinite presence of God"). Thus, next to rational justification of the absolute space hypothesis which came from Newton's belief that the laws of motion would be void unless they were referred to some specific (absolute) system (the justification was first noted by Mach in his critical analysis of Newton's famous pail experiment), Newton also presented the theological justification of the hypothesis.

In another work, *De aere et aether* (written in the 1670s), he discussed the hypothesis of "action at a distance."

There is no doubt that in his parascientific studies Newton was under a strong influence of neo-Platonic metaphysics and More's mysticism.

III. On the Interrelationship Between Newton's Methodology and Different Areas of His Study

The fact that Newton was engaged in three distinct areas of study leads to the conclusion that his methods must also have been of various types, regardless of the degree of his awareness of how much those areas were dissimilar.[4] There must be differences in the approach to problems of a strictly scientific nature (Newton's first field of interest), to questions of "normal" scholarship (his second area of activity) and, lastly, to para- (or even pseudo-) scientific studies of the third type which hardly ever comply with any scholarly standards being closer to mystical, theological or theosophical hypotheses. Thus, it seems only plausible to claim that depending on the field, he had to apply three distinct types of methods, i.e. strictly scientific, scholarly and para- (or pseudo-) scientific.

Newton's unpublished works provide enough evidence that he was fully aware of at least one of those distinctions, namely that between general scientific methodology and para- (pseudo-) scientific methods. It is not difficult to distinguish between those of his writings which are of a "phenomenalistic" character and the remaining ones which deal with "philosophical" questions of the nature of the investigated reality. The latter are often claimed to belong to works pursuing the ether hypothesis. Newton's aims and approach to the subject-matter are clearly dependent on the nature of study and differ considerably from one type to the other.

The phenomenalistic works (e.g. *Principia* or *Optics*) which continue the scientific traditions (in both of the above senses) are marked by mathematical rigour. They contain very few unjustified hypotheses and heuristic premises, in accordance with the principle *Hypotheses non fingo* (with emphasis on the last word). Instead, they are full of detailed descriptions of various experiments and their results. They are not intended to provide ultimate explanations of, for instance, the nature of gravitation or of absolute space; neither do they introduce base hypotheses.

Unlike the former, the works of the "philosophical" type (such as *De gravitatione* and *De aere et aether*) are devoid of any mathematical discussions and calculations, neither do they contain any analyses of experiments. They introduce, in that place, many hypotheses and suppositions as to the final causes and tend to present the ultimate interpretation of the facts of nature, mainly theological in nature. Newton's conceptions of both "material" and "immaterial" ether are deeply embedded in his metaphysical speculations.

It is worth noting that the above mentioned fundamental differences between the two types of work are best documented in his writings on absolute space. While the former studies deal with the purely physical problems, analyzing the functions which the absolute space would perform

---

[4] It seems possible to think of a higher number of those areas. One could question, for instance, grouping together (i.e. as "parascientific studies") such dissimilar activities as alchemy and theosophical speculations. While the former, in spite of being often inspired by metaphysics, were basically of experimental character, the latter had little in common with experimental study.

in the physical world, the latter focus on the theosophical questions attempting to determine the "divine" nature of the absolute space.

For Newton, who could not forsake the idea of searching for the ultimate theological interpretation of natural phenomena because he was deeply convinced that the utmost rationality of nature comes from God's rational project for the world, it must have been extremely difficult to remain confined in his "phenomenalistic" works to problems of a strictly scientific nature.

For that reason, the discipline which he was able to impose upon himself in his scientific studies is one more evidence of his greatness, not only of his intellectual genius but also of his personality.

HEGEL AND CONTEMPORARY NATURAL SCIENCES*

I. Introduction

My first thesis is the claim that Hegel's philosophy enables us to understand better both the nature and the development of natural phenomena, and above all, the nature and the development of natural sciences. This seems to be a less trivial thesis than the one concerning Hegel's influence on the understanding of social facts and social sciences. While the latter claim that of the impact that Hegel's philosophy exerted upon the development of the humanities, social sciences and philosophy cannot be questioned on any grounds, one could reasonably object to the claim about the influence of that philosophy on the development of natural sciences. Accordingly, one could incorrectly assume that it has no significance for modern natural sciences whatsoever.

II. Hegel's Philosophy of Nature and Contemporary Natural Sciences

1. The second thesis that I would like to defend is that although from the point of view of natural sciences the most interesting part of Hegel's system is not his philosophy of nature but his general theory of the world (Hegel's logic), yet his philosophy of nature also contains many ideas which are close enough to contemporary theories advanced by natural sciences. Moreover, it seems that Hegel's philosophy of nature is more interesting for present-day natural sciences than it was in his own times. This is due to the fact that the significance of the Hegelian conception of nature was not so much its contribution to his contemporary syntheses in natural sciences as in its anticipation of the future achievements of natural sciences and, even more importantly, of their future predicaments and quandaries.

2. The above follows from the fact that although Hegel was well acquainted with the contemporary state of natural sciences and mathematics, the knowledge played little role in the work which led to his

---

* First published in H. Kimmerle, W. Lefèvre, R.W. Meyer (eds.) (1989), *Hegel – Jahrbuch 1989*. Germinal Verlag Giessen, pp. 241-247. This text is reprinted with the kind permition of Germinal Verlag Giessen.

system, unlike his knowledge of the human being and of society. One could go even further, assuming that Hegel's studies of nature, although they were the starting point of his inquiry, were, at least in the case of inanimate nature, of no significance for the process of formation of his developed system.

This state of affairs is best evidenced in the artificiality of many of his "scientific" constructions and conceptions, as well as in a poor adjustment of the "scientific part" of his system to its "logical part." It goes without saying that Hegel would not have been a man of genius, not even an insightful researcher, had he not been aware of the precariousness of the "scientific foundations" of his system. That is why he constantly emphasized that nature (in particular, inanimate nature) which is being-for-other of the Notion, is the least adequate corporeality of the idea, realization of the concept, embodiment of the subject or of mind. In his opinion, one cannot learn much from nature about the dialectic either because nature does not undergo temporal progressive development (it has no history), and it is free (or almost free) of the mental concepts which are characteristic of the Notion. Besides, nature is too abstract to embody the dialectic. The dialectic is best manifested in concreteness, i.e. not in abstract immediate order but in concrete results, in the totality of qualifications, which are rich in mediations and cancelled assumptions.

3. The problems that Hegel raised in natural sciences had hardly preoccupied scientists until the beginning of the 20th century. Instead, it had been common to point out various inadequacies of this system of nature. Although the inadequacies are now better seen than ever before, the scientist's attitude to Hegel's philosophy, including his philosophy of nature, has been undergoing substantial revision.

Thus, two things have been found in Hegel's philosophy which could not have been noticed earlier. First, it has become evident that in some of his suggestions and hypotheses he definitely went ahead of his contemporaries, thus anticipating certain future findings of natural sciences. Second, in his philosophy of nature he raised the questions which are now the central issues of natural sciences, even though they were of no concern to his contemporaries. The latter point clearly indicates that the predicaments and quandaries of Hegel's philosophy of nature have nowadays become commonly acknowledged predicaments of modern science.

4. The above points will be illustrated below with some examples from physics. Hegel worked out a beautiful and a rather insightful conception of the unity of matter, motion, time and space. Many of his ideas contradicted Newton's physics but they bear comparison to modern physics, and may even be well appreciated from within it, especially from the point of view of the general theory of relativity. A weakness of his conception – which

was noticed by Riemann, who also encountered it in Kant – lay in that the philosophers, like many of their contemporaries, did not distinguish between such properties of time and space as infinity and boundlessness. That, in turn, is believed to account for the inadequacy of one of Kant's antinomies. According to Riemann, the paralogical deduction error which was committed by Kant and Hegel, was that having indicated that time and space are boundless they proceeded with further reasoning as if they had proven their infiniteness, which is far more difficult to decide and to prove.

In this connection, it is worth mentioning that Hegel's conception of unity of matter, motion, time and space, if placed within his mechanics, (which seems a plausible decision), shows that even in mechanics he is in favour of the totality theory, thus assuming an antimechanistic approach, against conceiving of phenomena as sums of elements.

5. Another of Hegel's interesting proposals put forward within his philosophy of nature is the aforementioned global approach to the investigated facts, conceiving of nature as an organic totality, which is already present in his mechanics and which finds full expression in higher regions of science, such as inorganic physics (and chemistry).

Today, complex (systemic) methods are widely applied not only in the so-called interdisciplinary and complex studies, such as cybernetics, the general system theory or synergetics but also in numerous branches of hardcore physics. For instance, the theory of elementary particles, field theories, astronomy and cosmology have well assimilated the concepts that are characteristic of the mentioned type of study, such as group theory, symmetry theory, matrix and system.

6. The third attractive issue of Hegel's philosophy of nature is his critical analysis of the concept of force, playing a crucial and much overestimated role in the natural sciences of his times, first of all in chemistry and biology but also in Newton's physics. Although it is still a disputable matter whether physics can do without that concept altogether (the first coherent conception of physics without the category of force was put forward by Heinrich Hertz), it is unquestionably true that the category plays a far lesser role in the physics of today than ever before. Its previous functions have largely been taken over by such concepts as energy, action and field, which are moved to the fore in modern physics. It seems, in, fact that all branches of physics can do without the concept except for two in which the terms used are "mechanical forces" and "electromotive forces."

7. There is a general agreement now as regards another point of Hegel's critical comments on the hypothesis of specific substances ("ponderable and imponderable") such as calorific matter, phlogiston, self-articulating sound, electric and magnetic fluids, and others. Thanks to both his

intuition and his knowledge, Hegel was clearly aware of a fictitious and "tautological" character of explanations provided, based on the assumed existence of those substances or of some hidden forces. In view of present-day discoveries, they were, indeed, mostly *ad hoc* explanations.

These are examples of the hypotheses and analyses included in Hegel's philosophy of nature which were ahead of their times, and whose significance became evident only from the perspective of modern physics.

8. I will now discuss two questions posed by Hegel whose importance is, likewise, noticed by modern physics. The first is the problem of relationship between the subjective and the objective. As we know, also in philosophy of nature Hegel stressed the unity of the opposites. I will not attempt, however, to summarize in this place his original and insightful, albeit extremely difficult, conception of objectivity.

As far as physics is concerned, it is only the scientists of today preoccupied with questions of microphysics that have reached the understanding of how difficult it is to separate the subjective from the objective, for instance in the *psi*-function of quantum-mechanical description. As we know, such description consists of two interrelated components: it is a description of an "objective" state of a microobject (or to be precise, of its potential transformations), and at the same time it is a meta-description of our "subjective" knowledge of that state. Consequently, the ideal of pure objective knowledge, which was typical of classical physics, has undergone radical change; it turned out to be principally unrealistic, which fact was so difficult to accept for Planck and Einstein. I will return to the question later, in connection with the subject-object relationship, the issue which is close albeit not identical (at least for Hegel) with the one discussed above.

9. The second issue raised by Hegel in his philosophy of nature which nowadays draws much attention of scientists (not only in natural sciences) is that of reduction. Hegel's stand in the matter is known to be definitely antireductionist; he argued against mechanistic theory as well as against other forms of reduction in science, such as reduction in chemistry and the so-called quantitative reduction which placed emphasis on quantitative differences between three spheres of nature but disregarded the existence of qualitative differences. In his antireductionist approach, Hegel assumes that each level is a specific whole and therefore any object placed at a given level must be described in relation to that level, as an element of its integral whole. Reduction in science remains an open question nowadays; so far, all reduction programmes (e.g. a reduction of physics to mechanics or to electromagnetics, of chemistry to physics, of biology to physics and chemistry, or in the entirely different department of science, of mathematics to logic), have been implemented only fragmentarily.

There is no doubt that Hegel exaggerated in his antireductionist vein, thus not allowing to use physical and chemical methods outside physics and chemistry respectively, for instance in studies of life phenomena.

10. In order to properly evaluate Hegel's attitude to sciences, it should be pointed out that he was well aware of his departure from the findings of mathematics, formal logic and natural sciences of his times. It follows from that the incompatibility was not a result of Hegel's insufficient knowledge of the sciences but rather of the latter's conscious attempts to "adjust" their results to the requirements of the logic he assumed. In the above light, all the more surprising is Hegel's criticism of Newton which in its essential part seems to have missed the point. That was the case of Newton's optics, mechanics, gravitation theory, of his differential and integral calculi, in other words, of Newton's all major achievements in physics and mathematics. Next to that, however, there were also, in Hegel's critique, a few justified points, in particular those concerning the more metaphysical sphere of Newton's work. Thus, he was quite right in rejecting Newton's substantive conception of the absolute space, which eventually led Hegel himself to defending the idea of superiority of the relative space over the absolute one.

Nonetheless, the person who fully appreciated Newton's genius and his contribution to science, was by no means Hegel but Kant. In my opinion, Hegel's false image of Newton was the immediate cause of his distorted picture of the contemporary natural sciences. Needless to say, his attitude is not difficult to explain. Namely, he must have held Newton responsible for excessive mathematization of theoretical physics which, in turn, gave rise – as he notes in *The Philosophy of Nature* – to "incredible metaphysics" that contradicted both experience and the Notion. He also criticized Newton for his "barbarian application of categories" which after him became a habit in physics. As regards optics, he additionally blamed him for using technical devices in order to eliminate the subject from the optical process, thus transforming optics into a purely objective science like mechanics.

I will now pass on to Hegel's logic, namely to those of its elements that seem of great interest for present-day natural sciences.

III. Hegel's Logic and Contemporary Natural Sciences

In Hegel's *Logic* there are three conceptions which are extremely interesting from the point of view of modern natural sciences.

1. The first one is the conception of universal progressive evolution, the evolution of ideas from abstractness to concreteness, which is accompanied by cancelling of the old notional determinations replaced by the new ones. Its result is a gradual emergence of thc Absolute Idea which

all the time becomes more and more concrete, individualized and enriched in contents. The conception had its equivalent in the biology but not in the physics of his times. The first physical theory of irreversible processes, i.e. thermodynamics, was to appear only towards the end of Hegel's life. Moreover, according to the latter theory, especially to its second principle of growing enthropy, the time "arrow" was not determined by progressive development but by a tendency to degradation and dispersion of energy. Seen from that perspective, it was not the progressive development but rather the regression that seemed to dominate the world. Therefore, it was hard to understand why there were so many highly organized and ordered things in the world, and where they came from.

Only the present-day nonlinear thermodynamics of open systems which are far from equilibrium, synergetics, Eigen's theory of hypercycles, and other theories of self-organization have placed the theory of progress in the centre of the study of material system evolution. It has been proved that self-organization processes occur in a natural way in a great many of unbalanced systems, which Prigogine calls dissipative systems while Haken refers to them as synergetic systems.

2. Another interesting conception of Hegel's *Logic* which met with unexpected response on the part of modern science is his famous theory of mediation and of the unity of opposite notional determinations, which stands out as a true jewel and the core of Hegel's dialectic. His highly sophisticated considerations which from a common-sense perspective, or even from the point of view of formal logic, might seem mere juggling with words, or an exercise in rhetoric, have turned out to have surprisingly close equivalents in many scientific studies of today. The statements found in the works of modern scientists are often quite similar to Hegel's, both in their form and in substance; they usually discuss, however, pairs of the opposite categories that are different from Hegel's.

I will illustrate the above with a pair of interrelated concepts which is the focus of synergetics and other theories of self-organization, i.e. order and disorder (or chaos). In classical thermodynamics and in information theory, the concepts seem to have stood in obvious opposition to one another, in other words, they were complementary and mutually exclusive. Enthropy stood for disorder in those theories whereas negative enthropy (or the potential information of a given system) was the measure of order. Decrease in disorder always meant increase in order, and vice versa.

The model has become radically more complicated – much in a "Hegelian" vein – when the sciences proceeded to search for explanation of qualitative change, in particular when they attempted to explain the process of self-organization of various complex systems.[1] It was, for instance, when normal radiation becomes laser radiation, when there appears superconductivity and superfluidity, when laminar flow of liquid becomes turbulent flow, as happens in water that is boiling over. An attempt to understand simple everyday phenomena, as the latter one, has led to the conclusion that the once assumed chaos of boiling over is, in fact, a manifestation of both order and disorder, or to be precise, of various kinds of order and disorder. In addition, the order and chaos are so splendidly well intermingled in that process, they infiltrate each other to such an extent that they cannot be separated, or even analyzed independently of each other. It turns out, therefore, that they are so much conditioned by each other that they can only co-occur as the elements of one and the same process: a given kind of order is always accompanied by its corresponding variety of disorder, and vice versa. Thus, we can repeat after Hegel that the process will take place only as long as order and disorder penetrate each other.

So as to understand the above problem better, a number of concepts of order (and their respective notions of disorder) have been introduced which are related to various "reference systems." In particular, there are orders which are relativized to various types of space, e.g. the order of normal (three-dimensional) space, of phase space, of momentum space and velocity space. The old concept of order (so-called structural order), used earlier by physicists and philosophers, has turned out to be a "purely" static notion which was based on the intuition that what does not change is ordered, and that any change results in chaos. On the other hand, the new concepts of order which are now introduced and analyzed are more and more clearly of a dynamic character. They refer not to the structural elements of matter but to the events and processes in which those elements participate. Thus, they concern the organization of the processes in time (i.e. they are temporal or spatiotemporal, and not purely spatial orders), and they adjust various functions of the system to one another. That dynamic order, in particular the functional one, is often contrasted with the static structural order. For instance, the biological order of life processes differs from a typical physical order (e.g. the one occurring in crystals) in that it is a functional order, a type of dynamic order.

It was the study of the unbalanced systems, which are the main points of interest of nonlinear thermodynamics, that led to the conclusion that the

---

[1] They are certain mathematical and physical theories, as well as some interdisciplinary studies exceeding the boundaries of physics. As for mathematics, one can list here the theory of dynamic systems, the bifurcation theory (which from the point of view of dialectics, is currently the most interesting part of dynamic system theory), the catastrophe theory (which, in turn, according to some authors belongs to bifurcation theory). The physical and neighbouring disciplines include nonlinear thermodynamics of unbalanced systems, macromolecule self-organization theory and synergetics.

stronger the correlations between the elements of a given system, the higher the degree of the dynamic order of that system. Accordingly, the whole in which the order is manifested is no longer of a synchronic but rather of a dynamic character of the Hegelian type. Such wholes are, first of all, all processes of self-organization in which each new stage of ordering, of one type or another, is reached thanks to modifications of the previous systems of interactions of the system parts, and due to an increase in the number of correlations within the system. In addition, the increased amount of correlations causes the former old correlations to be "backgrounded" within the system while the new ones are moved to the fore in their place, which changes the hierarchical order of the system. All this is more reminiscent of Hegel's ideas than of any other philosophers.

The above-mentioned focussing on the dynamic moment, even though very important, is not the most interesting similarity between Hegel's and the present-day studies of self-organizing and ordering processes. The most significant of all is the fact that, as mentioned, all interesting phenomena, processes and objects studied by modern science occur due to mutual adjustment of the elements of order and of chaos. That is a property of any turbulent motion, of all laser radiation processes, and of all superconductivity and superfluidity phenomena which cannot occur otherwise but in a state of, so to say, ordered chaos.

The very process of biological life is, likewise, a combination of order and disorder. The rigid order (as Hegel would say, "pure order") and rigid determination, excluding elements of fluctuation, spontaneity, indeterminacy and the statistical factor, preclude any basic life processes (see Careri 1982; Prigogine and Stengers 1984).

One could say that life (in the biological sense) is an art of the successful introduction of the elements of disorder into the ordered structures, or of the disharmonious components into the harmonious whole.

By analogy, one can conclude that certain 20th-century attempts to impose from above the rigid social order and stiff economic rules, based completely on central economic planning and devoid of any elements of free market, must inevitably lead to failure, for much the same reason. Namely, factors of spontaneity, statistical variation, randomness, indeterminacy, even of chaos, are absolutely necessary for any structure to survive.

As regards phenomena of nature, the conclusions are well accounted for by synergetics which explicated a decisive role of fluctuation, i.e. of a statistical factor, in phase transitions in which there appear qualitative changes, i.e. structural changes. Thus, a given fluctuation, which first brings about a moment of chaos to the ordered system, once it has spread all over the system, decides about the new type of order; it is, therefore, both order and disorder at the same time, it is the transition from the former to the latter.

3. I will now pass on to the third component of Hegel's logic which has found reflection in modern physics and in the philosophy of physics. What I have in mind is an eternal question of all philosophy, i.e. the problem of the subject-object relation. Kant seems to have been the first one who – due to his transcendental philosophy – presented the problem in a new and up-to-date light, which is especially valid for the present-day discussions on the subject held in modern physics (see Siemek 1977). For that reason, physicists quote Kant rather than Hegel in this connection. There is no doubt, however, that Hegel had his own opinion and presented some important conclusions in the matter.

Since I am not a historian of philosophy, I will not elaborate on the sophistication and subtleties of the differences between Hegel and Kant. I will merely say a few words on the relevance of the problem to modern physics.

It seems that any study, once it has gained sufficient insight into its subject matter, becomes aware of a subjective nature of the object of its inquiry, and of its own objective nature. No wonder that a discipline as sophisticated and well-developed as modern microphysics has come to similar conclusions in this respect as those reached earlier by transcendental philosophy (both Heisenberg and Weizsäcker write about it).

However, the problem that the subject-object relation poses for microphysics is more complex than was predicted in the philosophical theories. There are three new important factors which contribute to that.

The first difficulty comes from the existence of two distinct levels of reality, namely the macrolevel and the microlevel, and from that microphysics must take into account the fact that the subject – her conceptual, linguistic, sensual and technical faculties – are situated at one of them (i.e. at the macrolevel) whereas its object of study (i.e. microobjects) – at the other.

The second problem results from the corpuscular-wave dualism and from Bohr's principle of complementarity which make it impossible to enclose the results of microphysics in one coherent model, thus necessitating the construction of two complementary and mutually exclusive models.

The third difficulty is caused by the indeterminacy of the microworld (the microphysical reality) which is accounted for by Heisenberg's uncertainty principle and manifests itself in the violation of the determinism principle which states that "given the same conditions, the same things will occur."

The circumstances discussed above create a completely new and extremely difficult situation in scientific knowledge in the understanding of which the writings of such philosophers as Kant, Fichte and Hegel are certainly of great assistance, but which calls primarily for new, original philosophical (perhaps also logical) conceptions that would provide its adequate explanation.

IV. Conclusion

I have managed to cover only a few of the themes of Hegel's philosophy which are relevant to contemporary science. They are far more numerous and all of them are worth studying.

In concluding this essay, I would like to express my profound conviction that Hegel's works continue to be worth reading and, moreover, that Hegel belongs to those few people who will be read and studied as long as the human race exists.

Part VI
Some Problems of the Theory of Reality

UNITY OR VARIETY OF NATURE?*

I. Differentiation in the Sphere of Phenomena

There is a widespread conviction, which seems to have started with Aristotle, that the actual world is extremely varied (e.g. divided into individuals and species), and that the unity is rather a feature that is characteristic of our knowledge which is due to the process of generalization.[1] According to this conviction, the main function of rational thinking consists in identifying that which is varied and thus overcoming the resistance that is offered by empirically differentiated reality. The role of "the principle of identification" in our thinking, i.e. rational interpretation of reality was emphasized in particular by French theoreticians of knowledge such as Boutroux and Meyerson (cf. e.g. Meyerson 1957).

Assuming the evolutionary point of view that is characteristic of modern science, and using it (along with the vast majority of cosmologists) for the study of the Universe as a whole we can admit that the enormous variety of phenomena and things that is observed at present is a result of a long process of development, lasting billions of years. According to the standard cosmological model, as result of the Big Bang, the Universe which arose is so homogeneous that even the four fundamental physical interactions present in the world today (gravitational, weak, electromagnetic and strong) constituted a homogeneous universal interaction which is sometimes called supergravitation. Only in the course of further evolution, when the world crossed the Planck threshold and from the era of quantum gravitation entered the hadron era (which occurred at the $t = 10^{-43}$ sec time after the Big Bang), did its evolution lead – through breaking symmetry – to the

---

* Translation of "Jedność czy różnorodność natury?," published in A. Szołtysek (ed.) (1994), *Rozważania o filozofii a recentiori* [Considerations a Recentiori on Philosophy]. Katowice: Wydawnictwo Uniwersytetu Śląskiego, pp. 34-42.
[1] This view is in obvious contradistinction to the well-known standpoint adopted of the Eleates, according to which "being is uniform and unchanging," and phenomena discovered "by the blind manner of sensory perceptions" "are full of contradictions and are mutually exclusive," and as such, "are not a true picture of being." Cf. Bańka (1992, p. 10).

distinction of the universal interaction, gravitation interaction and, next, to strong interaction.

Lowering the temperature further led to splitting the electroweak interaction into electromagnetic interaction and weak interaction as a result of which we have the present Universe which is characterized by four fundamental physical interactions. Further changes led – under conditions of stars and galaxies being formed – to the rise of ever heavier elements and ever more complex chemical compounds until the moment when the physico-chemical evolution brought about the rise of organic compounds. Thus, conditions were created for the rise of life and, along with it, conditions for the biological evolution (on the Earth, this occurred about four billion years ago), which in turn ended in the appearance of human being (ca. 2 million years ago) and of the cultural evolution of human communities.

Apart from evolution and its consequences, another document of great variety of the world on the level of phenomena is the fact of the existence of several or perhaps even over a dozen thousand separate scientific disciplines, the majority of which study specific phenomena or types of phenomena.

So there seems to be no doubt that the unity of the world on the level of phenomena is not a characteristic feature of the present stage of evolution of the Universe but that it occurred in the early stages of its evolution. By the way, it is worth noting that the unity of the Universe may also be the consequence of a considerable reduction of the variety that is predicted according to the principle of growth of enthropy, i.e. the equalization of energetic potentials – in further phases of transformations, extremely distant from our times, occurring on the global scale. Experiencing of the variety of phenomena is not, however, the only kind of experience which we have by contacting the world. Equally strong is the feeling of the unity of the reality which is gained through the observation of the mutual convertibility of natural phenomena. This convertibility concerns, after all, not only phenomena that can be observed directly (i.e. by the naked eye) but also, and even most of all, elementary particles, i.e. the lowest and, so far, relatively well-known components of reality. Mutual transformability of elementary particles and of other objects of nature constitutes an expression of the so-called genetic unity of the world (cf. Z. Nowak 1990).

Apart from genetic unity, an attributive unity and a nomological unity of the world are also distinguished (see Eilstein 1961). Attributive unity is the mutual properties which can be ascribed to all the recognized and – as we may think – all existing material objects, while nomological unity is the universal laws which govern nature.

However, if the genetic unity of the world is manifested in forms that are partly accessible to observation by means of the senses (mutual transformability of phenomena) then attributive unity, and especially the nomological one we are informed about only rather in the process of study of the mechanisms of phenomena, i.e. deeper levels of reality which constitute the object of rational (intellectual) cognition.

II. A Dispute Over the Unity of the World and the Levels of Reality

The fact that the variety of reality is manifested on its surface, and the unity is fully manifested only on its deeper levels causes the dispute over the unity of nature to be entangled in a controversy between phenomenalism and essentialism about whether there exist various levels of reality, or whether it is only a one-level one. An extremist phenomenalist does not perceive unity apart from variety; one could say that s/he does not see the wood for the trees. However, not every essentialist may easily agree with the thesis on the unity of nature. If s/he is at the same time an antirealist in understanding the scientific theory, then the lack of agreement to the possibility of cognition of the so-called theoretical objects will make it impossible for him/her to give a positive answer to the question of whether it is possible to reach cognitively these levels of reality in which (possibly) the unity of nature is manifested. Therefore, it is only an essentialist taking a realistic point of view in understanding the scientific theory and its subject matter who can justly assume a thesis on the unity of nature.

A thesis on the multitude of levels of reality is not confirmed in such a versatile manner as the thesis on the variety of phenomena. These are not only very few disciplines – such as physics (which comprises microphysics, macrophysics and astronomy with cosmology), chemistry and biology (including microbiology) and certain similar sciences – whose subject matter are several (at least two) levels of reality. A vast majority of sciences, on the other hand, are restricted to the study of only one level of reality, which is usually a macrolevel. Since evolution (development understood as breaking symmetry, leading to the growth of the variety of phenomena) occurs mainly on the macrolevel, it is not strange that we can observe the greatest variety of phenomena just on the macrolevel. And exploration of the microlevel and the cosmic level indicates a considerably larger homogeneity of objects and processes occurring on these levels.

As a result, according to the data from modern science, the greatest variety of objects and phenomena occurs at the present stage of existence of the Universe,[2] and this happens just at the macrolevel, or – one could say – in the time and space "surrounding" that is closest to us. The further

---

[2] This stage in the standard cosmological model is usually called the galactic era; the galactic era is the fifth consecutive one and was preceded by the era of quantum cosmology, the hadron era, the lepton era and the radiant era.

we reach "in depth" (the lower and lower levels of the structure of matter) and "in width" (the higher and higher structural levels of nature) and when we look into the further past or future of the Universe, the more homogeneous the world seems to be. If it is really so, then the Universe is "a flower which blossoms only once," never to repeat this fantastic variety of forms which we are able to see. In order to explain the circumstances due to which it is just we who witness this unique fact of the blossoming of the Universe, we can refer to the anthropic principle or, in this case, to the circumstance that an observer could appear only at the phase of full blossom since s/he him/herself is its product.

III. The Problem of the Infinity of the World "in Depth" and "in Width"

The existence of many levels of reality opens not only the perspective of the unity of nature into the deeper levels of reality but also into a perspective of still another differentiation of nature, which may be called "vertical." It consists in the dissimilarity and uniqueness of particular "storeys" of the structure of matter: if the number of different levels of reality was infinite then we could say that there is a new kind of infinite variety, i.e. an infinite variety of structural levels of matter.

As far as infinity "in depth" is concerned, until now physicist's considerations have been generally restricted to quarks and preons which, due to the so-called mass defect and other peculiarities of the phenomena of the microworld, complicate the problem of the relation of the part to the whole (as e.g. quarks have a greater mass than hadrons of which the latter consist). A prevailing view which seems to be held by physicists is that even though there are no absolutely elementary particles (i.e. those devoid of internal degrees of freedom), matter is finite in "depth," i.e. the number of levels of reality lower than the microlevel is finite. Obviously, there still remains the problem of infinity spreading in the "opposite" direction – the problem of "in-width" infinity (is the number of levels of reality rising over the macrolevel infinite?). Although the hypotheses of the material and spatial infinity of the Universe cannot be precluded at the present level of development of science, cosmologists do not in principle use concepts of levels higher than the cosmic or megacosmic one. Therefore, it cannot be precluded that the number of fundamentally different levels of reality is limited to three (microlevel, macrolevel and cosmic level) or to five (if submicrolevel, studied by picophysics, and megacosmic level, studied by megaphysics were added).

Let us now look closer at the three aspects of unity of the world mentioned above: the genetic, the attributive and the nomological one.

IV. The Genetic Unity of the Universe

The genetic unity of the Universe, consisting in the mutual transformability of objects and phenomena which occur in the world, and in the formation of higher, better organized structures from the lower ones may be viewed on all levels of the structure of matter. For instance, the evidence is the change in the state of aggregation of various chemical substances, the occurrence of chemical reactions, leading to a transformation of some chemical substances into others, transmutations of chemical elements. However, most significantly, it is observed, as I have already emphasized, in the fact of the mutual transformability of elementary particles. All transformations allowed by the laws of physics are possible in this field, in particular by the laws of conservation which determine the conditions of transformability and the rules of transformation.

It is worth noting that Democritus (and other representatives of speculative atomistics), by assuming the eternity and indestructibility of atoms, considered them all as the completely separate bricks of matter, which precluded genetic unity of the world on the level of matter which he thought to be an elementary one. Full proof of the genetic unity of the world is not possible without showing the mutual transformability on the lowest level of the structure of the matter. This results from the fact that transformability of objects on the structural levels of the matter higher than the elementary one may not shake the identity of the components of these objects, while the mutual transformability of elementary objects may already consist only in transformations occurring in the same objects, and not, for example, in shifting some of their component parts (since there are no such parts). Thus, until we reach the elementary level of reality (assuming that such a level exists at all), the genetic unity of the world will constitute a programme to be carried out, and not a fact established by science.

Unification theories which are a present being developed in physics, such as the theory of great unification (whose aim is to give conditions under which strong interaction becomes indistinguishable from the electroweak interaction) and the theory of superunification of all interactions, including gravitation, aim at showing that under conditions of enormous pressures and temperatures prevailing in the early stages of the evolution of the Universe both the division into hadrons and leptons and the division into bosons and fermions did not occur. Thus, the disappearance of differences between fundamental physical interactions is accompanied by the disappearance of differences between elementary particles and their components.

What is more, together with the disappearance of differences between elementary physical objects (particles) also the difference between matter and interaction disappears. This is because both matter and interaction represent certain physical microobjects; namely, matter occurs in the form of fermions (i.e. leptons and hadrons), and interaction occurs in the form of carrying bosons.[3] So, when fermions become indistinguishable from bosons any difference between matter and interaction disappears.

V. The Attributive Unity of the World

A thesis claiming that only those objects which have common properties may interact with each other seems to be justified. Thus, the basis of the gravitational interaction of physical objects is that they have (gravitational) mass, the foundation of electromagnetic interaction is having an electric charge, and the like. Therefore, assuming that in the Universe there are no objects which are in absolute isolation (that, to put it differently, the Universe does not consist of parts that are totally independent of one another), we may conclude that all its components (concretes), independent of their kind and magnitude, include certain common properties – let us call them attributes of matter. Most often, such properties are included within the attributes of matter as motion, time, space (or space-time) and determination (subject to laws, including causality). Some scientists, especially physicists, also include mass and energy, since physics does not know objects that would be devoid of these two properties, although there are objects which do not have the so-called rest mass, e.g. photons (light quanta); but they have the relativist mass connected with their motion, since every object having energy also has mass that is proportional to it.

Certainly, it is not possible to preclude a possibility that certain (or even all) properties that are today considered as attributes of matter belong only to certain subsets of material objects. In particular, one may think that such apparently indispensable and fundamental properties as time (lasting in time) and space (extension in space) characterize only objects of "the medium level" and do not concern sufficiently large objects (e.g. megacosmic ones) or sufficiently small ones (e.g. submicroscopic ones).[4] It seems to be more probable, however that not so much time and space themselves but rather some of their properties (e.g. continuity of time and space, anisotropy of time and three-dimensionality of space) are different from what we now assume or that they do not belong to them in certain scales or under certain conditions.

---

[3] Interaction between matter (fermions) consists in the fact that fermions mutually transfer to one another (i.e. send and absorb) virtual bosons.

[4] What is particularly doubtful is the possibility of time and space location of virtual particles. Cf. e.g. Z. Nowak (1990).

VI. The Nomological Unity of the Universe

The name of universal laws belongs to laws that are non-emptily fulfilled in the whole domain of reality, in all the spatial and temporal areas of the physical Universe (we assume that the absolute vacuum does not exist), or occurring ("acting") under all conditions (and in this sense "unconditional," conditionally unlimited) (cf. Such 1972, pp. 279-281). Laws of this kind concern all levels of reality in their whole range, or refer to all objects of the physical Universe, regardless of their quality and magnitude (p. 283).

Only rather few laws that have so far been discovered by physics may claim the name of universal laws in the sense given here. One such law is considered to be (since the time of its formulation by Newton) the law of universal gravitation (at present in Einstein's formulation within the general theory of relativity) and some universal (of an unrestricted range of application) laws of conservation, e.g. the law of the conservation of mass and energy ($E = mc^2 = constans$).

The universal status of the laws of conservation may additionally raise serious doubts from the point of view of both fundamental theories of contemporary physics, i.e. the general theory of relativity (GTR) and quantum mechanics (QM). Within the framework of GTR it is difficult to give the laws of conservation the generally covariant form, without which no law of physics can be considered to be properly formulated. On the other hand, the trouble within QM is that the law of the conservation of mass and energy is undermined in the processes of the rise of the so-called virtual particles. However, one may doubt whether the undermining is real, since energy (and mass) of the real elementary particles disappears in the process of the rise of virtual particles in a degree and at the time which is within the limits of indeterminacy, determined by Heisenberg's principle of indeterminacy.

There is no doubt that the nomological unity of the world does not reach as far as it was assumed by mechanistic materialists, according to whom "man is submitted to the same laws as is the other part of nature" (Holbach), which led to the conclusion that all laws are actually universal and there are no peculiar laws concerning only some fragments of nature.

VII. Conclusion

While the predominantly descriptive sciences, such as experimental physics, experimental chemistry, astronomy, geology, zoology, botany, microbiology, geography and history, reveal ever newer aspects of phenomena, theoretical sciences, especially theoretical physics and its extensions, i.e. astrophysics and cosmology, formulate theories which discover new and ever deeper manifestations of the unity of the world.

What has not been achieved by classical, uniform theories of the field can be achieved by theories of unification based on the quantum theory of the field. Especially the creation of a complete theory of superunification would constitute a breakthrough, since it would have crucial significance for solidifying the unity of the world in all the three aspects. As it is, this theory aims at showing that, first, all elementary particles, both those which represent matter (fermions) and those responsible for interactions (bosons), are indistinguishable from one another under conditions of extremely high temperatures, and second that all the four fundamental interactions are (at low temperatures) manifestations of a uniform superinteraction (supergravitation). Obtaining the two results would lead to the conclusion that:

– all elementary objects which represent both matter (fermions, that is leptons and quarks) as well as interaction (bosons), are mutually transformable (genetic unity);

– all of them undergo certain universal laws, established by theoretical physics (nomological unity).

Relativist cosmology, whose outcome is a standard model of the evolution of the Universe, leads to the conclusion that a state of extremely high temperatures actually dominated the early stages of the existence of the Universe. In such a state, the Unity of the world occurred in an as if pure form, which was not contaminated by the variety of phenomena that are the result of the developments, which really broke the early initial symmetry of the world.

THE PLACE OF PROCESSES IN THE STRUCTURE OF REALITY*

I. Introduction

It is usual to include within the basic ontological categories that characterize the structure of reality the categories of a thing, an event and a process. Three trends take their names from them. In a different way, they give answer to the question of what the ontological category is to which the basic components of reality belong. According to reism, the world ultimately consists of things, according to eventism – of events, and according to processualism – of processes (Krajewski 1987, pp. 450-451).

In this essay, I shall use a somewhat different terminology, which is used by Roman Ingarden in *Spór o istnienie świata* [A Debate over the Existence of the World]. Ingarden divides all the individual objects "into two great classes: (1) temporal objects, and (2) atemporal objects (which in particular include ideal objects)" (Ingarden 1987a, p. 188). Among the individual objects, he subsequently defines three basic types: "(1) objects lasting in time, in particular – things, (2) processes, (3) events" (p. 189). Apart from things, Ingarden includes within objects lasting in time also living creatures and human beings (pp. 207, 220-221).

Ingarden tends to think that the three presented categories of temporal objects are mutually irreducible to one another and that in the real world, there are both objects lasting in time, processes and, finally, events. However, this does not mean that, according to him, they are existentially (i.e. from the point of view of being) equivalent: objects lasting in time constitute an indispensable basis of being ("carriers") both of processes and events (which are, furthermore, not independent existentially with respect to processes and, in this sense, are ontologically distinguished) (Ingarden 1987a, pp. 190, 209-210).

This viewpoint is close to the natural, common view according to which the world consists of things in which various processes take place and various events happen.

---

* Translation of "Miejsce procesów w strukturze rzeczywistości," published in W. Heller (ed.) (1996), *Świat jako proces* [The World as a Process]. Poznań: Wydawnictwo Naukowe IF UAM, pp. 9-16.

The aim of this essay is to outline an answer to the following questions: (1) do there exist all three kinds of objects that are temporal, and (2) what is the place and role of processes in the structure of reality?

II. Objects Lasting in Time, Processes, Events

Definitions of objects lasting in time, of processes and of events, adopted in this essay, come from Ingarden. Objects lasting in time and processes differ mainly from events in that their existence always encompasses some time interval, while events are "momental-point" phenomena (or "quantum" phenomena) whose existence is restricted to one moment (or quantum) of time.

On the other hand, the main difference occurring between objects lasting in time and processes "concerns the *manner* in which an object lasting in time may survive particular moments" (Ingarden 1987a, p. 208). Namely, an object lasting in time (for the sake of brevity, hereinafter referred to as an "object" or "thing"), *"remains identically the same* in newer and newer moments in which it exists" (p. 208). It does not mean that it remains (in these moments) the same, since certain changes may occur in it, and thus new properties may appear thanks to the processes occurring in its "inside" or thanks to the events which occur in it (p. 208). On the other hand, a process does not exist from the first moment of its existence as a fully formed object but happens in time in the sense that "just its current" phase is transferred into a completely new phase, although not delimited from the former one that it is constantly prolonged into it (p. 208).

A process – as distinct from an object (a thing) – is characterized by "a specific two-sidedness of the structure and mode of existence" (Ingarden 1987a, p. 209). A process is something which "on the one hand is a whole of always growing phases, and on the other, an occurring subject of properties in time" (Ingarden 1987b, Part I, p. 407).

> Each definite process, as a subject of properties, which is constituted in the occurrence of phases also has the important characteristic, which is specific to it, that the phases of which the whole which is growing in its course is composed *elapse in a continuous manner. This constant passing of phases constitutes their* specific *manner of existence*. It is importantly connected with the temporality of the process and is distinguished by the following elements: (1) *one* and *only one* phase is always current; (2) the ever newer phase becomes current; (3) the current phase loses in a *continuous* manner its topicality and just because of this a new, just coming phase becomes current; (4) at the moment when the current phase begins, there no longer exist (to be more precise: are not current) the phases preceding it but they already *existed*; on the other hand phases which come later do not yet exist but they *will* only exist (will be current); (5) at the moment when the last phase gains its topicality, the process has already *passed*. But not every process has to have its last phase (Ingarden 1987a, pp. 194-195).

In turn, an event is an "occurrence, more precisely: coming into existence of a certain state of things or a certain object situation" (Ingarden 1987a, p. 189). Events are characterized just by the fact that
they do not last. They occur and so they cease to exist. They are the ends, results of processes or their beginnings (sometimes their "intersection"). The processes as if lead to them, and, closing the process, they are already something totally new in relation to it. However, an event is not constituted by a state of things itself created by a certain course, because, as it is, a state of things may last shorter or longer (Ingarden 1987a, p. 189).

An *event* itself is a *happening* of some state, its *coming into being* (Ingarden 1987a, pp. 189-190).

Ingarden remarks that the "existence of an event" does not have to be "punctual."
Whether it is really so still depends on an answer to the question of whether moments are only "points" in a one-dimensional time continuum as mathematical-geometrical conception of time demands or, conversely, they are particular time quanta which mark their separateness in the course of time, and at the same time they are neither a time *point* nor time *segment*. What is advisable here is that one should be particularly cautious and that one should still wait a while before resolving this problem. Anyway, an event does not exceed the span of *one concrete now*. (Ingarden 1987a, p. 190).

According to the above definitions – things, processes and events are mutually exclusive, i.e. no element of the structure of reality can fall under more than one of these definitions.

III. Do All Three Kinds of Temporal Objects Exist?

Three points of view on the question of the nature of basic components of reality represent an extreme, reductionist approach: each of them recognizes the existence of only those objects that fall under one of the ontological categories.

What speaks in favour of reism, which is supported by the common experience and also has a long and still vivid tradition in philosophy includes the fact that without objects lasting in time both processes and events seem to be "suspended in a vacuum," i.e. devoid of their natural carriers.

However, the autonomous dynamism (and in particular energetism), which allows for the existence of a "motion without that which moves" and "changes without that which changes" (the substance of the world is "energy" deprived of any carrier, either material or spiritual) has its own proponents not only among philosophers but also natural scientists, especially physicists. This trend has found some support; the support came from many as five physical theories and research areas and programmes connected with them.

(1) The first one is classical electrodynamics which initiated field conceptions in physics. The field, being the structure of an infinite number of degrees of freedom, spreading with light velocity does not have

distinctly marked carriers of motion. Indeed, there are field quanta; however, according to "uniform" conceptions of field, field quanta are only certain of its singularities, constituting the condensation of energy.

(2) The next theory is the so-called Wheeler's geometrodynamics, according to which all material structures may be reduced to the spatio-temporal structure which abounds in variable properties (whereby, strictly speaking, Wheeler's geometrodynamics is not so much a ready-made theory but more of a research programme which tends to reduce any physical properties of objects to spatiotemporal, chronogeometric characteristics).

(3) The third theory is quantum mechanics (and microphysics in general) with its wave-corpuscle dualism. According to the Copenhagen interpretation of this theory, a microobject is in itself neither a corpuscle nor a wave, that is exists only "potentially" and becomes a particle or a wave only in contact with a macroinstrument of a given kind, with the reservation that its full (complete) description requires (according to Bohr's complementarity principle hat is taken into account) both kinds of properties: those of corpuscles and those of waves. Therefore, from the point of view of the principle of the visuality – it is a bi-model one. At the same time, a microobject, even revealing (in a definite experimental situation, i.e. in contact with a macroinstrument of a proper kind) its corpuscular properties, is not a particle in the full (classical) sense of the word, since – according to Heisenberg's principle of indeterminacy – its location and momentum (constituting two physical attributes of being a corpuscle) are not precisely determined at the same time. Consequently, microobjects (microphenomena) have indeed their own corpuscular aspect; however, it is not distinctly marked, which is a problem for reism.

(4) The next support for autonomous dynamism (being in discord with reism) comes from quantum electrodynamics, according to which the moving microobject is surrounded by a cluster of quickly rising and vanishing virtual corpuscles, which in addition also rise in the process of the fluctuation of a vacuum. These are corpuscles which for their existence in the form of ordinary elementary corpuscles perceived in experience, must "supplement" a certain "shortage" of energy which characterizes them. Therefore, their existence as corpuscles (carriers of features, motions and the like) is even more troublesome than the existence of ordinary elementary corpuscles.

(5) Finally, the special relativity theory, by means of its fundamental result contained in the Einstein mass-energy equivalence law ($E = mc^2$), provided a new argument to the adherents of energetism who proclaim either a thesis that mass and energy are mutually transformable, or a thesis on the identity of mass and energy, clearly formulated by Eddington. According to Eddington, the difference between mass and energy is purely subjective and comes from the fact that both physicists and chemists dealt

with this property, with the exception that in physics it occurred under the name of "energy" and obtained a different system of measuring units (erg and the like) than that in chemistry (gram and the like) where it received the name of "mass." According to Eddington, the unification of the system of measuring units transforms Einstein's formula into a simple identity $E = m$, which points to the identity of mass and energy and which makes the law of conservation of mass (formulated by chemists) and the law (principle) of the conservation of energy (discovered by physicists) one and the same law.

None of the above-mentioned arguments for autonomous dynamism is indisputable and has a full power of proof. Rather, they indicate only that the reduction of the basic components of the world to things (or, to put it more precisely, to objects existing in time) is not possible to conduct and that – at least on the grounds of physics – both processes and events have a full right of existence.

IV. The Role of Processes in the Natural and Social Realm

The modern stage of the development of sciences of all types is characterized by a dynamic approach to the spheres of reality under study. It reveals the role of processes in the structure of the studied objects.

An exact analysis of the mechanical motion (Newton) and changes in general (Leibniz) led to the creation of differential and integral calculus, on the basis of which mathematical analysis was developed which has provided means for the exact quantitative description of dynamic aspects of physical phenomena.

Physics had studied static systems in the antiquity (Archimedes' statics) and then shifted to the study of systems in motions (Galileo-Newton dynamics). At the beginning, it focused only on the study of reversible processes (classical mechanics, Faraday-Maxwell electrodynamics), to perceive in the 19th century the importance of irreversible processes, determining the time "arrow" (classical thermodynamics of Carnot-Clausius).

At the same time, the ever more pronounced paradigm of an evolutionary interpretation of nature, initiated by Kant and Laplace (cosmogonic hypothesis of the rise of the solar system), which through observing transformations occurring in the Earth's crust (Lyell) led to the formulation of the theory of evolution of species (Darwin).

This trend encompassed philosophy and the humanities. The cult of what is eternal and unchanging, which is so characteristic of the ancient (Plato, Aristotle) as well as mediaeval philosophy and science (St. Augustine, St. Thomas), gives way to the perception of the rank of that which is the result of a long-lasting evolution of nature (Lamarck, Darwin) and social progress (Hegel, Comte, Marx).

The processual interpretation of reality became a distinguishing factor of the modern history and other sciences of society as well. This is attested by Hannah Arendt:

> In the same way and due to the same reasons interest has been shifted from things to processes, as a result of which things were soon to become almost accidental by-products of processes. . . . Since the 17th century, the main subject of attention of all scientific studies, both natural and historical, were processes (Arendt 1994, p. 74).

At present the importance of processes is perceived even more distinctly than in modern science and philosophy. This is so both because of science and technology. Two fields of physics contributed to it in particular: microphysics and the (nonlinear) thermodynamics of open systems that are far from equilibrium (Prigogine). The mutual transformability of elementary particles and the arguments that were cited previously for autonomous dynamism demonstrated that the processual character of reality is visible on a microlevel yet more distinctly than on the macroscopic level. This is emphasized by one of the best-known proponents of processualism, A.N. Whitehead in his book *Process and Reality* (Whitehead 1929). On the other hand, the thermodynamics of dissipative systems (Prigogine) and synergetics (Haken) showed that processes of self-organization are characteristic of all systems that are far from equilibrium, both those that are living systems and those which are "purely" physical and chemical systems.

In turn, the rank of processes in modern technology arises from the fact that apart from traditional technology, based mainly on mechanical processing of manufactured goods, a powerful field emerged, composed of various technologies using physical and, especially, chemical processes for the production of new materials and processing the means for living.

The negative aspect of the processual approach, especially of starting a number of technological processes by humans, is obviously the environmental pollution. Also, for the first time in history, the human being became capable of creating the processes the starting of which on a full scale would mean self-annihilation.

SCIENCE AND TECHNOLOGY AND THE CURRENT TRENDS IN THE DEVELOPMENT OF CULTURE*

1. As is well-known, the adaptation of animate nature to conditions of existence occurs in humans in a form that is different from that which occurs in other living beings. Humans adapt their surroundings actively according to their needs, while the other species in general adapt themselves in a rather passive way to the conditions in which they live. Undoubtedly, science and technology support and even substantially reinforce this trend of active transformation of living conditions of humans, according to their needs. Science, and especially modern experimental science has been involved in the process of transforming nature and creating an artificial environment of human existence. Scientific experiment may be recognized as the kind of activity which from the cognitive point of view prepares the human for active interference with the course of events in nature.

2. This points to another trend in the development of the human being, strengthened by science – a trend that is peculiar to the human being since the very beginning, and even observed already in its embryonic form also in animals: the cognitive trend aiming at the most adequate reflection of what occurs.

Science contributes primarily to the independence of that trend, which was initially completely subordinated to the first trend – that of the transformation of reality. In view of the two basic functions of science: the practical one, consisting in supplying people with premises for predicting as a *sine qua non* condition of the efficient activity, and the theoretical one, aiming at explanation of phenomena, the cognitive trend in science is always present – in one way or another – in connection with the trend transforming the reality. It is valid even in reference to spheres that are seemingly separated from practical life, such as religion or philosophy, which aim at change because of their axiological and normative dimension. However, in this case they do not aim at changing the surrounding world but at changing the human being him/herself, i.e. at the formation of human personality.

---

* Translation of "Nauka i technika a trendy rozwojowe obecne w kulturze," published in J. Such, J. Wiśniewski (eds.) (1996), *Kulturowe uwarunkowania wiedzy* [Cultural Conditions of Knowledge]. Poznań: Wydawnictwo Naukowe IF UAM, pp. 9-12.

In science, strengthening the cognitive trend, which is so specific to the human race and its growing independence, occurred not so much thanks to the formation of experimental sciences but rather due to the division of scientific research into fundamental and applied studies. Since then theoretical sciences forced the practical tasks out in the sense that immediately only the purely cognitive goals are their guiding principle, rather than the applications of the knowledge obtained. Thus, the care for the development of the trend of transforming reality first of all became a part of practical and in particular technical sciences. Due to the fact that all knowledge – and, in any case, scientific knowledge – may be, and actually is, as proved by the history of science and technology, reforged into practical skills; the two trends discussed so far cannot stand in opposition, i.e. they cannot be mutually exclusive.

3. The third trend that is characteristic of human nature will be termed the contemplative trend. It consists in the human being's focus on him/herself in delineating his/her place in the world and in an attempt at self-improvement.

It is my conviction that science and technology have given up this trend. It is clearly visible if we compare the highly contemplative Asian cultures and a number of other cultures with the European culture, which is the only one based on science and technology.

Science and technology have been coping very well with the surrounding natural world for three centuries but they still poorly cope with the society, and even worse with the human being – his/her spiritual life and personality. This helplessness concerns both the above dimensions. No wonder then that the most interesting and the most thorough knowledge on people still seems to come not from scientists but from writers and artists, philosophers and theologians. It is also they who, together with social workers and politicians, and not scientists and technicians, have the most say in the sphere of transformations concerning society and human nature. The so-called social engineering, consisting in the application of scientific knowledge to the study of social and personal transformations, if it exists at all, it is still in its infancy.

The human as a social and spiritual being is still located within the domain of religion, philosophy, ethics and art, i.e. those fields that are well-developed in every culture, which has achieved a definite level of maturity, so also in European culture. However, in the latter they have been partly ousted and dominated by science and technology, which caused the contemplative trend to remain in the shadow of the transformational and cognitive trends.

4. On the example of fate of the three above-mentioned main trends, determining directions of the development of the human and society, it can be seen that science and technology fundamentally influence people's endeavours and aspirations. On the one hand, they essentially strengthen

two of them. On the other hand, they distinctly weaken the human tendency that is so visible in the ancients, e.g. in Socrates, to focus on him/herself, and not on the outside world. This tendency is now quickly supplanted in Asian and other cultures which have also developed science and technology. It can be supposed that in future, humankind as a whole will go in the direction set by the Euro-American culture and civilization.

Since the scientific-technological culture poses very concrete and, by now, already well-recognized threats, the process of combining various cultures into one – the culture of all human beings, quickly developing on the basis of science and technology is undoubtedly a dangerous phenomenon. To put it graphically, it can easily lead us to conquer Mars but at the same time to destroy our cradle, which is the Earth.

It would be a Utopia to try to "free" European culture from science and technology and, at the same time, from the heritage which they bring with them. However, it is perhaps not a Utopia to strengthen considerably those spheres of spiritual culture of which the contemplative trend consists and which determine the axiological dimension of human life.

The conviction that human culture is ultimately bound to become axiologically sterilized, and such of its spheres as religion, ethics, philosophy and art are doomed to decay, does not have to be true.

However, it is still a fact that the main distinguishing factor of the scientific-technological culture – such as has so far been shaped – is that it provides the ever more powerful means, making human activities more and more efficient but does not provide a distinct compass, which would indicate how to use these means and in which direction to go.

On the other hand, the ascientific and atechnological cultures are less involved in taking care of the means, and are more concentrated on goals that are worth carrying out. Thus, the contemplative component results in culture acquiring a clearly axiological dimension, which is shadowed in the scientific-technological culture by the prevalence of the transformational and cognitive components.

It is surely some tragic decree of fate that the scientific-technological culture, of which we are heirs, lacks a distinct guidance just at the moment when it is capable of annihilating itself, and with it all life on the Earth. In this situation, it seems that when humankind has ever more powerful means of destruction at its disposal, it is possible to avoid self-destruction only when at least one of the two (non-mutually-exclusive) conditions is fulfilled. The first is that people develop some common, universal systems of values, i.e. such systems which will in any case not differ much with respect to the fundamental, highest values. The second one is that such a level of tolerance is obtained which will allow the co-existence of human societies, regardless of the differences in their ways leading to the future.

HEGEL'S CATEGORY OF TOTALITY
AND HIS CONCEPT OF STATE*

I. State as an Organized Whole

In his *Encyclopaedia of the Philosophical Sciences* Hegel writes:

> The state as a living spirit exists without fail only as a certain organized whole subdivided into particular functions which, stemming from *one* term of rational will (although not yet known as a term), produce it constantly as their result. A *political system* is the very fragmentation of the *state authority*. It contains a description of how the rational will, as far as in individuals it is the general will only *in itself*, reaches on the one hand awareness and self-understanding and *finds* itself, while on the other hand through the actions of the government and its particular branches it is embodied in reality and kept in it, as well as defended from accidental subjectivity of both the government and individuals. It is an existing *justice* as the reality of *freedom* in which all its rational descriptions are developed (Hegel 1990, p. 528).

If the form of the government is the fragmentation of the state's authority, then the state's binding agent is the government which is after all a "general part of the political system." That is because, not much further on, we read:

> The living totality, the maintaining, i.e. constantly producing the state in general and its political system is *the government*. The forming of *family* and *classes* of the civic society is necessary in the natural sense. The government is a *general* part of the political system, i.e. the part that has the conscious purpose of keeping the other parts but at the same time it takes into consideration and fulfils the general goals of the whole, goals which stand above the designations of family and civic society. The structure of government encompasses also its fragmentation into powers, in accordance with how their specific qualities are defined on the basis of the term but in its subjectivity they penetrate each other creating a *real* unity (p. 532).

Thus the government "realizes general goals of the whole, those which stand above the designations of family and civic society" and which are the society's "natural structure."

II. State as a Fragmented Whole

State, as seen by Hegel, is not however, a totalitarian state, despite what Popper thinks (see Popper 1993b, pp. 9-87). This is evident from the very

---

* First published in A. Arndt, K. Bal, H. Ottmann (eds.). *Hegel – Jahrbuch 2000. Hegels Ästhetik. Die Kunst der Politik – Die Politik der Kunst*, Zweiter Teil. Akademia Verlag, pp. 154-156. This text is reprinted with the kind permission of Akademia Verlag.

fact of the fragmentation of the state authority, which Hegel calls the political system. This is because the political system defines how the rational will

is embodied in reality through activities of the government and its specific branches, and kept in reality as well as defended from accidental subjectivity of both the government and individuals. The political system is an existing *justice* as the reality of *freedom* in which all its rational descriptions are developed (Hegel 1990, p. 528).

This fragmentation of state authority means

a division of the state's functions" into "legislative power, the administration of justice, or the judicature, Civil Service, police, etc. – and at the same time its *division* between different organs of power, which are equipped in rights appropriate to their functions and for that purpose and reason have independence in their functioning, while remaining under higher supervision (p. 535).

That Hegel's state is not a totalitarian one is manifested also by the fact that

it witnesses a participation in state matters of a *greater number of people*, who altogether make up the general state (§ 528) as long as they make the matter of general goals the actual definition of their particular life, with the further condition that in order to individually take part in that matter they must have the education and the talent to do so (p. 535).

The "higher supervision" over the legislative, judicial and executive powers mentioned above is, according to Hegel, performed by the monarch. Hegel writes:

In governing as a limited totality (1) *subjectivity* exists as . . . deciding the will of the state, its highest peak, similarly to an all-penetrating unity – *the princely* power of governing. In the perfect form of state, in which all moments of notion have reached their free existence, this subjectivity is not the so-called *moral person* or a certain decision-making *coming from the majority of votes* – a form in which the unity of the deciding will does not have a *real* existence – but as a real individuality it is the will of *one* deciding individual – it is *the monarchy*. This is why the political system of a *developed* mind is the monarchic system. All the other systems belong to lower forms of development and self-realization of the mind (pp. 533-534).

Thus, the best political system according to Hegel is not the democratic system (because "the participation of all in all matters, taken for itself stands in opposition to the principle of *division* of powers, i.e. the developed freedom of moments of notion"; p. 534), nor the totalitarian system but the monarchic system. Because the real unity of the formed variety resulting from the division of powers

is only the individuality of the monarch – existing in one person subjectivity of abstract, ultimate determining. . . . That subjectivity being a moment of abstract determining in general defines itself further, on the one hand, so that the monarch's name appears as an external bond and a sanction consecrating all that takes place in governing, and on the other hand, as a simple self-reference it has in itself the attribute of *directness*, and thus *nature*, and because of that the assignment of individuals to the honour of princely power is determined by *hereditary* means (p. 534).

The third, and perhaps most crucial indication that imputing totalitarianism to Hegel is wide of the mark, is the role which, according to him, is played by such categories as freedom and equality, and especially the former, in determining the political system of a state. On this matter Hegel writes:

Freedom and *equality* are simple categories, which are often treated as the essence of what should be the basic definition and the basic goal and result of a political system. It is true in principle but in these formulations there is also a fallacy stemming mostly from the fact that they are totally abstract. It is they which fixed in the form of abstraction do not allow the constitution of what is concrete, i.e. the fragmentation of the state, or *political system* and government at all, or destroy them. Together with the state is created inequality, the difference between the governing power and the governed, superior authority is created, along with offices, administration etc. The consequent principle of equality rejects all differences and in this way does not allow the existence of any statehood. ... As for freedom, it is usually taken partially in the *negative* sense in opposition to alien anarchy and lawless treatment, and partially in the *positive* sense, as *subjective* freedom. However, this freedom is given a very wide range, so that it encompasses both license and acts furthering one's own particular ends, as well as the right to own opinion on general matters and to caring for them and participating in them (pp. 528-529).

Noticing this antinomy (the impossibility of joint fulfilment) of freedom and equality Hegel writes:

it is in reality the high level of development and education in modern states which creates the highest concrete *inequality* of individuals, while on the other hand through deeper wisdom of laws and of guarantees of legal state it realizes an even more entrenched freedom, one which it can allow and tolerate. Even the superficial differentiation contained in the words freedom and equality indicates that the former sends towards inequality (p. 530).

The fourth piece of evidence against the truthfulness of attributing totalitarianism to Hegel is the role that he ascribes to constitution.

The *guarantee* of constitution – writes Hegel – i.e. the necessity of laws being rational and their realization ensured, is contained in the spirit of the whole nation – namely in the definiteness according to which a nation has the self-knowledge of its reason (religion is that consciousness in its absolute substantiateness) followed by *real organization*, corresponding to that spirit, as an *expansion* of that principle. Constitution assumes as its preceding premise that very consciousness of the spirit and vice versa, the spirit takes that constitution for granted, as the real spirit alone has awareness of its principles only so long as they exist for it as existing (p. 531).

The above evidence clearly shows that the assertion formulated by Popper and accusing Hegel of being – next to Plato and Marx – one of the three main representatives of the current in social philosophy called historicism is an accusation which is totally unfounded. At the same time, the suspicion that Hegel was guilty of totalitarianism and of a negative attitude towards the possibility of democratic reforms loses its validity (cf. Popper 1993a, pp. 21-22 and 31, and Popper 1993b, pp. 35-87).

Contrary to what Popper believes, "Hegel's holism and his organic theory of state" are far from meaning that Hegel was a promulgator of totalitarian political system – a totalitarian state. According to Hegel's scheme of history the lowest "first political form we see in history" is the despotic form. In *Lectures on the Philosophy of History* Hegel writes:

The East knew ... only that Individual is free; the Greeks and the Romans that some are free; the German world knows that everybody is free. The first philosophical form which we can see in history is *despotism*, followed by *democracy* and *aristocracy*, followed by *monarchy* (Hegel 1958a, p. 155).

III. Conclusion

As we have seen, the fact that Hegel understands the state as an organized but at the same time fragmented whole does not provide grounds to accuse him of totalitarianism. This is because in Hegel's dialectics "the domination of the whole over the parts," "formulating all the partial phenomena as moments of the whole" does not stand in the way of seeing the significant role of differentiation and fragmentation of the whole.

Also, his famous concept of history as a process of becoming aware of one's freedom so that each nation and each age reach only a certain limited awareness of freedom because of which they must yield the palm to other nations, which are more deeply and better aware of the freedom, is a testimony to Hegel's and his philosophy's general antitotalitarian attitude (see Such 1992, pp. 95-100).

Hegel, a theoretician of the civic society, clearly saw its role in the proper functioning of the state. As demonstrated by later history of the European civilization, civic society, which Hegel perceives as among others "a system integrating people," has remained of elementary importance in the process of warranting civic freedoms and, at the same time, in consolidation of democracy.

THE BEING OF BEINGS IN HEIDEGGER'S
SEIN UND ZEIT

I. Introduction

In this essay, I intend to focus my attention on the kinds of beings discussed in Heidegger's *Sein und Zeit*, and the modes of their Being without engaging in a consideration of the genetic relations between particular beings. I will place special emphasis on the differences between the particular beings, and in a similar way between the modes of their Being. I do not deny, however, that the former set of differences (between the kinds of beings) can be reduced in some way to the latter type of differences (i.e. between the modes of Being of beings).

I will develop my reflections as a person whose major interest is the philosophy of natural sciences, especially the philosophy of physics, astronomy and cosmology, and therefore a person who is not interested mainly in existence (*Dasein*) and its Being but in nature and its Being. Accordingly, this is not the existential and ontological viewpoint of Heidegger but a view which Heidegger would probably call a (purely) ontic view or approach.

Moreover, I believe that the reflections presented in *Sein und Zeit* are of great interest also to those who do not concentrate on humans but who deal with the world of nature and technology partly created by humans and which surrounds them, and who approach this world from the perspective of natural scientists.

I also think that such a starting point for my reflections is in some respect justified by the fact that – unlike Heidegger– I shall not begin from the Being of beings but from the beings themselves.

II. Kinds of Beings in Sein und Zeit

Heidegger's ontology, a proposed fundamental ontology of human beings (that is, an existential analytics of *Dasein*), was intended as a general ontology which – unlike the ontology of Jaspers, which is consciously and entirely concentrated upon the analysis of human existence – would not confine itself to the ontology of human Being. For the latter, human existence is only a starting point, whereas the goal is the Being of all beings, and therefore being as a whole.

In the last section (§83) of *Sein und Zeit* Heidegger writes: "The seemingly clear difference between the Being of an existing *Dasein* and the Being of a being other than *Dasein* (presence for example) is merely a point of departure for ontological problems and not something that would satisfy philosophy" (Heidegger, p. 609).

Although Heidegger in his work did not arrive at the "idea" of Being in general, and likewise, at the "idea" of universal being, "on his way" he outlined the kinds of Being and beings which could compose such "ideas."

The point of departure for Heidegger – and I should think for existentialism in general – is the discrimination between Human being (*Dasein*) and non-human being (*Natur*). Human being, as primordially and authentically historical, is also called history, and the non-human being is called nature. In non-human being, the being of things present-at-hand (real), and the being of things ready-to-hand (tools) are distinguished. That which is (in the broad sense of this word) can "exist" in three manners: *existiere* (*Dasein*), be ready-at-hand (*Zuhandensein*), (equipment/*das Zuhandene*), or be present-at-hand (things/*das Vorhandene*).

Human being (*Dasein*) is an entity that is ontologically distinguished both in the sense that – as our own being – it is to us the only being that is entirely concrete, and through the fact that certain characteristics of other beings and their modes of Being (e.g. temporality and historicality) are derivative of the characteristics (existentials/*die Existenzialen*) of *Dasein* and its mode of Being.

Moreover, compared to *Dasein*, not only ideas but also tools and things are abstracts. What are some of other features of Human being?

The third feature of Human being is the fact that it is the being of creatures who are conscious and always have some attitude towards their own existence. They are distinguished by consciousness of their existence and understanding of their Being. I constitute a being which is in its substance different from the being of all things, because I am able to say about myself: I am, states Heidegger. The essence of *Dasein* is its existence, i.e. Being through various manners of "taking up" one's Being.

The fourth feature of Human being is the fact that it is the being of entities full of care about their Being and, to a certain extent, making decisions about it. Humans – being "thrown into life" – indeed have no influence on their births and coming to existence but they can either accept or condemn this existence.

The fifth distinguishing feature of Human being is that it is not an isolated being but a being existing-in-the-world, connected with the world, which means (1) connected with other people, and (2) connected with things, particularly with things ready-to-hand, that is tools.

The sixth distinguishing feature of Human being is that humans not only exist-in-the-world but also know it, hence are conscious not only of their own Being but also of the Being of others (i.e. the Being of other people and the Being of the world of things: ready-to-hand and present-at-hand).

Finally, the seventh important distinguishing feature of Human being is that it is a being that is authentically (in a certain primordial sense) temporal, and therefore, also historical.

A certain fundamental unitary characteristics of *Dasein*, determining its attitude towards other beings, is its ability to transcend. Transcendence is the fundamental character of *Dasein*. It means "transcending being," because of which we can only secondarily discuss the existence or reality of beings other than *Dasein* as other beings. Through transcendence, *Dasein* is able to relate both to its own being, and to other beings. Since *Dasein* transcends, and so exists-in-the-world, it is able to relate to particular beings, including its own (Baran, p. XIX).

Unlike *Dasein*, things (both ready-at-hand and the merely present-at-hand) are characterized by: (1) the unconsciousness of their existence (Being), (2) the inability to determine their existence (Being), (3) the inability to know the world, and (4) only derivative temporality and historicality (possessed in nature and technology merely through their relation with humans that is to say through the fact of Being-in-the world).

If characterizing the beings, human and natural, separately from characterizing their Being (Being of beings), presents considerable difficulties, the problem is even greater when, in a similar manner, one tries to establish the differences between things present-at-hand and things ready-to-hand. Things ready-to-hand (tools) differ from things merely-present-at-hand (i.e. phenomena and objects of nature), above all, by the fact that they possess "equipmental constitution" called "assignment" (*Verweisung*). Individual "assignments" are "serviceability-for," "harmfulness," "applicability" etc. Serviceability "for" and applicability "for-the-purpose-of" always constitute a possible concretization of the reference. What distinguishes tools (e.g. hammering with a hammer or "indicating" with a sign) is not, however, a feature (or any other definiteness) of things. Something that is ready-to-hand always has its suitability and unsuitability (Heidegger, p. 118). However, serviceability (reference) as equipmental constitution is not merely the suitability of some being but a condition pertaining to Being, about which the being can be characterized by suitabilities. Beings are discovered because, as entities, they are assigned to something. It is involvement (*Bewandtnis*) that is the nature of Being of the ready-to-hand. The fact that it is endowed with this involvement makes up the ontological description of Being of this being but is not an ontic statement on being, as Heidegger remarks (Heidegger, p. 119). In this way, the ontic characteristics of a ready-to-hand being evolve into an ontological characteristics of the Being of that being.

Further on, Heidegger emphasizes the unitary nature of involvement in which a given tool is involved. At the same time, the involvement whole (*Bewandtnisganzheit*), which constitutes for example the readiness-to-hand of something ready-to-hand in a workshop, is "prior" to the individual tool.

The impossibility of characterizing a tool as a being without continuing on to a characterization of the mode of its Being, causes me to confine my discussion, in the following paragraph, to the difference between the modes of Being of beings to the Being of human being (*Dasein*) and non-human being (nature).

III. Being of Beings: Human (Dasein) and Non-Human (Natur)

The degree of the difference in Heiddegger's view between the Being of *Dasein* and *Natur*, is manifested by the fact, among others that in order to characterize each of these kinds of Being he uses a separate set of notions, hence a different terminology.

When discussing the Being of *Dasein*, Heidegger does not use the term "structure," which would suggest an "occurrent" nature of being. Instead, he prefers to use the term "constitution-formulation" (*Verfassung*). What he means is the constitution-formulation of Being of Dasein (*Seinverfassung*). In this way, the moments of *Dasein's* structure are seemingly resolved into its "modes of Being." *Dasein*, however, is each time related to its "modes of Being." Each of its structures is a kind of "formulation" (*Verfassung*), "taking up" of Being (Baran 1994, p XVII). The universal form of Being of Dasein that distinguishes it from the Being of nature, is its existence. This consists in various ways of "taking up" one's Being but first of all, in understanding one's Being. *Dasein's* understanding of its own Being, of the fact that it itself is, being the fundamental moment of *Dasein's* Being, its elementary assignment to Being, makes other modes of its Being possible, or in other words its assignments (*Verweisung*) to Being. In "experiencing" by *Dasein* that "it is" (i.e. its "throwness"), lies the fundamental "affectedness" (*Befindlichkeit*) of *Dasein*, as a result of which the world appears to Dasein as the world, and therefore an intraworldly being (Baran 1994, p. XVIII).

Although Being "needs" *Dasein* in order to "realize" itself, the founding role of understanding of Being, and so the assignment of *Dasein* to Being, results in the fact that being-there (*Da-sein*), *Dasein* simply means Being (Baran 1994, p. XVIII).

I will not engage in the discussion the specific modes of Being of *Dasein* underlying the collective word care (*Sorge*). I will only mention that one of the important features of the Being of *Dasein* is that all the modes of its Being are divided into authentic and inauthentic, and that the inauthentic Being is always founded on the authentic Being (Baran, p. XIX).

What, in turn, can be said about the Being of a being other than *Dasein*: the Being of nature (in a broad meaning of the word including tools)? This being exists merely in a certain derivative sense, because anything that "Is" is primordially *Dasein*. For "to be" in the primordial sense means to exist, and to exist means to "project" one's possibilities. Nevertheless, the non-human being – the being of things and tools – is a real being in the sense that it is present-at-hand and ready-to-hand.

IV. Concluding Remarks

It is very interesting that Heidegger's approach to the relation of Human being to non-human entities (natural) – an approach undoubtedly inspired by the transcendental philosophy initiated by Kant and Fichte – markedly corresponds to the natural sciences and their philosophy. This correspondence lies in the conviction that physical reality yields to purely objective description, as realized by classic physics, exclusively on the level of macrocosm. Therefore, the description of reality on other levels, i.e. on the microcosm level, on one hand, and on the cosmic level, on the other – inevitably entails visible subjective elements, and this suggests that even the world of nature is in its existence dependent on the conscious observer.

In quantum physics, it is manifest in the fact that description of a microobject is never a pure, objective description of the state of the microobject but a simultaneous (inseparable) metadescription of our knowledge pertaining to this state. This is one of the most vital differences between classical physics and quantum physics; due to this difference, physics (and, at any rate, the physics of the microcosm) was compelled to renounce the ideal of an entirely objective description, which was the guiding principle for classical physics. As is widely known, scientists, even of such a high calibre as Planck or Einstein were unable to accept this renunciation.

This means that a microobject described by quantum mechanics cannot be isolated (even mentally) from either the situation in which its placed or from the "factors" which are at least partly mental, such as observational and measuring procedures, and the state of knowledge of the observer. The fact of the inadmissibility of such isolation is stressed by the notion of observational situation (*Beobachtungssituation*) introduced by Heisenberg (see Heisenberg 1959, p. 82).

The notion of an observational situation presupposes the inseparability of the object and the subject of observation, that is their unity. Only viewing them as a unity makes it possible to include in the description the knowledge that the subject has about the object.

Hence, the observer appears twice in reflections on the interpretation of quantum mechanics: first in connection with the macro measuring instrument manipulated by the observer, and second with to the role of the knowledge, which participates in the quantum mechanical description.

Consequently, the first function of the observer consists in the "physical effect" (hence it may be replaced by the analogical function of a measuring instrument), the second function consists in the "mental effect" (the influence of knowledge).

On the other hand, the state of affairs discussed here is encountered in contemporary cosmology, in the area of anthropic cosmology, according to which the fact of human existence correlates with the fact of the existence of the Universe in some important way.

REFERENCES

Ajdukiewicz, K. (1960). *Język i poznanie* [Language and Knowledge], vol. 1. Warszawa: PWN.

Ajdukiewicz, K. (1965). *Logika pragmatyczna* [Pragmatic Logic]. Warszawa: PWN.

Amsterdamski S. (1965). *Mechanika kwantowa a materializm* [Quantum Mechanics and Materialism]. *Posłowie* [Afterword] in Heisenberg (1965).

Amsterdamski, S. (1973). *Między doświadczeniem a metafizyką. Z filozoficznych zagadnień rozwoju nauki* [Between Experiment and Metaphysics. Some of the Philosophical Problems of the Development of Science]. Warszawa: PWN.

Arendt, H. (1994). *Między czasem minionym a przyszłym* [Between the Past and the Future]. Vol. I. Warszawa: Aletheia.

Aristotle (1968). *Fizyka* [Physics]. Warszawa: PWN.

Atkatz, D., H. Pagels (1982). Origin of the Universe as a Quantum Tunnelling Event. *Physical Review*, **D 25**.

Augustynek, Z. (1964). Determinizm fizyczny [Physical Determinism]. In: S. Amsterdamski, Z. Augustynek, W. Mejbaum, *Prawo, Konieczność, Prawdopodobieństwo* [Law, Necessity, Probability]. Warszawa: PWN, pp. 125-221.

Banaszak, G. and Kmita, J. (1991). *Społeczno-regulacyjna teoria kultury* [Socio-regulative Theory of Culture]. Warszawa: Wydawnictwo Instytutu Kultury.

Bańka, J. (1992). *Medytacje parmenidiańskie o pierwszej filozofii. Recentywizm i pannyngeneza* [Parmenidian Meditations on the First Philosophy. Recentivism and Pannyngenesis]. Katowice: Wydawnictwo Uniwersytetu Śląskiego.

Baran, B. (1994). Wprowadzenie [Introduction]. In: Heidegger (1994), pp. XI-XXXI.

Barrow, J.D. (1994). *The Origin of the Universe*. London: Orion Publishing Group Ltd.

Bażański, St., M. Demiański (1967). Zasada Macha [Mach's Principle]. *Studia Filozoficzne*, **2**, 145-160.

Bohr, N. (1913). On the Constitution of Atoms and Molecules. *Philosophical Magazine*, **26**, 1-25; 476-502; 857-875.

Bohr, N. (1920). Über die Linienspektren der Elemente. *Zeitschrift für Physik*, **2**, 423-469.

Bohr, N. (1923). *Über die Qauntetheories der Linienspektren*. Braunschweig.

Bohr, N. (1924). On the Application of the Quantum Theory to Atomic Structure. In: *Proceedings of the Cambridge Philosophical Society*, Part I. Cambridge: Cambridge Univ/Press.

Bohr, N. (1934). *Atomic Theory and Description of Nature*. Cambridge: Cambridge Univ/Press.

Braithwaite, R.B. (1954). *Scientific Explanation*. Cambridge: Cambridge Univ/Press.

Bridgman, P.W. (1958). *The Logic of Modern Physics*. New York: Macmillan.

Brout, R., F. Englert, E. Gunzig (1976). The Creation of the Universe as a Quantum Phenomenon. *Annals of Physics*, **115**, 78-106.

Bunge, M. (1967). *Intuicja i nauka* [Intuition and Science], Moskwa: Progress.

Bunge, M. (1977). Quantum Physics and Measurement. *International Journal of Quantum Chemistry*, Vol. **XII**, Suppl. I, 1-13.

Butterfield, H. (1958). *The Origins of Modern Science 1300-1800*. London: G. Bell and Sons Ltd.

Cackowski, Z. (1972). Działanie, praktyka, poznanie [Action, Practice, Knowledge]. *Studia Filozoficzne*, **1**, 95-118.

424

Careri, G. (1982). *Ordine e disordine nella materia.* Lattera.

Davis, P.L. (1983). *God and the New Physics.* New York: Simon and Schuster.

Drat-Ruszczak, Z. (1987). Twórczość w nauce [Creativity in Science]. In: Cackowski, Z. *et al.* (eds.), *Filozofia a nauka. Zarys encyklopedyczny* [Philosophy and Science. An Encyclopedic Survey]. Wrocław: Ossolineum, pp. 732-734.

Drees, W.B. (1990). Theology and Cosmology beyond the Big Bang Theory. In: J. Fennema (ed.), *Science and Religion: One World – Changing Perspectives on Reality.* Dordrecht: Kluwer.

Duhem, P. (1911). *La théorie physique. Son objet et sa structure.* Paris: Rivière.

Eilstein, H. (ed.) (1961). *Jedność materialna świata* [The Material Unity of the Universe]. Warszawa: PWN.

Eilstein, H., M. Przełęcki (eds.) (1966). *Teoria i doświadczenie* [Theory and Experience], Warszawa: PWN.

Einstein, A. (1917). Kosmologische Betrachtungen zur allgemeinen Relativitätstheorie, *Sitzungbericht. preuss. Akad. Wiss.*, 142-152.

Einstein, A. (1933). *On the Method of Theoretical Physics.* Oxford: Bell and Sons Ltd.

Einstein, A. (1934). *Mein Weltbild.* Amsterdam: Querido.

Einstein, A. (1965). *Fizika i rialnost* [Physics and Reality]. Moskva: Progress.

Einstein, A. (1978a). Science and Religion. In: *Ideas and Opinions.* New York: Dell Publishing.

Einstein, A. (1978b). The Fundamentals of Theoretical Physics. In: *Ideas and Opinions.* New York: Dell Publishing.

Einstein, A., L. Infeld, B. Hoffmann (1938). Gravitational Equations and the Problem of Motion. I. *Annals of Mathematics,* **39**.

Einstein, A., L. Infeld, B. Hoffmann (1940). Gravitational Equations and the Problem of Motion. II. *Annals of Mathematics,* **41**.

Engels, F. (1953). *Dialektyka przyrody* [Dialectics of Nature]. Warszawa: KiW.

Feyerabend, P.K. (1962). Explanation, Reduction and Empiricism. In: H. Feigl, G. Maxwell (eds.), *Minnesota Studies in the Philosophy of Science,* vol. III. Minneapolis: University of Minnesota Press, pp. 29-97.

Feyerabend, P.K. (1963). How to be a Good Empiricist – A Plea for Tolerance in Matters Epistemological. In: Baumrin B. (ed.), *Philosophy of Science. The Delaware Seminar,* vol. II. New York: Wiley, pp. 3-39.

Feyerabend, P.K. (1962). Problems of Microphysics. In: *Frontiers of Science and Philosophy.* Pittsburgh: University of Pittsburgh Press.

Fock, V.A. (1939). Sur le mouvement des masses finies d'aprés la théorie de gravitation einsteinienne. *Journal of Physics (USSR),* **1**.

Fock, W.A. (1969). Kvantovaja fizika i filosofskije probliemy [Quantum Physics and Philosophical Problems]. In: *Lenin i sovriemiennoe jestiestvoznanije* [Lenin and the Contemporary Natural Sciences]. Moskva: Progress.

Galileo, G. (1930). *Rozmowy i dowodzenia matematyczne dotyczące dwóch nowych umiejętności w zakresie mechaniki i ruchów lokalnych* [Colloquies and Mathematical Proofs with Respect to the Two New Skills Concerning Mechanics and Local Movements]. Warszawa: Wydawnictwo Kasy im. Mianowskiego.

Galileo, G. (1962). *Dialog o dwóch najważniejszych układach świata Ptolemeuszowm i Kopernikowym* [Dialogue on the Two Most Important World Systems, the Ptolemaic and the Copernican]. Warszawa: PWN.

Geymonat, L. (1966). *Filozofia a filozofia nauki* [Philosophy and the Philosophy of Science]. Warszawa: PWN.

Giedymin, J. (1964). *Problemy, założenia, rozstrzygnięcia* [Problems, Assumptions, Solutions]. Poznań: PWN.

Giedymin, J. (1966). O teoretycznym sensie tzw. terminów i zdań obserwacyjnych [On the Theoretical Sense of the so-called Observational Sentences]. In: H. Eilstein, M. Przełęcki (eds.), *Teoria i doświadczenie* [Theory and Experience]. Warszawa: PWN, pp. 91-110.

425

Gordon. M. (1966). *Poznanie prawomocne a wiedza o świecie* [The Legitimatic Cognition and Knowledge about the World]. Warszawa: PWN.

Grünbaum, A. (1973). *Philosophical Problems of Space and Time.* Dordrecht, Boston: D. Reidel.

Griszczuk, L.P., J.B. Zeldowicz (1982). Complete Cosmological Theories. In: M.J. Duff, C.J. Isham (eds.), *The Quantum Theory of Space and Time.* Cambridge: Cambridge University Press.

Hartle, J.B. and S.W. Hawking (1983). Wave Function of the Universe, *Physical Review,* **D 28**, 2960-2975.

Hawking, S.W. (1988). *A Brief History of Time: From the Big Bang to Black Holes.* Toronto, New York: Bantam Books.

Hawking, S. (1990). *Krótka historia czasu* (Polish translation of *A Brief History of Time*). Warszawa: „Alfa".

Hawking, S.W., R. Penrose (1970). The Singularities of Gravitational Collapse and Cosmology. *Procedings of the Royal Society,* **A 314**, London, 529-548.

Hawking, S.W. and R. Penrose (1996). *The Nature of Space and Time.* Princeton: Princeton University Press.

Hegel, G.W.F. (1929). *Sämtliche Werke.* H. Glockner, Bd. VIII. Stuttgart: Friedrich Frommann Verlag.

Hegel, G.W.F. (1958a). *Wykłady z filozofii dziejów* [Lectures on the Philosophy of History]. Volume I, Warszawa: PWN.

Hegel, G.W.F. (1958b). *Wykłady z filozofii dziejów* [Lectures on the Philosophy of History]. Volume II, Warszawa: PWN.

Hegel, G.W.F. (1965). *Sämtliche Werke.* H. Glockner, Bd. XVII. Stuttgart – Bad Cannstatt: Friedrich Frommann Verlag.

Hegel, G.W.F. (1990). *Encyklopedia nauk filozoficznych* [The Encyclopaedia of the Philosophical Sciences]. Warszawa: PWN.

Heidegger, M. (1994). *Bycie i Czas* (Polish translation of *Sein und Zeit*). Warszawa: PWN.

Heisenberg, W. (1959). *Physics and Philosophy.* London: George Allen and Unwin (Ruskin House).

Heisenberg, W. (1965). *Fizyka a filozofia* [Physics and Philosophy]. Warszawa: KiW.

Heisenberg, W. (1979). *Ponad granicami* [Beyond the Limits]. Warszawa: PWN.

Heisenberg, W. (1987). *Część i Całość* [Part and Whole]. Warszawa: PIW.

Heller, M. (1978). Uwagi o metodologii kosmologii [Remarks on the Methodology of Cosmology]. *Roczniki Filozoficzne,* vol. **26**, 3.

Heller, M. (1994). *Wszechświat u schyłku stulecia* [Universe at the Turn of the Century]. Kraków: Znak.

Heller, M. (1995). *Nauka i wyobraźnia* [Science and Imagination]. Kraków: Znak.

Hempel, C. (1979). Scientific Rationality: Analytic vs. Pragmatic Perspectives. In: T.F. Geraets (ed.), *Rationality Today.* Ottawa.

Hempel, C.G. (1968). *Podstawy nauk przyrodniczych* [Foundations of Natural Sciences]. Warszawa: PWN.

Horton, R. (1992). Tradycyjna myśl afrykańska a nauka zachodnia [Traditional African Thought and Western Science]. In: E. Mokrzycki (ed.), *Racjonalność i styl myślenia* [Rationality and the Style of Thinking]. Warszawa: Wydawnictwo IFiS PAN, pp. 396-453.

Hubble, E. (1958). *The Realm of the Nebulae.* New York: Dover Publishing.

Ingarden, R. (1987a). *Spór o istnienie świata* [A Debate over the Existence of the World]. Vol. I. Ontologia egzystencjalna [Existential Ontology]. Warszawa: PWN.

Ingarden, R. (1987b). *Spór o istnienie świata* [A Debate over the Existence of the World]. Vol. II. Ontologia formalna, Część 1. Forma i istota [Formal Ontology, Part 1. Form and Essence]. Warszawa: PWN.

Jevons, W.S. (1960a). *Zasady nauki* [Principles of Science], vol. I. Warszawa: PWN.

Jevons, W.S. (1960b). *Zasady nauki* [Principles of Science], vol. II. Warszawa: PWN.

426

Kant, I. (1957). *Krytyka czystego rozumu* (Polish translation of *The Critique of Pure Reason*). Warszawa: PWN.

Kedrov, B.M. (1965). *Przedmiot i więź wzajemna nauk przyrodniczych* [The Subject and Mutual Bond of Natural Sciences]. Warszawa: PWN.

Kmita, J. (1966). Uwagi na marginesie problemu sensu empirycznego terminów teoretycznych [Side Notes on the Problem of Empirical Sense of Theoretical Terms]. In: H. Eilstein, M. Przełęcki (eds.), *Teoria i doświadczenie* [Theory and Experience]. Warszawa: PWN, pp. 177-205.

Kmita, J. (1967). Potoczny okres warunkowy [The Common Conditional]. *Studia Metodologiczne*, 3, 33-44.

Kmita, J. (1971). C. Lévi-Straussa propozycje metodologiczne [C. Lévi-Strauss's Methodological Suggestions]. *Studia Filozoficzne*, 3, 127-130.

Kmita, J. (1971). *Z metodologicznych problemów interpretacji humanistycznej* [Methodological Problems of the Humanistic Interpretation]. Warszawa: PWN.

Kmita, J. (1972). Metodologia jako dyscyplina humanistyczna [Methodology as a Humanist Discipline]. *Studia Filozoficzne*, 1, 43-63.

Kmita, J. (1973a). Interpretacja Humanistyczna a wyjaśnianie funkcjonalne [Humanistic Interpretation and Functional Explanation]. In: J. Kmita (ed.), *Elementy marksistowskiej metodologii humanistyki* [Elements of Marxist Methodology of the Humanities]. Poznań: Wydawnictwo Poznańskie, pp 206-227.

Kmita, J. (1973b). Dyrektywa wyjaśniania funkcjonalno-genetycznego [The Directive of the Functional-Genetic Explanation]. In: J. Kmita (ed.), *Elementy marksistowskiej metodologii humanistyki* [Elements of the Marxist Methodology of the Humanities]. Poznań: Wydawnictwo Poznańskie, pp. 237-254.

Kmita, J. (1976). *Szkice z teorii poznania naukowego* [Essays in the Theory of Scientific Cognition]. Warszawa: PWN.

Kmita, J. (1979). Historyczny charakter epistemologii marksistowskiej [The Historical Nature of Marxist Epistemology]. In: Z. Cackowski and J. Kmita (eds.), *Społeczny kontekst poznania* [The Social Context of Knowledge]. Wrocław: Ossolineum, pp. 81-117.

Kmita, J. (1985). *Kultura i poznanie* [Culture and Knowledge]. Warszawa: PWN.

Kmita, J. (1987a). Rozwój poznania naukowego [The Development of Scientific Cognition]. In: Z. Cackowski *et al.* (eds.), *Filozofia a nauka. Zarys encyklopedyczny* [Philosophy and Science. An Encyclopedic Survey]. Wrocław: Ossolineum, pp. 598-606.

Kmita, J. (1987b). Indywidualizm i antyindywidualizm metodologiczny [Methodological Individualism and Anti-individualism]. In: Z. Cackowski *et al.* (eds.), *Filozofia a nauka. Zarys encyklopedyczny* [Philosophy and Science. An Encyclopedic Survey]. Wrocław: Ossolineum, pp. 235-244.

Kotarbiński, T. (1961). *Elementy teorii poznania, logiki formalnej i metodologii nauk.* [Preliminaries to Epistemology, Formal Logic and Methodology]. Wrocław-Warszawa-Kraków: Ossolineum.

Krajewski, W. (1974). Kopernik i Galileusz Versus Arystoteles – nowa metoda naukowa przeciw dogmatyzmowi i wąskiemu empiryzmowi [Copernicus and Galileo against Aristotle. New Scientific Method against Dogmatism and „Narrow Empiricism"]. *Studia Metodologiczne*, 12, 3-22.

Krajewski, W. (1977). *Correspondence Principle and Growth of Science. Episteme*, 4. Dordrecht/Boston: D. Reidel.

Krajewski, W. (1987). Ontologia [Ontology]. In: Z Cackowski *et al* (eds.), *Filozofia a nauka. Zarys encyklopedyczny* [Philosophy and Science. An Encyclopedic Survey]. Wrocław: Ossolineum, pp. 444-451.

Kuhn, T.S. (1962). *The Structure of Scientific Revolutions*. Chicago: The University of Chicago Press.

Kuhn, T.S. (1977). *The Essential Tension*. Chicago: The University of Chicago Press.

Kuhn, T.S. (1985). Tradycje matematyczne a tradycje eksperymentalne w rozwoju nauk fizycznych (Mathematical Traditions and Experimental Traditions in the Development

427

of Physical Sciences). In: *Dwa bieguny* (Polish translation of *The Essential Tension*). Warszawa: PIW, pp. 67-112.

Lakatos, I. (1970). Falsification and the Methodology of Scientific Research Programmes. In: *Criticism and the Growth of Knowledge*. Cambridge: Cambridge University Press, pp. 91-195.

Lakatos, I. (1976). Proofs and Refutations. In: J. Worrall, E. Zahar (eds.), *The Logic of Mathematical Discovery*. Cambridge: Cambridge University Press.

Lakatos, I. (1978). History of Science and its Rational Reconstructions. In: Worrall J., G. Curie (eds.), *The Methodology of Scientific Research Programmes. Philosophical Papers*, vol. I. Cambridge: Cambridge University Press.

Lange, O. (1962). *Całość i rozwój w świetle cybernetyki* [The Whole and Development in the Light of Cybernetics]. Warszawa: PWN.

Laue, N. (1957). *Historia fizyki* [The History of Physics]. Warszawa: PWN.

Leibniz, G.W. (1981). *New Essays on Human Understanding*. Cambridge: Cambridge University Press.

Lewenstam, A. (1974). Zasada korespondencji u Bohra [The Principle of Correspondence in Bohr's Works]. In: W. Krajewski, W. Mejbaum, J. Such (eds.), *Zasada korespondencji w fizyce a rozwój nauki* [The Principle of Correspondence in Physics and the Development of Science]. Warszawa: PWN, pp. 21-36.

Łubnicki, N. (1967). Armin Teske (1910-1967). *Studia filozoficzne*, **3**.

Malewski, A., J. Topolski (1960). *Studia z metodologii historii* [Studies in the Methodology of History]. Warszawa-Wrocław: Ossolineum.

Manteuffel, R. (1985). Wiedza naukowa i zdroworozsądkowa [Scientific and Common Knowledge). In: B. Suchodolski, J. Kubin (eds.), *Nauka w kulturze ogólnej* [Science within the General Kulture]. Wrocław: Ossolineum.

Markov, M.A. (1976). *O prirodie matierii* [On the Nature of Matter]. Moskva: Progress.

Matuszewski, R. (1980). Zasady, prawa i hipotezy mechaniki Galileusza oraz problem ich empirycznego sprawdzania [Principles, Laws and Hypotheses of Galileo's Mechanics and the Problem of Their Empirical Testing]. In: J. Such (ed.), *O swoistości uzasadniania w różnych naukach* [On the Peculiarity of Justification in Various Sciences]. Poznań: Wydawnictwo Naukowe UAM, pp. 51-74.

Mejbaum, W. (1960). *Posłowie do pracy H. Reichenbacha Powstanie filozofii naukowej [Afterword to H. Reichenbach's The Origin of Scientific Philosophy]*. Warszawa: PWN.

Mejbaum, W. (1964). Prawa i sformułowania [Laws and Formulations]. In: S. Amsterdamski, Z. Augustynek, W. Mejbaum, *Prawo, konieczność, prawdopodobieństwo* [Law, Necessity and Probability]. Warszawa: KiW, pp. 225-256.

Mejbaum, W. (1966). *O twierdzeniach bazowych* [On Basic Statements]. In: H. Eilstein, M. Przełęcki (eds.), *Teoria i doświadczenie* [Theory and Experience]. Warszawa: PWN, pp. 111-129.

Meyerson, E. (1957). *Identité et réalité*. Paris.

Mierkulov, I.P. (1971). Probliema semioticheskoy prostoty v logikie nauki [The Problem of Semiotic Simplicity in the Logic of Science]. *Voprosy Filosofii*, **6**.

Mill, J.St. (1947). *A System of Logic Ratiocinative and Inductive*. Vol. I. London-New York-Toronto: Longmans, Greens and Co.

Moszczeńska, W. (1960). *Wstęp do badań historycznych* [An Introduction to Historical Studies]. Warszawa: PWN.

Nagel, E. (1961). *The Structure of Science*. London: Routledge and Kegan Paul.

Neumann, C. (1896). *Allgemeine Untersuchungen über das Newtonische Prinzip der Ferrwirkungen*. Leipzig.

Neumann, J. (1932). *Die mathematische Grundlagen der Quantenmechanik*. Berlin: Springer.

Niedźwiecki, W. (1974). Teoria, korespondencja, zasada korespondencji [Theory, Correspondence, Principle of Correspondence]. In: W. Krajewski, W. Mejbaum, J. Such (eds.), *Zasada korespondencji w fizyce a rozwój nauki* [The Principle of

428

Correspondence in Physics and the Development of Science]. Warszawa: PWN, pp. 349-445.

Nowak, L. (1970). O zasadzie abstrakcji i stopniowej konkretyzacji [On the Principle of Abstraction and Gradual Concretization]. In: *Założenia metodologiczne "Kapitału" Marksa* [Methodological Assumptions of Marx's "Capital"]. Warszawa: KiW, pp. 123-218.

Nowak, L. (1971). *U podstaw marksowskiej metodologii nauk* [Foundations of the Marxian Methodology of Science]. Warszawa: PWN.

Nowak, L. (1972). Theories, Idealization and Measurement. *Philosophy of Science*, 39, 4, 533-547.

Nowak, L. (1973). Popperowska koncepcja praw i sprawdzania [Popper's Conception of Laws and Testing]. In: J. Kmita (ed.), *Elementy marksistowskiej metodologii humanistyki* [Elements of the Marxist Methodology of the Humanities]. Poznań: Wydawnictwo Poznańskie, pp. 303-304.

Nowak, L. (1974). *Zasady marksistowskiej filozofii nauki. Próba systematycznej rekonstrukcji* [Principles of the Marxist Philosophy of Science. An Attempt a Systematic Reconstruction]. Warszawa: PWN.

Nowak, L. (2000). Galileo – Newton's Model of Free Fall. In: I. Nowakowa, L. Nowak, *Idealization X: The Richness of Idealization, (Poznań Studies in the Philosophy of the Sciences and the Humanities*, vol. 69), Amsterdam – Atlanta: Rodopi, 18-27.

Nowak, Z.M. (1990). *Jedność materialna świata w świetle wzajemnej przekształcalności cząstek elementarnych* [Material Unity of the World in the Light of Mutual Transformability of Elementary Particles]. Opole: Wydawnictwo WSP.

Nowakowa, I. (1972). *Zasada korespondencji w fizyce* [The Principle of Correspondence in Physics]. Ph.D. thesis. Dept. of Philosophy. Adam Mickiewicz University, Poznań.

Nowakowa, I. (1975a). *Dialektyczna korespondencja a rozwój nauki* [The Dialectical Correspondence and the Development of Science]. Warszawa/Poznań: PWN.

Nowakowa, I. (1975b). Idealization and the Problem of the Correspondence. *Poznań Studies in the Philosophy of the Sciences and the Humanities*, 1, 65-70.

Nowakowa, I. (1994). *Idealization V: The Dynamics of Idealizations (Poznań Studies in the Philosophy of the Sciences and the Humanities*, vol. 34). Amsterdam-Atlanta: Rodopi.

Pakszys, E. (1980). Redukcjonistyczne wyjaśnianie i sprawdzanie w biologii [Reductionist Explanation and Testing in Biology]. In: J. Such (ed.), *O swoistości uzasadniania w różnych naukach* [On the Peculiarity of Justification in Various Sciences]. Poznań: Wydawnictwo Naukowe UAM, pp. 107-112.

Pałubicka, A. (1977). *Orientacje epistemologiczne a rozwój nauki* [Epistemological Orientations and the Development of Science]. Warszawa-Poznań: PWN.

Pałubicka, A. (1992). Kantowskie formy oglądu zmysłowego a kulturowy charakter doświadczenia [Kantian Forms of Sensory Perception and the Cultural Character of Experience]. *Nowa Krytyka*, 2, 63-76.

Pap, A. (1955). *Erkenntnististheories*. Wien.

Pelc, J., M. Przełęcki, K. Szaniawski (1957). *Prawa nauki* [Laws of Science]. Warszawa: PWN.

Piaget, J. (1972). *The Principles of Genetic Epistemology*. London.

Piaget, J. (1977). *Psychologia i epistemologia* [Psychology and Epistemology]. Warszawa: PWN.

Planck, M. (1970). *Jedność fizycznego obrazu świata* [The Unity of the Physical Picture of the World]. Warszawa: Książka i Wiedza.

Popowicz, M.W. (1969). *Ob universalnosti logiki* [On the Universality of Logic]. *Voprosy Filosofii*, 7.

Popper, K.R. (1959). *The Logic of Scientific Discovery*. New York: Basic Books Inc.

Popper, K.R. (1972). *Objective Knowledge*. Oxford: At the Clarendon Press.

Popper, K.R. (1993a). *Społeczeństwo otwarte i jego wrogowie* (Polish translation of *The Open Society and its Enemies*). Volume 1. Urok Platona (The Charm of Plato). Warszawa: PWN.

429

Popper, K.R. (1993b). *Społeczeństwo otwarte i jego wrogowie* [Polish translation of *The Open Society and its Enemies*). Volume 2. Wysoka fala proroctw: Hegel, Marks i następstwa (The High Tide of Prophecies: Hegel, Marx and the Consequences). Warszawa: PWN.

Prigogine, I., I. Stengers (1984). *Order out of Chaos*. London: Heinemann.

Przełęcki, M. (1966). Interpretacja systemów aksjomatycznych [Interpretation of Axiomatic Systems]. In: T Pawłowski (ed.), *Logiczna teoria nauki* [The Logical Theory of Science]. Warszawa: PWN.

Rudnicki, K. (1995). *The Cosmological Principles*. Kraków: Uniwersytet Jagielloński.

Russell, B. (1943). *An Inquiry into Meaning and Truth*. London: G. Allen and Unwin.

Russell, B. (1949). *ABC of Relativity*. London: Peace Foundation.

Rutkiewicz, M.N. (1965). Razwitije, progriess i zakony dialektiki [Development, Progress and the Laws of Dialectics]. *Voprosy Filosofii*, **8**, 25-34.

Schlesinger, G. (1963). *Method in Physical Science*. London.

Siemek, M. (1977). *Idea transcendentalizmu u Fichtego i Kanta* [The Idea of Transcendentalism in Fichte and Kant]. Warszawa: PWN.

Siemek, M. (1980). Lukács i Hegel: Problemy filozoficznej samowiedzy marksizmu (II) [Lukács and Hegel: Problems of the Philosophical Self-Knowledge of Marxism (II)]. *Studia Filozoficzne*, **7**, 4-18.

Skarżyński, G. (1969). O paradoksach kosmologicznych [On Cosmological Paradoxes]. *Studia Filozoficzne*, **8**.

Such, J. (1967). Johna Stuarta Milla koncepcja uniwersalności oraz niezawodności praw nauki [John Stuart Mill's Conception of Universality and Reliability of Scientific Laws]. In: W. Krajewski (ed.), *Pojęcie prawa nauki w XIX wieku* [The Notion of the Law of Science in the 19th Century]. Warszawa: PWN, pp. 29-70.

Such, J. (1972). *O uniwersalnosci praw nauki* [On the Universality of the Laws of Science]. Warszawa: KiW.

Such, J. (1975a). *Czy istnienie experimentum crucis? Problemy sprawdzania, praw i teorii naukowych* [Is there Experimentum Crucis? Problems of the Verification of Laws and Theories in Science], Warszawa: PWN.

Such, J. (1975b). *Problemy weryfikacji wiedzy* [Problems of Verification of Knowledge]. Warszawa: PWN.

Such, J. (1978). Idealization and Concretization in Natural Sciences. *Poznań Studies in the Philosophy of the Sciences and the Humanities*, 1-4, vol. IV, 49-74.

Such, J. (1982). Are there Definitively Falsifying Procedures in Science? In: W. Krajewski (ed.), *Polish Essays in the Philosophy of the Natural Sciences, (Boston Studies in the Philosophy of Science)*, vol. 68. Dordrecht/Boston/London, D. Reidel Publishing Company, 113-126.

Such, J. (1992). *Dialektyczne wizje świata* [Dialectic Visions of the World]. Warszawa-Poznań: PWN.

Such, J. (1994). Przedmiot i metoda kosmologii [The Subject Matter and Method of Cosmology]. In: J. Such, J. Szymański and A. Szczuciński (eds.), *Swoistość metod badawczych a przedmiot nauk szczegółowych* [The Peculiarity of Research Methods and the Subject of Special Sciences]. Poznań: Wielkopolska Agencja Wydawnicza, pp. 7-15.

Swieżawski, S. (1966). *Zagadnienie historii filozofii* [The Issue of the History of Philosophy]. Warszawa: PWN.

Szczuciński, A. (1986). Rozwój fizyki a rozwój astrofizyki [The Development of Physics and the Development of Astrophysics]. In: J. Such, E. Pakszys (eds.), *Szkice o rozwoju nauki* [Essays on the Development of Science]. Poznań: Wydawnictwo Naukowe IF UAM, pp. 61-68.

Tamm, I.J. (1961). Problemy cząsteczek elementarnych [Problems of Elementary Particles]. In: *Mezony, grawitacja, antymateria* [Mesons, Gravitation, Antimatter]. Warszawa: PWN.

Tryon, E.P. (1973). Is the Universe a Vacuum Fluctuation? *Nature*, **246**.

430

Vilenkin, A. (1982). Creation of the Universe from Nothing. *Physical Letters*, **B 117**.

Watkins, J. (1984). *Science and Scepticism*. Princeton, New Jersey: Princeton University Press.

Weinberg, S. (1992). *Dreams of a Final Theory*. New York: Pantheon Books.

Weingarten, D.H. (1996). Kwarki z komputera [Quarks from the Computer]. *Świat Nauki*, **3**.

Weizsäcker, C.F. (1971). *Die Einheit der Natur*. München: Carl Hanser Verlag.

Weizsäcker, C.F. (1978). *Jedność przyrody* [The Unity of Nature]. Warszawa: PIW.

Whitehead, A.N. (1929). *Process and Reality*. Cambridge: The University Press.

Zamiara, K. (1974). *Metodologiczne znaczenie sporu o status poznawczy teorii* [Methodological Significance of the Dispute over the Cognitive Status of a Theory], Warszawa: PWN.

Zamiara, K. (1977). Epistemologia genetyczna J. Piageta a społeczny rozwój nauki [Genetic Epistemology of J. Piaget and the Social Development of Science]. In: Z. Cackowski, J. Kmita (eds.), *Społeczny kontekst poznania* [Social Context of Cognition]. Wrocław: Ossolineum.

Zinovev, A. (1976) *Logika nauki* (Polish translation of *The Logic of Science*). Warszawa: PWN

Zinovev, A. (1970). O logikie micromira [On the Logic of the Microworld]. *Voprosy Filosofii*, **2**.

Zotov, A.F. (1969). Problemy postroyeniya i interpretatsyi obshchkih fizicheskih tieorii [Problems of the Construction and Interpretation of General Physical Theories]. *Voprosy Filosofii*, **7**.

Życiński, J. (1993). *Granice racjonalności* [Boundaries of Rationality]. Warszawa: PWN.

POZNAŃ STUDIES IN THE PHILOSOPHY OF THE SCIENCES AND THE HUMANITIES

Contents of back issues

VOLUME 1 (1975)

Main topics:
The Method of Humanistic Interpretation; The Method of Idealization; The Reconstruction of Some Marxist Theories.
(sold out)

VOLUME 2 (1976)

Main topics:
Idealizational Concept of Science; Categorial Interpretation of Dialectics.
(sold out)

VOLUME 3 (1977)

Main topic:
Aspects of the Production of Scientific Knowledge.
(sold out)

VOLUME 4 (1978)

Main topic:
Aspects of the Growth of Science.
(sold out)

VOLUME 5 (1979)

Main topic:
Methodological Problems of Historical Research.
(sold out)

VOLUME 6 (1982)

SOCIAL CLASSES ACTION & HISTORICAL MATERIALISM

Main topics:
On Classes; On Action; The Adaptive Interpretation of Historical Materialism; Contributions to Historical Materialism.
(sold out)

VOLUME 7 (1982)

DIALECTICAL LOGICS FOR THE POLITICAL SCIENCE
(Edited by Hayward R. Alker, Jr.)

VOLUME 8 (1985)

CONSCIOUSNESS: METHODOLOGICAL AND PSYCHOLOGICAL APPROACHES
(Edited by Jerzy Brzeziński)

VOLUME 9 (1986)

THEORIES OF IDEOLOGY AND IDEOLOGY OF THEORIES
(Edited by Piotr Buczkowski and Andrzej Klawiter)

VOLUME 10 (1987)

WHAT IS CLOSER-TO-THE-TRUTH?
A PARADE OF APPROACHES TO TRUTHLIKENESS
(Edited by Theo A.F. Kuipers)

VOLUME 11 (1988)

NORMATIVE STRUCTURES OF THE SOCIAL WORLD
(Edited by Giuliano di Bernardo)

VOLUME 12 (1987)

POLISH CONTRIBUTIONS TO THE THEORY AND PHILOSOPHY OF LAW
(Edited by Zygmunt Ziembiński)

VOLUME 20 (1990)

Jürgen Ritsert
MODELS AND CONCEPTS OF IDEOLOGY

VOLUME 21 (1991)

PROBABILITY AND RATIONALITY
STUDIES ON L. JONATHAN COHEN'S PHILOSOPHY OF SCIENCE
(Edited by Ellery Eells and Tomasz Maruszewski)

VOLUME 22 (1991)

THE SOCIAL HORIZON OF KNOWLEDGE
(Edited by Piotr Buczkowski)

VOLUME 23 (1991)

ETHICAL DIMENSIONS OF LEGAL THEORY
(Edited by Wojciech Sadurski)

VOLUME 24 (1991)

ADVANCES IN SCIENTIFIC PHILOSOPHY
ESSAYS IN HONOUR OF PAUL WEINGARTNER ON THE OCCASION OF
THE 60TH ANNIVERSARY OF HIS BIRTHDAY
(Edited by Gerhard Schurz and Georg J.W. Dorn)

VOLUME 25 (1992)

IDEALIZATION III: APPROXIMATION AND TRUTH
(Edited by Jerzy Brzeziński and Leszek Nowak)

VOLUME 26 (1992)

IDEALIZATION IV: INTELLIGIBILITY IN SCIENCE
(Edited by Craig Dilworth)

VOLUME 27 (1992)

Ryszard Stachowski

THE MATHEMATICAL SOUL.
AN ANTIQUE PROTOTYPE OF THE MODERN MATEMATISATION OF
PSYCHOLOGY

VOLUME 40 (1995)

THE HERITAGE OF KAZIMIERZ AJDUKIEWICZ
(Edited by Vito Sinisi and Jan Woleński)

VOLUME 41 (1994)

HISTORIOGRAPHY BETWEEN MODERNISM AND POSTMODERNISM.
CONTRIBUTIONS TO THE METHODOLOGY OF
THE HISTORICAL RESEARCH
(Edited by Jerzy Topolski)

VOLUME 42 (1995)

IDEALIZATION VII: IDEALIZATION, STRUCTURALISM,
AND APPROXIMATION
(Edited by Martti Kuokkanen)

Idealization, Approximation and Counterfactuals in the Structuralist Framework – T.A.F. Kuipers, *The Refined Structure of Theories*; C.U. Moulines and R. Straub, *Approximation and Idealization from the Structuralist Point of View*; I.A. Kieseppä, *A Note on the Structuralist Account of Approximation*; C.U. Moulines and R. Straub, *A Reply to Kieseppä*; W. Balzer and G. Zoubek, *Structuralist Aspects of Idealization*; A. Ibarra and T. Mormann, *Counterfactual Deformation and Idealization in a Structuralist Framework*; I.A. Kieseppä, *Assessing the Structuralist Theory of Verisimilitude*. **Idealization, Approxima-tion and Theory Formation** – L. Nowak, *Remarks on the Nature of Galileo's Methodological Revolution*; I. Niiniluoto, *Approximation in Applied Science*; E. Heise, P. Gerjets and R. Westermann, *Idealized Action Phases. A Concise Rubicon Theory*; K.G. Troitzsch, *Modelling, Simulation, and Structuralism*; V. Rantala and T. Vadén, *Idealization in Cognitive Science. A Study in Counterfactual Correspondence*; M. Sintonen and M. Kiikeri, *Idealization in Evolutionary Biology*; T. Tuomivaara, *On Idealizations in Ecology*; M. Kuokkanen and M. Häyry, *Early Utilitarianism and Its Idealizations from a Systematic Point of View*. **Idealization, Approximation and Measurement** – R. Westermann, *Measurement-Theoretical Idealizations and Empirical Research Practice*; U. Konerding, *Probability as an Idealization of Relative Frequency. A Case Study by Means of the BTL-Model*; R. Suck and J. Wienöbst, *The Empirical Claim of Probality Statements, Idealized Bernoulli Experiments and their Approximate Version*; P.J. Lahti, *Idealizations in Quantum Theory of Measurement*.

VOLUME 43 (1995)

Witold Marciszewski and Roman Murawski
MECHANIZATION OF REASONING IN A HISTORICAL PERSPECTIVE

Chapter 1: *From the Mechanization of Reasoning to a Study of Human Intelligence*; Chapter 2: *The Formalization of Arguments in the Middle Ages*; Chapter 3: *Leibniz's Idea of Mechanical Reasoning at the Historical Background*; Chapter 4: *Between Leibniz and Boole: Towards the Algebraization of Logic*; Chapter 5: *The English Algebra of Logic in the 19th Century*; Chapter 6: *The 20th Century Way to Formalization and Mechanization*; Chapter 7: *Mechanized Deduction Systems*.

VOLUME 44 (1995)

THEORIES AND MODELS IN SCIENTIFIC PROCESSES
(Edited by William Herfel, Władysław Krajewski,
Ilkka Niiniluoto and Ryszard Wójcicki)

Introduction; **Part 1. Models in Scientific Processes** – J. Agassi, *Why there is no Theory of Models?*; M. Czarnocka, *Models and Symbolic Nature of Knowledge*; A. Grobler, *The Representational and the Non-Representational in Models of Scientific Theories*; S. Hartmann, *Models as a Tool for the Theory Construction: Some Strategies of Preliminary Physics*; W. Herfel, *Nonlinear Dynamical Models as Concrete Construction*; E. Kałuszyñska, *Styles of Thinking*; S. Psillos, *The Cognitive Interplay Between Theories and Models: the Case of 19th Century Optics.* **Part 2. Tools of Science** – N.D. Cartwright, T. Shomar, M. Suarez, *The Tool-Box of Science*; J. Echeverria, *The Four Contexts of Scienctific Activity*; K. Havas, *Continuity and Change; Kinds of Negation in Scientific Progress*; M. Kaiser, *The Independence of Scientific Phenomena*; W. Krajewski, *Scientific Meta-Philosophy*; I. Niiniluoto, *The Emergence of Scientific Specialties: Six Models*; L. Nowak, *Antirealism, (Supra-) Realism and Idealization*; R.M. Nugayev, *Classic, Modern and Postmodern Scientific Unification*; V. Rantala, *Translation and Scientific Change*; G. Schurz, *Theories and Their Applications* – *a Case of Nonmonotonic Reasoning*; W. Strawiñski, *The Unity of Science Today*; V. Torosian, *Are the Ethics and Logic of Science Compatible?* **Part 3. Unsharp Approaches in Science** – E.W. Adams, *Problems and Prospects in a Theory of Inexact First-Order Theories*; W. Balzer, G. Zoubek, *On the Comparision of Approximative Empirical Claims*; G. Cattaneo, M. Luisa Dalla Chiara, R. Giuntini, *The Unsharp Approaches to Quantum Theory*; T.A.F. Kuipers, *Falsification Versus Effcient Truth Approximation*; B. Lauth, *Limiting Decidability and Probability*; J. Pykacz, *Many-Valued Logics in Foundations of Quantum Mechanics*; R.R. Zapatrin, *Logico-Algebraic Approach to Spacetime Quantization.*

VOLUME 45 (1995)

COGNITIVE PATTERNS IN SCIENCE AND COMMON SENSE.
GRONINGEN STUDIES IN PHILOSOPHY OF SCIENCE,
LOGIC, AND EPISTEMOLOGY
(Edited by Theo A.F. Kuipers and Anne Ruth Mackor)

L. Nowak, *Foreword*; **General Introduction** – T.A.F. Kuipers and A.R. Mackor, *Cognitive Studies of Science and Common Sense*; **Part I: Conceptual Analysis in Service of Various Research Programmes** – H. Zandvoort, *Concepts of Interdisciplinarity and Environmental Science*; R. Vos, *The Logic and Epistemology of the Concept of Drug and Disease Profile*; R.C. Looijen, *On the Distinction Between Habitat and Niché, and Some Implications for Species' Differentiation*; G.J. Stavenga, *Cognition, Irreversibility and the Direction of Time*; R. Dalitz, *Knowledge, Gender and Social Bias*; **Part II: The Logic of the Evaluation of Arguments, Hypotheses, Rules, and Interesting Theorems** – E.G.W. Krabbe, *Can We Ever Pin One Down to a Formal Fallacy?*; T.A.F. Kuipers, *Explicating the Falsificationist and Instrumentalist Methodology by Decomposing the Hypothetico-Deductive Method*; A. Keupink, *Causal Modelling and Misspecification: Theory and Econometric Historical*

Practice; M.C.W. Janssen and Y.-H. Tan, *Default Reasoning and Some Applications in Economics*; B. Hamminga, *Interesting Theorems in Economics*; **Part III: Three Challenges to the Truth Approximation Programme** – S.D. Zwart, *A Hidden Variable in the Discussion About 'Language Dependency' of Truthlikeness*; H. Hettema and T.A.F. Kuipers, *Sommerfeld's Atombau: A Case Study in Potential Truth Approximation*; R. Festa, *Verisimilitude, Disorder, and Optimum Prior Probabilities*; **Part IV: Explicating Psychological Intuitions** – A.R. Mackor, *Intentional Psychology is a Biological Discipline*; J. Peijnenburg, *Hempel's Rationality. On the Empty Nature of Being a Rational Agent*; L. Guichard, *The Causal Efficacy of Propositional Attitudes*; M. ter Hark, *Connectionism, Behaviourism and the Language of Thought.*

VOLUME 46 (1996)

POLITICAL DIALOGUE: THEORIES AND PRACTICE
(Edited by Stephen L. Esquith)

Introduction – S.L. Esquith, *Political Dialogue and Political Virtue*. **Part I: The Modern Clasics** – A.J. Damico, *Reason's Reach: Liberal Tolerance and Political Discourse*; T.R. Machan, *Individualism and Political Dialogue*; R. Kukla, *The Coupling of Human Souls: Rousseau and the Problem of Gender Relations*; D.F. Koch, *Dialogue: An Essay in the Instrumentalist Tradition*. **Part II: Toward a Democratic Synthesis** – E. Simpson, *Forms of Political Thinking and the Persistence of Practical Philosophy*; J.B. Sauer, *Discoursee, Consensus, and Value: Conversations about the Intelligible Relation Between the Private and Public Spheres*; M. Kingwell, *Phronesis and Political Dialogue*, R.T. Peterson, *Democracy and Intellectual Mediation – After Liberalism and Socialism*. **Part III: Dialogue in Practice** – S. Rohr Scaff, L.A. Scaff, *Political Dialogue in the New Germany: The Burdens of Culture and an Asymmetrical Past*; J.H. Read, *Participation, Power, and Democracy*; S.E. Bennett, B. Fisher, D. Resnick, *Speaking of Politics in the United States: Who Talks to Whom, Why, and Why Not*; J. Forester, *Beyond Dialogue to Transformative Learning: How Deliberative Rituals Encourage Political Judgement in Community Planning Processes*; A. Fatić, *Retribution in Democracy.*

VOLUME 47 (1996)

EPISTEMOLOGY AND HISTORY. HUMANITIES AS A PHILOSOPHICAL PROBLEM AND JERZY KMITA'S APPROACH TO IT
(Edited by Anna Zeidler-Janiszewska)

A. Zeidler-Janiszewska, *Preface*. **Humanistic Knowledge** – K.O. Apel, *The Hermeneutic Dimension of Social Science and its Normative Foundation*; M. Czerwiński, *Jerzy Kmita's Epistemology*; L. Witkowski, *The Frankfurt School and Structuralism in Jerzy Kmita's Analysis*; A. Szahaj, *Between Modernism and Postmodernism: Jerzy Kmita's Epistemology*; A. Grzegorczyk, *Non-Cartesian Coordinates in the Contemporary Humanities*; A. Pałubicka, *Pragmatist Holism as an Expression of Another Disenchantment of the World*; J. Sójka, *Who is Afraid of Scientism?*; P. Ozdowski, *The Broken Bonds with the World*; J. Such, *Types of Determination vs. the Development of Science in Historical Epistemology*; P. Zeidler, *Some Issues of Historical Epistemology in the Light of the Structuralist Philosophy of Science*; M. Buchowski, *Via Media: On the Consequences of Historical Epistemology for the Problem of Rationality*; B. Kotowa, *Humanistic Valuation and Some Social Functions of the Humanities*. **On Explanation and Humanistic Interpretation** – T.A.F. Kuipers, *Explanation by Intentional, Functional, and Causal Specification*; E. Świderski, *The Interpretational Paradigm in the Philosophy of the Human Sciences*; L. Nowak, *On the*

Limits of the Rationalistic Paradigm; F. Coniglione, *Humanistic Interpretation between Hempel and Popper*; Z. Ziembiński, *Historical Interpretation vs. the Adaptive Interpretation of a Legal Text*; W. Mejbaum, *Explaining Social Phenomena*; M. Ziółkowski, *The Functional Theory of Culture and Sociology*; K. Zamiara, *Jerzy Kmita's Social-Regulational Theory of Culture and the Category of Subject*; J. Brzeziński, *Theory and Social Practice. One or Two Psychologies?*; Z. Kwieciński, *Decahedron of Education (Components and aspects). The Need for a Comprehensive Approach*; J. Paśniczek, *The Relational vs. Directional Concept of Intentionality*. **The Historical Dimension of Culture and its Studies** – J. Margolis, *The Declension of Progressivism*; J. Topolski, *Historians Look at Historical Truth*; T. Jerzak-Gierszewska, *Three Types of the Theories of Religion and Magic*; H. Paetzold, *Mythos und Moderne in der Philosophie der symbolishen Formen Ernst Cassirers*; M. Siemek, *Sozialphilosophische Aspekte der Übersetzbarkeit*. **Problems of Artistic Practice and Its Interpretation** – S. Morawski, *Theses on the 20th Century Crisis of Art and Culture*; A. Erjavec, *The Perception of Science in Modernist and Postmodernist Artistic Practice*, G. Dziamski, *The Avant-garde and Contemporary Artistic Consciousness*; H. Orłowski, *Generationszugehörigkeit und Selbsterfahrung von (deutschen) Schriftstellern*; T. Kostyrko, *The "Transhistoricity" of the Structure of Work of Art and the Process of Value Transmission in Culture*; G. Banaszak, *Musical Culture as a Configuration of Subcultures*; A. Zeidler-Janiszewska, *The Problem of the Applicability of Humanistic Interpretation in the Light of Contemporary Artistic Practice*; J. Kmita, *Towards Cultural Relativism "with a Small 'r' "*; The Bibliography of Jerzy Kmita.

VOLUME 48 (1996)

THE SOCIAL PHILOSOPHY OF ERNEST GELLNER
(Edited by John A. Hall and Ian Jarvie)

J.A. Hall, I. Jarvie, *Preface*; J.A. Hall, I. Jarvie, *The Life and Times of Ernest Gellner*. **Part 1: Intelectual Background** – J. Musil, *The Prague Roots of Ernest Gellner's Thinking*; Ch. Hahn, *Gellner and Malinowski: Words and Things in Central Europe*; T. Dragadze, *Ernest Gellner in Soviet East*. **Part 2: Nations and Nationalism** – B. O'Leary, *On the Nature of Nationalism: An Appraisal of Ernest Gellner's Writings on Nationalism*; K. Minogue, *Ernest Gellner and the Dangers of Theorising Nationalism*; A.D. Smith, *History and Modernity: Reflection on the Theory of Nationalism*; M. Mann, *The Emergence of Modern European Nationalism*; N. Stagardt, *Gellner's Nationalism: The Spirit of Modernisation?* **Part 3: Patterns of Development** – P. Burke, *Reflections on the History of Encyclopaedias*; A. MacFarlane, *Ernest Gellner and the Escape to Modernity*; R. Dore, *Soverein Individuals*; S. Eisenstadt, *Japan: Non Axial Modernity*; M. Ferro, *L'Indépendance Telescopée: De la Décolonisation a L'Impérialisme Multinational*. **Part 4: Islam** – A. Hammoudi, *Segmentarity, Social stratification, Political Power and Sainthood: Reflections on Gellner's Theses*; H. Munson, Jr., *Rethinking Gellner's Segmentary Analysis of Morocco's AitᶜAtta*; J. Baechler, *Sur le charisme*; Ch. Lindholm, *Despotism and Democracy: State and Society in PreModern Middle East*; H. Munson, Jr. *Muslim and Jew in Morocco: Reflections on the Distinction between Belief and Behaviour*; T. Asad, *The Idea of an Anthropology of Islam*. **Part 5: Science and Disenchantment** – P. Anderson, *Science, Politics, Enchantment*; R. Schroeder, *From the Big Divide to the Rubber Cage: Gellner's Conception of Science and Technology*; J. Davis, *Irrationality in Social Life*. **Part 6: Relativism and Universals** – J. Skorupski, *The Post-Modern Hume: Ernest Gellner's 'Enlightenment Fundamentalism'*; J. Wettersten, *Ernest Gellner: A Wittgensteinian Rationalist*; I. Jarvie, *Gellner's Positivism*; R. Boudon, *Relativising Relativism: When Sociology Refutes the Sociology of Science*; R. Aya, *The Empiricist Exorcist*. **Part 7: Philosophy of History** – W. McNeill, *A Swang Song for British Liberalism?*; A. Park, *Gellner and the Long Trends of History*; E. Leone, *Marx, Gellner, Power*; R. Langlois, *Coercion, Cognition and Production: Gellner's Challenge to*

Historical Materialism and Post-Modernism; E. Gellner, *Reply to my Critics*; I. Jarvie, *Complete Bibliography of Gellner's Work*.

VOLUME 49 (1996)

THE SIGNIFICANCE OF POPPER'S THOUGHT
(Edited by Stefan Amsterdamski)

Karl Popper's Three Worlds – J. Watkins, *World 1, World 2 and the Theory of Evolution*; A. Grobler, *World 3 and the Cunning of Reason*. **The Scientific Method as Ethics** – J. Agassi, *Towards Honest Public Relations of Science*; S. Amsterdamski, *Between Relativism and Absolutism: the Popperian Ideal of Knowledge*. **The Open Society and its Prospects** – E. Gellner, *Karl Popper – The Thinker and the Man*; J. Woleński, *Popper on Prophecies and Predictions*.

VOLUME 50 (1996)

THE IDEA OF THE UNIVERSITY
(Edited by Jerzy Brzeziński and Leszek Nowak)

Introduction – K. Twardowski, *The Majesty of the University*. **I** – Z. Ziembiński, *What Can be Saved of the Idea of the University?*; L. Kołakowski, *What Are Universities for?*; L. Gumański, *The Ideal University and Reality*; Z. Bauman, *The Present Crisis of the Universities*. **II** – K. Ajdukiewicz, *On Freedom of Science*; H. Samsonowicz, *Universities and Democracy*; J. Topolski, *The Commonwealth of Scholars and New Conceptions of Truth*; K. Szaniawski, *Plus ratio quam vis*. **III** – L. Koj, *Science, Teaching and Values*; K. Szaniawski, *The Ethics of Scientific Criticism*; J. Brzeziński, *Ethical Problems of Research Work of Psychologists*. **IV** – J. Goćkowski, *Tradition in Science*; J. Kmita, *Is a "Creative Man of Knowledge" Needed in University Teaching?*; L. Nowak, *The Personality of Researchers and the Necessity of Schools in Science*. **Recapitulation** – J. Brzeziński, *Reflections on the University*.

VOLUME 51 (1997)

KNOWLEDGE AND INQUIRY: ESSAYS ON JAAKKO HINTIKKA'S EPISTEMOLOGY AND PHILOSOPHY OF SCIENCE
(Edited by Matti Sintonen)

M. Sintonen, *From the Science of Logic to the Logic of Science*. **I: Historical Perspectives** – Z. Bechler, *Hintikka on Plentitude in Aristotle*; M.-L. Kakkuri-Knuuttila, *What Can the Sciences of Man Learn from Aristotle?*; M. Kusch, *Theories of Questions in German-Speaking Philosophy Around the Turn of the Century*; N.-E. Sahlin, *'He is no Good for My Work': On the Philosophical Relations between Ramsey and Wittgenstein*. **II: Formal Tools: Induction, Observation and Identifiability** – T.A.F. Kuipers, *The Carnap-Hintikka Programme in Inductive Logic*; I. Levi, *Caution and Nonmonotic Inference*; I. Niiniluoto, *Inductive Logic, Atomism, and Observational Error*; A. Mutanen, *Theory of Identifiability*. **III: Questions in Inquiry: The Interrogative Model** – S. Bromberger, *Natural Kinds and Questions*; S.A. Kleiner, *The Structure of Inquiry in Developmental Biology*; A. Wiśniewski, *Some Foundational Concepts of Erotetic Semantics*; J. Woleński, *Science and Games*. **IV: Growth of Knowledge: Explanation and Discovery** – M. Sintonen, *Explanation: The Fifth Decade*; E. Weber, *Scientific Explanation and the Interrogative Model of Inquiry*; G. Gebhard, *Scientific Discovery, Induction, and the Multi-Level*

Character of Scientific Inquiry; M. Kiikeri, *On the Logical Structure of Learning Models*. **V: Jaakko Hintikka: Replies.**

VOLUME 52 (1997)

Helena Eilstein
LIFE CONTEMPLATIVE, LIFE PRACTICAL.
AN ESSAY ON FATALISM

Preface. **Chapter One: Oldcomb and Newcomb** – 1. *In the King Comb's Chamber of Game*; 2. *The Newcombian Predicaments.* **Chapter Two: Ananke** – 1. *Fatalism: What It Is Not?*; 2. *Fatalism: What Is It?*; 3. *Fatalism and* a priori *Arguments*; 4. *Fatalism and 'Internal' Experience*; 5. *Determinism, Indeterminism and Fatalism*; 6. *Transientism, Eternism and Fatalism*; 7. *Fatalism: What It Does Not Imply?.* **Chapter Three: Fated Freedom** – 1. *More on Libertarianism*; 2. *On the Deterministic Concept of Freedom*; 3. *Moral Self and Responsibility in the Light of Probabilism*; 4. *Fated Freedom.* **Chapter Four: The Virus of Fatalism** – 1. *Fatalism and Problems of Cognition*; 2. *The Virus of Fatalism: Why Mostly Harmless?*

VOLUME 53 (1997)

Dimitri Ginev
A PASSAGE TO THE HERMENEUTIC PHILOSOPHY OF SCIENCE

Preface; **Introduction**: *Topics in the Hermeneutic Philosophy of Science*; **Chapter 0**: *On the Limits of the Rational Reconstruction of Scientific Knowledge*; **Chapter 1**: *On the Hermeneutic Nature of Group Rationality*; **Chapter 2**: *Towards a Hermeneutic Theory of Progressive Change in Scientific Development*; **Chapter 3**: *Beyond Naturalism and Traditionalism*; **Chapter 4**: *A Critical Note on Normative Naturalism*; **Chapter 5**: *Micro and Macrohermeneutics of Science*; *Concluding Remarks.*

VOLUME 54 (1997)

IN ITINERE. EUROPEAN CITIES AND THE BIRTH OF MODERN
SCIENTIFIC PHILOSOPHY
(Edited by Roberto Poli)

Introduction – R. Poli, *In itinere. Pictures from Central-European Philosophy.* **Stages of the Tour** – K. Schuhmann, *Philosophy and Art in Munich around the Turn of the Century*; M. Libardi, *In itinere: Vienna 1870-1918*; W. Baumgartner, *Nineteenth-Century Würzburg: The Development of the Scientific Approach to Philosophy*; L. Dappiano, *Cambridge and the Austrian Connection*; J. Sebestik, *Prague Mosaic. Encounters with the Prague Philosophers*; J.J. Jadacki, *Warsaw: The Rise and Decline of Modern Scientific Philosophy in the Capital City of Poland*; J. Woleński, *Lvov*; L. Albertazzi, *Science and the Avant-Garde in Early Nineteenth-Century Florence*; F. Minazzi, *The Presence of Phenomenology in Milan between the Two World Wars. The Contribution of Antonio Banfi and Giulio Preti.*

VOLUME 55 (1997)

REALISM AND QUANTUM PHYSICS
(Edited by Evandro Agazzi)

Introduction – E. Agazzi. **Part One: Philosophical Considerations** – P. Horwich, *Realism and Truth*, E. Agazzi, *On the Criteria for Establishing the Ontological Status of Different Entities*; A. Baltas, *Constraints and Resistance: Stating a Case for Negative Realism*; M. Paty, *Predicate of Existence and Predicability for a Theoretical Object in Physics*. **Part Two: Observability and Hidden Entities** – F. Bonsak, *Atoms: Lessons of a History*; A. Cordero, *Arguing for Hidden Realities*; B. d'Espagnat, *On the Difficulties that Attributing Existence to "Hidden" Quantities May Raise*; M. Pauri, *The Quantum, Space-Time and Observation*. **Part Three: Applications to Quantum Physics** – D. Albert, *On the Phenomenology of Quantum-Mechanical Superpositions*; G.C. Ghiraldi, *Realism and Quantum Mechanics*; M. Crozon, *Experimental Evidence of Quark Structure Inside Hadrons*.

VOLUME 56 (1997)

IDEALIZATION VIII: MODELLING IN PSYCHOLOGY
(Edited by Jerzy Brzeziński, Bodo Krause and Tomasz Maruszewski)

Part I: Philosophical and Methodological Problems of Cognition Process – J. Wane, *Idealizing the Cartesian-Newtonian Paradigm as Reality: The Impact of New-Paradigm Physics on Psychological Theory*; E. Hornowska, *Operationalization of Psychological Magnitudes. Assumptions-Structure-Consequences*; T. Bachmann, *Creating Analogies – On Aspects of the Mapping Process between Knowledge Domains*; H. Schaub, *Modelling Action Regulation*. **Part II: The Structure of Ideal Learning Process** – S. Ohlson, J.J. Jewett, *Ideal Adaptive Agents and the Learning Curve*; B. Krause, *Towards a Theory of Cognitive Learning*; B. Krause, U. Gauger, *Learning and Use of Invariances: Experiments and Network Simulation*; M. Friedrich, *"Reaction Time" in the Neural Network Module ART 1*. **Part III: Control Processes in Memory** – J. Tzelgov, V. Yehene, M. Naveh-Benjamin, *From Memory to Automaticity and Vice Versa: On the Relation between Memory and Automaticity*; H. Hagendorf, S. Fisher, B. Sá, *The Function of Working Memory in Coordination of Mental Transformations*; L. Nowak, *On Common-Sense and (Para-)Idealization*; I. Nowakowa, *On the Problem of Induction. Toward an Idealizational Paraphrase*.

VOLUME 57 (1997)

EUPHONY AND LOGOS
ESSAYS IN HONOUR OF MARIA STEFFEN-BATÓG AND TADEUSZ BATÓG
(Edited by Roman Murawski, Jerzy Pogonowski)

Preface. **Scientific Works of Maria Steffen-Batóg and Tadeusz Batóg** – *List of Publications of Maria Steffen-Batóg*; *List of Publications of Tadeusz Batóg*; J. Pogonowski, *On the Scientific Works of Maria Steffen-Batóg*; Jerzy Pogonowski, *On the Scientific Works of Tadeusz Batóg*; W. Lapis, *How Should Sounds Be Phonemicized?*; P. Nowakowski, *On Applications of Algorithms for Phonetic Transcription in Linguistic Research*; J. Pogonowski, *Tadeusz Batóg's Phonological Systems*. **Mathematical Logic** – W. Buszkowski, *Incomplete Information Systems and Kleene 3-valued Logic*; M. Kandulski, *Categorial Grammars with Structural Rules*; M. Kołowska-Gawiejnowicz, *Labelled*

Deductive Systems for the Lambek Calculus; R. Murawski, *Satisfaction Classes – a Survey*; K. Świrydowicz, *A New Approach to Dyadic Deontic Logic and the Normative Consequence Relation*; W. Zielonka, *More about the Axiomatics of the Lambek Calculus*. **Theoretical Linguistics** – J.J. Jadacki, *Troubles with Categorial Interpretation of Natural Language*; M. Karpiński, *Conversational Devices in Human-Computer Communication Using WIMP UI*; W. Maciejewski, *Qualitative Orientation and Gramatical Categories*; Z. Vetulani, *A System of Computer Understanding of Text*; A. Wójcik, *The Formal Development of van Sandt's Presupposition Theory*; W. Zabrocki, *Psychologism in Noam Chomsky's Theory (Tentative Critical Remarks)*; R. Zuber, *Defining Presupposition without Negation*. **Philosophy of Language and Methodology of Sciences** – J. Kmita, *Philosophical Antifundamentalism*; A. Luchowska, *Peirce and Quine: Two Views on Meaning*; S. Wiertlewski, *Method According to Feyerabend*; J. Woleński, *Wittgenstein and Ordinary Language*; K. Zamiara, *Context of Discovery – Context of Justification and the Problem of Psychologism*.

VOLUME 58 (1997)

THE POSTMODERNIST CRITIQUE OF THE PROJECT OF ENLIGHTENMENT
(Edited by Sven-Eric Liedman)

S.-E. Liedman, *Introduction*; R. Wokler, *The Enlightenment Project and its Critics*; M. Benedikt, *Die Gegenwartsbedeutung von Kants aufklärender Akzeptanz und Zurückweisung des Modells der Naturwissenschaft für zwischenmenschliche Verhältnisse: Verfehlte Beziehungen der Geisterwelt Swedenborgs*; S.-E. Liedman, *The Crucial Role of Ethics in Different Types of Enlightenment (Condorcet and Kant)*; S. Dahlstedt, *Forms of the Ineffable: From Kant to Lyotard*; P. Magnus Johansson, *On the Enlightenment in Psycho-Analysis*; E. Lundgren-Gothlin, *Ethics, Feminism and Postmodernism: Seyla Benhabib and Simone de Beauvoir*; E. Kiss, *Gibt es ein Projekt der Aufklärung und wenn ja, wie viele (Aufklärung vor dem Horizont der Postmoderne)*; E. Kennedy, *Enlightenment Anticipations of Postmodernist Epistemology*; L. Nowak, *On Postmodernist Philosophy: An Attempt to Identify its Historical Sense*; M. Castillo, *The Dilemmas of Postmodern Individualism*.

VOLUME 59 (1997)

BEYOND ORIENTALISM. THE WORK OF WILHELM HALBFASS AND ITS IMPACT ON INDIAN AND CROSS-CULTURAL STUDIES
(Edited by Eli Franco and Karin Preisendanz)

E. Franco, K. Preisendanz, *Introduction and Editorial Essay on Wilhelm Halbfass*; Publications by Wilhelm Halbfass; W. Halbfass, *Research and Reflections: Responses to my Respondents. Beyond Orientalism? Reflections on a Current Theme.* **Part I: Cross-Cultural Encounter and Dialogue** – F. X. Clooney, SJ, *Wilhelm Halbfass and the Openness of the Comparative Project*; F. Dallmayr, *Exit from Orientalism: Comments on Wilhelm Halbfass*; S. D. Serebriany, *Some Marginal Notes on India and Europe*; R. Sen (née Mookerjee), *Some Reflections on India and Europe: An Essay in Understanding*; K. Karttunen, *Greeks and Indian Wisdom*; D. Killingley, *Mlecchas, Yavanas and Heathens: Interacting Xenologies in Early Nineteenth-Century Calcutta*; W. Halbfass, *Research and Reflections: Responses to my Respondents. Cross-Cultural Encounter and Dialogue.* **Part II: Issues of Comparative Philosophy** – J. Nath Mohanty, *Between Indology and Indian Philosophy;* J. S. O'Leary, *Heidegger and Indian Philosophy*; S. Rao, *"Subordinate" or "Supreme"? The Nature of Reason in India and the West*; R. Ivekoviæ, *The Politics of Comparative Philosophy*; B.-A. Scharfstein, *The Three Philosophical Traditions*; W. Halbfass, *Research and Reflections: Responses to my Respondents. Issues of Comparative Philosophy.* **Part III: Topics in Classical Indian Philosophy** – J. E. M. Houben, *Bhartṛhari's Perspectivism: The Vṛtti and*

Bhartṛhari's Perspectivism in the First kāṇ ḍa of the Vākyapadīya; J. Bronkhorst, *Philosophy and the Vedic Exegesis in the Mīmāṃsā*; J. Taber, *The Significance of Kumārila's Philosophy*; K. Harikai, *Kumārila's Acceptance and Modification of Categories of the Vaiœeṣika School*; V. Lysenko, *The Vaiœeṣika Notions of ākāsa and diœ from the Perspective of Indian Ideas of Space*; B. M. Perry, *Early Nyāya and Hindu Orthodoxy: ānvīkṣikī and adhikāra*; W. Halbfass, *Research and Reflections: Responses to my Respondents: Topics in Classical Indian Philosophy*. **Part IV: Developments and Attitudes in Neo-Hinduism** – A. O. Fort, *Jīvanmukti and Social Service in Advaita and Neo-Vedānta*; S. Elkman, *Religious Plurality and Swami Vivekananda*. **Part V: Indian Religion, Past and Present** – M. Hara, *A Note on dharmasya sūkṣmā gatiḥ*; A. Wezler, *The Story of Aṇī-Māṇḍavya as told in the Mahābhārata: Its significance for Indian Legal and Religious History*; Y. Grinshpon, *Experience and Observation in Traditional and Modern Pātañjala Yoga*; F. J. Korom, *Language Belief and Experience in Bengali Folk Religion*; W. Halbfass, *Research and Reflection: Responses to my Respondents. Developments and Attitudes in Neo-Hinduism; Indian religion, Past and Present.*

VOLUME 60 (1998)

MARX'S THEORIES TODAY
(Edited by Ryszard Panasiuk and Leszek Nowak)

R. Panasiuk, *Introduction*; **Part I: On Dialectics and Ontology** – S.-E. Liedman, *Engels and the Laws of Dialectics*; R. Panasiuk, *On Dialectics in Marxism Again*; R. Albritton, *The Unique Ontology of Capital*; R. Washner, *It is not Singularity that Governs the Nature of Things. The Principle of Isolated Individual and Its Negation in Marx's Doctoral Thesis*; **Part II: On Historical Materialism and Social Theories** – Z. Cackowski, *The Continuing Validity of the Marxian Thought*; P. Casal, *From Unilineal to Universal Historical Materialism*; I. Hunt, *A Dialectical Interpretation and Resurrection of Historical Materialism*; W. Krajewski, *The Triumph of Historical Materialism*; L. Nowak, *The Adaptive Interpretation of Historical Materialism: A Survey. On a Contribution to Polish Analytical Marxism*; M. Kozłowski, *A New Look at Capitalism. Between the Decommunisation of Marx's and the Defeudalisation of Hegel's Visions of Capitalism*; F. Moseley, *An Empirical Appraisal of Marx's Economic Theory*; Ch. Bertram and A. Carling, *Stumbling into Revolution. Analytical Marxism, Rationality and Collective Action*; K. Graham, *Collectives, Classes and Revolutionary Potential in Marx*; U. Himmelstrand, *How to Become and Remain a Marxicising Sociologist. An Egocentric Report*; **Part III: On Axiology and the Socialist Project** – P. Kamolnick, *Visions of Social Justice in Marx: An Assessment of Recent Debates in Normative Philosophy*; W. Schmied-Kowarzik, *Karl Marx as a Philosopher of Human Emancipation*; H. J. Sandkühler, *Marx – Welche Rationalität? Epistemische Kontexte und Widersprüche der Transformation von Philosophie in Wissenschaft*; J. Kmita, *The Production of "Rational Reality" and the "Systemic Coercion"*; J. Bidet, *Metastructure and Socialism*; T. Andreani, *Vers une Issue Socialiste à la Crise du Capitalisme*; W. Becker, *The Bankruptsy of Marxism. About the Historical End of a World Philosophy*; D. Aleksandrowicz, *Myth, Eschatology and Social Reality in the Light of Marxist Philosophy.*

VOLUME 61 (1997)

REPRESENTATIONS OF SCIENTIFIC REALITY
CONTEMPORARY FORMAL PHILOSOPY OF SCIENCE IN SPAIN
(Edited by Andoni Ibarra and Thomas Mormann)

Introduction – A. Ibarra, T. Mormann, *The Long and Winding Road to Philosophy of Science in Spain.* **Part 1: Representation and Measurement** – A. Ibarra, T. Mormann, *Theories as Representations*; J. Garrido Garrido, *The Justification of Measurement*; O. Fernandez Prat, D. Quesada, *Spatial Representations and Their Physical Content*; J.A. Díez Calzada, *The Theory-Net of Interval Measurement Theory.* **Part 2: Truth, Rationality, and Method** – J.C. García-Bermejo Ochoa, *Realism and Truth Approximation in Economic Theory*; W.J.Gonzáles, *Rationality in Economics and Scientific Predictions*; J.P. Zamora Bonilla, *An Invitation to Methodonomics.* **Part 3: Logics, Semantics and Theoretical Structures** – J.L.Falguera, *A Basis for a Formal Semantics of Linguistic Formulations of Science*; A. Sobrino, E. Trillas, *Can Fuzzy Logic Help to Pose Some Problems in the Philosophy of Science?*; J. de Lorenzo, *Demonstrative Ways in Mathematical Doing*; M. Casanueva, *Genetics and Fertilization: A Good Marriage*; C.U. Moulines, *The Concept of Universe from a Metatheoretical Point of View.*

VOLUME 62 (1998)

IN THE WORLD OF SIGNS
(Edited by Jacek Juliusz Jadacki and Witold Strawiński)

Introduction. *How to Move in the World of Signs.* **Part I: Theoretical Semiotics** – A. Bogusławski, *Conditionals and Egocentric Mental Predicates*; W. Buszkowski, *On Families of Languages Generated by Categorial Grammar*; K.G. Havas, *Changing the World – Changing the Meaning. On the Meanings of the "Principle of Non-Contradiction"*; H. Hiż, *On Translation*; S. Marcus, *Imprecision, Between Variety and Uniformity: The Conjugate Pairs*; J. Peregrin, P. Sgall, *Meaning and "Propositional Attitudes"*; O.A. Wojtasiewicz, *Some Applications of Metric Space in Theoretical Linguistic.* **Part II: Methodology** – E. Agazzi, *Rationality and Certitude*; I. Bellert, *Human Reasoning and Artifical Intelligence. When Are Computers Dumb in Simulating Human Reasoning?*; T. Bigaj, *Analyticity and Existence in Mathematics*; G.B. Keene, *Taking up the Logical Slack in Natural Languages*; A. Kertész, *Interdisciplinarity and the Myth of Exactness*; J. Srzednicki, *Norm as the Basis of Form*; J.S. Stepanov, *"Cause" in the Light of Semiotics*; J.A. Wojciechowski, *The Development of Knowledge as a Moral Problem.* **Part III: History of Semiotics** – E. Albrecht, *Philosophy of Language, Logic and Semiotics*; G. Deledalle, *A Philosopher's Reply to Questions Concerning Peirce's Theory of Signs*; J. Deledalle-Rhodes, *The Transposition of Linguistic Sign in Peirce's Contributions to "The Nation"*; R. E. Innis, *From Feeling to Mind: A Note on Langer's Notion of Symbolic Projection*; R. Kevelson, *Peirce's Semiotics as Complex Inquiry: Conflicting Methods*; J. Kopania, *The Cartesian Alternative of Philosophical Thinking*; Xiankun Li, *Why Gonsung Long (Kungsun Lung) Said "White Horse Is Not Horse"*; L. Melazzo, *A Report on an Ancient Discussion*; Ding-fu Ni, *Semantic Thoughts of J. Stuart Mill and Chinese Characters*; I. Portis-Winner, *Lotman's Semiosphere: Some Comments*; J. Réthoré, *Another Close Look at the Interpretant*; E. Stankiewicz, *The Semiotic Turn of Breal's "Semantique"*; **Part IV: Linguistic** – K. Heger, *Passive and Other Voices Seen from an Onomasiological Point of View*; L.I. Komlószi, *The Semiotic System of Events, Intrinsic Temporal and Deictic Tense Relations in Natural Language. On the Conceptualization of Temporal Schemata*; W. M. Osadnik, E. Horodecka, *Polysystem Theory, Translation Theory and Semiotics*; A. Wierzbicka, *THINK – a Universal Human Concept and a Conceptual Primitive.* **Part V: Cultural Semiotics** – G. Bettetini,

Communication as a Videogame; W. Krysiński, *Joyce, Models, and Semiotics of Passions*; H. Książek-Konicka, *"Visual Thinking" in the Poetry of Julian Przyboś and Miron Białoszewski*; U. Niklas, *The Space of Metaphor*; M. C. Ruta, *Captivity as Event and Metaphor in Some of Cervantes' Writings*; E. Tarasti, *From Aestetics to Ethics: Semiotics Observations on the Moral Aspects of Art, Especially Music*; L. Tondl, *Is It Justified to Consider the Semiotics of Technological Artefacts?*; V. Voigt, *Poland, Finland and Hungary (A Tuatara's View)*; T.G. Winner, *Czech Poetism: A New View of Poetics Language*; J. Wrede, *Metaphorical Imagery – Ambiguity, Explicitness and Life*. **Part VI: Psycho-Socio-Semiotics** – E.M. Barth, *A Case Study in Empirical Logic and Semiotics. Fundamental Modes of Thought of Nazi Politician Vidkun Quisling, Based on Unpublished Drafts and Notebooks*; P. Bouissac, *Why Do Memmes Die?* W. Kalaga, *Threshold of Signification*, A. Podgórecki, *Do Social Sciences Evaporate?*

VOLUME 63 (1998)

IDEALIZATION IX: IDEALIZATION IN CONTEMPORARY PHYSICS
(Edited by Niall Shanks)

N. Shanks, *Introduction*; M. Bishop, *An Epistemological Role for Thought Experiments*; I. Nowak & L. Nowak, *"Models" and "Experiments" as Homogeneous Families of Notions*; S. French & J. Ladyman, *A Semantic Perspective on Idealization in Quantum Mechanics*; Ch. Liu, *Decoherence and Idealization in Quantum Measurement*; S. Hartmann, *Idealization in Quantum Field Theory*; R. F. Hendry, *Models and Approximations in Quantum Chemistry*; D. Howard, *Astride the Divided Line: Platonism, Empiricism, and Einstein's Epistemological Opportunism*; G. Gale, *Idealization in Cosmology: A Case Study*; A. Maidens, *Idealization, Heuristics and the Principle of Equivalence*; A. Rueger & D. Sharp, *Idealization and Stability:A Perspective from Nonlinear Dynamics*; D. L. Holt & R. G. Holt, *Towards a Very Old Account of Rationality in Experiment: Occult Practices in Chaotic Sonoluminescence*.

VOLUME 64 (1998)

PRAGMATIC IDEALISM. CRITICAL ESSAYS ON NICHOLAS RESCHER'S SYSTEM OF PRAGMATIC IDEALISM
(Edited by Axel Wüstehube and Michael Quante)

Introduction: A. Wüstehube, *Is Systematic Philosophy still Possible?*; T. Airaksinen, *Moral Facts and Objective Values*; L. Rodríguez Duplá, *Values and Reasons*; G. Gale, *Rescher on Evolution and the Intelligibility of Nature*; J Kekes, *The Nature of Philosophy*; P. Machamer, *Individual and Other-Person Morality: A Plea for an Emotional Response to Ethical Problems*; D. Marconi, *Opus Incertum*; M. Marsonet, *Scientific Realism and Pragmatic Idealism*; R. Martin, *Was Spinoza a Person?*; H. Pape, *Brute Facts, Real Minds and the Postulation of Reality: Resher on Idealism and the Ontological Neutrality of Experience*; J. C. Pitt, *Doing Philosophy: Rescher's Normative Methodology*; L. B. Puntel, *Is Truth "Ideal Coherence"?*; M. Quante, *Understanding Conceptual Schemes: Rescher's Quarrel with Davidson*; A. Siitonen, *The Ontology of Facts and Values*; M. Willaschek, *Skeptical Challenge and the Burden of Proof: On Rescher's Critique of Skepticism*; N. Rescher, *Responses*.

VOLUME 65 (1999)

THE TOTALITARIAN PARADIGM AFTER THE END OF COMMUNISM.
TOWARDS A THEORETICAL REASSESSMENT.
(Edited by Achim Siegel)

A. Siegel, **Introduction**: *The Changing Fortunes of the Totalitarian Paradigm in Communist Studies.* **On Recent Controversies Over The Concept Of Totalitarianism** – K. von Beyme, *The Concept of Totalitarianism – A Reassessment after the End of Communist Rule*; K. Mueller, *East European Studies, Neo-Totalitarianism and Social Science Theory*; L. Nowak, *A Conception that is Supposed to Correspond to the Totalitarian Approach to Realsocialism*; E. Nolte, *The Three Versions of the Theory of Totalitarianism and the Significance of the Historical-Genetic Version*; E. Jesse, *The Two Major Instances of Totalitarianism: Observations on the Interconnection between Soviet Communism and National Socialism.* **Classic Concept Of Totalitarianism: Reassessment And Reinterpretation** – J.P. Arnason, *Totalitarianism and Modernity: Franz Borkenau's "Totalitarian Enemy" as a Source of Sociological Theorizing on Totalitarianism*; A. Sölner, *Sigmund Neumann's "Permanent Revolution": A Forgotten Classic of Comparative Research into Modern Dictatorship*; F. Pohlmann, *The "Seeds of Destruction" in Totalitarian Systems. An Interpretation of the Unity in Hannah Arendt's Political Philosophy*; W.J. Patzelt, *Reality Construction under Totalitarianism: An Ethnomethodological Elaboration of Martin Draht's Concept of Totalitarianism*; A. Siegel, *Carl Joachim Friedrich's Concept of Totalitarianism: A Reinterpretation*; M.R. Thompson, *Neither Totalitarian nor Authoritarian: Post-Totalitarianism in Eastern Europe.*

VOLUME 66 (1999)

Leon Gumański

TO BE OR NOT TO BE? IS THAT THE QUESTION?
AND OTHER STUDIES IN ONTOLOGY, EPISTEMOLOGY AND LOGIC

Preface; *The Elements of a Judgment and Existence*; *Traditional Logic and Existential Presuppositions*; *To Be Or Not To Be? Is That The Question?*; *Some Remarks On Definitions; Logische und semantische Antinomien; A New Approach to Realistic Epistemology; Ausgewählte Probleme der deontischen Logik; An Attempt at the Definition of the Biological Concept of Homology; Similarity.*

VOLUME 67 (1999)

Kazimierz Twardowski

ON ACTIONS, PRODUCTS AND OTHER TOPICS IN PHILOSOPHY
(Edited by Johannes Brandl and Jan Woleński)

Introduction; Translator's Note; Self-Portrait (1926/91); Biographical Notes. **I. On Mind, Psychology, and Language:** *Psychology vs. Physiology and Philosophy (1897); On the Classification of Mental Phenomena (1898); The Essence of Concepts (1903/24); On Idio- and*

Allogenetic Theories of Judgment (1907); *Actions and Products (1912)*; *The Humanities and Psychology (1912/76)*; *On the Logic of Adjectives (1923/27)*. **II. On Truth and Knowledge:** *On So-Called Relative Truths (1900)*; A priori, *or Rational (Deductive) Sciences and* a posteriori, *or Empirical (Inductive) Sciences (1923)*; *Theory of Knowledge. A Lecture Course (1925/75)*. **III. On Philosophy:** *Franz Brentano and the History of Philosophy (1895)*; *The Historical Conception of Philosophy (1912)*; *On Clear and Unclear Philosophical Style (1920)*; *Symbolomania and Pragmatophobia (1921)*; *Address at the 25th Anniversary Session of the Polish Philosophical Society (1929/31)*; *On the Dignity of the University (1933)*. **Bibliography.**

VOLUME 68 (2000)

Tadeusz Czeżowski

KNOWLEDGE, SCIENCE AND VALUES. A PROGRAM FOR SCIENTIFIC PHILOSOPHY
(Edited by Leon Gumański)

L. Gumański, *Introduction.* **Part 1: Logic, Methodology and Theory of Science** – *Some Ancient Problems in Modern Form*; *On the Humanities*; *On the Method of Analytical Description*; *On the Problem of Induction*; *On Discussion and Discussing*; *On Logical Culture*; *On Hypotheses*; *On the Classification of Sentences and Propositional Functions*; *Proof*; *On Traditional Distinctions between Definitions*; *Deictic Definitions*; *Induction and Reasoning by Analogy*; *The Classification of Reasonings and its Consequences in the Theory of Science*; *On the so-called Direct Justification and Self-evidence*; *On the Unity of Science*; *Scientific Description*. **Part 2: The World of Human Values and Norms** – *On Happiness*; *How to Understand "the Meaning of Life" ?*; *How to Construct the Logic of Goods?*; *The Meaning and the Value of Life*; *Conflicts in Ethics*; *What are Values?*; *Ethics, Psychology and Logic*. **Part 3: Reality–Knowledge–World** – *Three Attitudes towards the World*; *On Two Views of the World*; *A Few Remarks on Rationalism and Empiricism*; *Identity and the Individual in Its Persistence*; *Sensory Cognition and Reality*; *Philosophy at the Crossroads*; *On Individuals and Existence*. J.J. Jadacki, *Trouble with Ontic Categories or Some Remarks on Tadeusz Czeżowski's Philosophical Views*; W. Mincer, *The Bibliography of Tadeusz Czeżowski.*

VOLUME 69 (2000)

Izabella Nowakowa, Leszek Nowak

THE RICHNESS OF IDEALIZATION

Preface; **Introduction** – *Science as a Caricature of Reality.* **Part I: THREE METHODO-LOGICAL REVOLUTIONS** – *1. The First Idealizational Revolution. Galileo's-Newton's Model of Free Fall*; *2. The Second Idealizational Revolution. Darwin's Theory of Natural Selection*; *3. The Third Idealizational Revolution. Marx's Theory of Reproduction.* **Part II: THE METHOD OF IDEALIZATION** – *4. The Idealizational Approach to Science: A New Survey*; *5. On the Concept of Dialectical Correspondence*; *6. On Inner Concretization. A Certain Generalization of the Notions of Concretization and Dialectical Correspondence*; *7. Concretization in Qualitative Contexts*; *8. Law and Theory: Some Expansions*; *9. On Multiplicity of Idealization.* **Part III: EXPLANATIONS AND APPLICATIONS** – *10. The Ontology of the Idealizational Theory*; *11. Creativity in Theory-building*; *12. Discovery and Correspondence*; *13. The Problem of Induction. Toward an Idealizational Paraphrase*; *14. "Model(s) and "Experiment(s). An Analysis of Two Homogeneous Families of Notions*; *15. On Theories, Half-Theories, One-fourth-Theories, etc.*; *16. On Explanation and Its Fallacies*; *17. Testability and Fuzziness*; *18. Constructing the Notion*; *19. On Economic Modeling*; *20. Ajdukiewicz, Chomsky and the Status of the Theory of Natural Language*; *21. Historical Narration*; *22. The Rational Legislator.* **Part IV: TRUTH AND IDEALIZATION** – *23.*

A Notion of Truth for Idealization; 24. *"Truth is a System": An Explication*; 25. *On the Concept of Adequacy of Laws*; 26. *Approximation and the Two Ideas of Truth*; 27. *On the Historicity of Knowledge*. **Part V: A GENERALIZATION OF IDEALIZATION** – 28. *Abstracts Are Not Our Constructs. The Mental Constructs Are Abstracts*; 29. *Metaphors and Deformation*; 30. *Realism, Supra-Realism and Idealization*. **REFERENCES** – *I. Writings on Idealization*; *II. Other Writings.*

VOLUME 70 (2000)

QUINE. NATURALIZED EPISTEMOLOGY, PERCEPTUAL KNOWLEDGE AND ONTOLOGY
(Edited by Lieven Decock and Leon Horsten)

Introduction. **Naturalized Epistemology** – T. Derksen, *Naturalistic Epistemology, Murder and Suicide? But what about the Promises!*; Ch. Hookway, *Naturalism and Rationality*; M. Gosselin, *Quine's Hypothetical Theory of Language Learning. A Comparison of Different Conceptual Schemes and Their Logic*. **The Nature of Perceptual Knowledge** – J. van Brakel, *Quine and Innate Similarity Spaces*; D. Koppelberg, *Quine and Davidson on the Structure of Empirical Knowledge*; E. Picardi, *Empathy and Charity*. **Ontology** – S. Laugier, *Quine: Indeterminacy, 'Robust Realism', and Truth*; R. Vergauwen, *Quine and Putnam on Conceptual Relativity and Reference: Theft or Honest Toil?*; I. Douven, *Empiricist Semantics and Indeterminism of Reference*; L. Decock, *Domestic Ontology and Ideology*; P. Gochet, *Canonical Notation, Predication and Ontology*.

VOLUME 71 (2000)

LOGIC, PROBABILITY AND SCIENCE
(Edited by Niall Shanks)

N. Shanks & R.R. Gardner, *Introduction*; C. Morgan, *Canonical Models and Probabilistic Semantics (Commentary by François Lepage; Reply by Morgan)*; F. Lepage, *A Many-Valued Probabilistic Logic (Commentary by Charles Morgan; Reply by Lepage)*; P. Rawling, *The Exchange Paradox, Finite Additivity, and the Principle of Dominance (Commentary by Robert R. Gardner; Reply by Rawling)*; S. Vineberg, *The Logical Status of Conditional and its Role in Confirmation (Commentary by Piers Rawling; Reply by Vineberg)*; D. Mayo, *Science, Error Statistics, and Arguing from Error (Commentary by Susan Vineberg; Reply by Mayo)*; M.N. Lance, *The Best is the Enemy of the Good: Bayesian Epistemology as a Case Study in Unhelpful Idealization (Commentary by Leszek Nowak; Reply by Lance)*; R.B. Gardner & M.C. Wooten, *An Application of Bayes' Theorem to Population Genetics (Commentary by Lynne Seymour; Reply by Gardner and Wooten)*; P.D. Johnson, Jr., *Another Look at Group Selection (Commentary by Niall Shanks; Reply by Johnson)*; C.F. Juhl, *Teleosemantics, Kripkenstein and Paradox (Commentary by Daniel Bonevac; Reply by Juhl)*; D. Bonevac, *Constitutive and Epistemic Principles (Commentary by Mark Lance; Reply by Bonevac)*; O. Bueno, *Empiricism, Mathematical Truth and Mathematical Knowledge (Commentary by Chuang Liu; Reply by Bueno)*; Ch. Liu, *Coins and Electrons: A Unified Understanding of Probabilistic Objects (Commentary by Steven French; Reply by Liu)*; A. Maidens, *Are Electrons Vague Objects? (Commentary by David Over; Reply by Maidens)*.

VOLUME 72 (2000)

ON COMPARING AND EVALUATING SCIENTIFIC THEORIES
(Edited by Adam Jonkisz and Leon Koj)

L. Koj, *Preface*; L. Koj, *Methodology and Values*; L. Koj, *Science as System*; A. Grobler, *Explanation and Epistemic Virtue*; P. Giza, *"Intelligent" Computer System and Theory Comparison*; H. Ogryzko-Wiewiórowski, *Methods of Social Choice of Scientific Theories*; K. Jodkowski, *Is the Causal Theory of Reference a Remedy for Ontological Incommensurability?*; W. Balzer, *On Approximative Reduction*; C. Ulises Moulines, *Is There Genuinely Scientific Progress?*; A. Jonkisz, *On Relative Progress in Science*.

VOLUME 73 (2000)

THE RATIONALITY OF THEISM
(Edited by Adolfo García de la Sierra)

Preface; A. García de la Sierra, *Introduction*; W. Redmond, *A Logic of Religious Faith and Development*; J.M. Bocheński, O.P., *The Five Ways*; M. Beuchot, *Saint Thomas' Third Way: Possibility and Necessity, Essence and Existence*; R. Swinburne, *Cosmological and Teleological Arguments*; A. García de la Sierra, *The Ontological Argument*; A. García de la Sierra, *Pascal's Wager*; A. García de la Sierra & A. Araujo, *The Experience of God in Moral Obligation*; A. Plantinga, *The onus probandi of Theism*; R. A. Clouser, *Is God Eternal?*; A. Tomasini, *The Presence and Absence of God*; L. Nowak, *On the Common Structure of Science and Religion*.

VOLUME 74 (2000)

POLISH PHILOSOPHERS OF SCIENCE AND NATURE IN THE 20th CENTURY
(Edited by Władysław Krajewski)

W. Krajewski, *Introduction*; **I. Philosophers** J. Woleński, *Tadeusz Kotarbiński – Reism and Science*; A. Jedynak, *Kazimierz Ajdukiewicz – From Radical Conventionalism to Radical Empiricism*; L. Gumański, *Tadeusz Czeżowski – Our Knowledge though Uncertain is Probable*; M. Tałasiewicz, *Jan Łukasiewicz – The Quest for the Form of Science*; I. Szumilewicz-Lachman, *Zygmunt Zawirski – The Notion of Time*; A. Jedynak, *Janina Hosiasson-Lindenbaumowa – The Logic of Induction*; T. Bigaj, *Joachim Metallmann – Causality, Determinism and Science*; J. Woleński, *Izydora Dąmbska – Between Conventionalism and Realism*; A. Koterski, *Henryk Mehlberg – The Reach of Science*; I. Nowakowa, *Adam Wiegner's Nonstandard Empiricism*; W. Krajewski, *Janina Kotarbińska – Logical Methodology and Semantic*; M. Tałasiewicz, *Maria Kokoszyńska-Lutmanowa – Methodology, Semantics, Truth*; T. Batóg, *Seweryna Łuszczewska-Romahnowa – Logic and Philosophy of Science*; M. Omyła, *Roman Suszko – From Diachronic Logic to Non-Fregean Logic*; J. Woleński, *Klemens Szaniawski – Rationality and Statistical Methods*; A. Jedynak, *Halina Mortimer – The Logic of Induction*; K. Zamiara, *Jerzy Giedymin – From the Logic of Science to the Theoretical History of Science*; J. M. Dołęga, *B. J. Gawecki – A Philosopher of the Natural Sciences*; A. Bronk, *Stanisław Kamiński – A Philosopher and Historian of Science*; Z. Hajduk, *Stanisław Mazierski – A Theorist of Natural Lawfulness*. **II. Scientists** W. Krajewski, *Marian Smoluchowski – A Forerunner of the Chaos Theory*; A. Motycka, *Czesław Białobrzeski's Conception of Science*; M Tempczyk, *Leopold Infeld – The Problem of Matter and Field*; M. Czarnocka, *Grzegorz Białkowski – Science and Its Subject*; J. Płazowski, *Jerzy Rayski –*

Physicist and Philosopher of Physics; J. Misiek, *Zygmunt Chyliński – Physics, Philosophy, Music*; W. Sady, *Ludwik Fleck – Thought Collectives and Thought Styles*. **II. General Surveys** K. Ajdukiewicz, *Logicist Anti-Irrationalism in Poland*; K. Szaniawski, *Philosophy of Science in Poland*; I. Nowakowa, *Main Orientations in the Contemporary Polish Philosophy of Science*.

VOLUME 75 (2000)

STRUCTURALIST KNOWLEDGE REPRESENTATION
PARADIGMATIC EXAMPLES
(Edited by Wolfgang Balzer, Joseph D. Sneed and C. Ulises Moulines)

W. Balzer, U. Moulines, *Introduction*; J. A Diez Calzada, *Structuralist Analysis of Theories of Fundamental Measurement*; A. García de la Sienra, P. Reyes, *The Theory of Finite Games in Extensive Form*; H. J. Burscheid, H. Struve, *The Theory of Stochastic Fairness – its Historical Development, Formulation and Justification*; W. Balzer, R. Mattessich, *Formalizing the Basis of Accounting*; W. Diederich, *A Reconstruction of Marxian Economics*; B. Hamminga, W. Balzer, *The Basic Structure of Neoclassical General Equilibrium Theory*; K. Manhart, *Balance Theories: Two Reconstructions and the Problem of Intended Applications*; R. Westermann, *Festinger's Theory of Cognitive Dissonance: A Structuralist Theory-Net*; R. Reisenzein, *Wundt's Three-Dimensional Theory of Emotion*; P. Lorenzano, *Classical Genetics and the Theory-Net of Genetics*; H. Hettema, T.A.F. Kuipers, *The Formalisation of the Periodic Table*; C. Ulises Moulines, *The Basic Core of Simple Equilibrium Thermodynamics*; T. Bartelborth, *An Axiomatization of Classical Electrodynamics*; *Author's Index*; *Subject Index*.

VOLUME 76 (2000)

EVENTS, FACTS AND THINGS
(Edited by J. Faye, U. Scheffler and Max Urchs)

J. Faye, U. Scheffler and M. Urchs, *Philosophical Entities: An Introduction*; J. Faye, *Facts as Truth Makers*; J. Persson, *Examining the Facts*; U. Scheffler and Y. Shramko, *The Logical Ontology of Negative Facts: On What is Not*; W. Stelzner, *The Impact of Negative Fact for the Imaginary Logic of N.A. Vasil'ev*; B. Rode Meinertsen, *Events, Facts and Causation*; U. Meixner, *Essential Conception of Events*; P. Stekeler-Weithofer, *Questions and Theses Concerning (Mental) Events and Causation*; E. Tegtmeier, *Events as Facts*; M. Urchs, *Events of Episystems*; J. Seibt, *The Dynamic Constitution of Things*; K. Trettin, *Tropes and Things*; D. von Wachter, *A World of Fields*; A. Bartels, *Quantum Field Theory: A Case for Event Ontologies?*; M. Dorato, *Facts, Events, Things and the Ontology of Physics*; M. Kuhlmann, *Processes as Objects of Quantum Field Theory: Consequences for the Interpretation of QFT*; J. Paśniczek, *Objects vs. Situations*; A. Siitonen, *Effects or Consequences of Action*; P. Needham, *Hot Stuff*; L. Bo Gundersen, *Goodman's Gruesome Modal Fallacy*; Th. Mormann, *Topological Representations of Mereological Systems*; U. Scheffer & M. Winkler, *Tools*.

VOLUME 77 (2003)

KNOWLEDGE AND FAITH
(Edited by J. J. Jadacki and Kordula Świętorzecka)

Editorial Note; J.J. Jadacki and K. Świętorzecka, *On Jan Salamucha's Life and Work*. **Part I. Logic and Theology** - *On the «Mechanization» of Thinking*; *On the Possibilities of a Strict formalization of the Domain of Analogical Notion*; *The Proof ex motu for the Existence of God. Logical Analysis of St. Thomas Aquinas' Arguments*. **Part II. History of Logic** - *The Propositional Logic in William Ockham*; *The Appearance of Antinomial Problems within Medieval Logic*; *From the History of Medieval Nominalism*. **Part III. Metaphysics and Ethics** - *From the History of One Word ("Essence")*; *The Structure of the Material World*; *Faith*; *The Relativity and Absoluteness of Catholic Ethics*; *The Problem of Force in Social Life*; *A Vision of Love*. **Comments and Discussions** – J. M. Bocheński, *J. Salamucha The Notion of Deduction in Aristotle and St. Thomas Aquinas*; J. M. Bocheński, *J. Salamucha, "The Proof ex motu for the Existence of God. Logical Analysis of St. Thomas Aquinas' Arguments"*; J.F. Drewnowski, *J. Salamucha, "The Proof ex motu for the Existence of God. Logical Analysis of St. Thomas Aquinas' Arguments"*; H. Scholz, *The Mathematical Logic and the Metaphysics*; H. Scholz, *J. Salamucha "The Appearance of Antinomial Problems within Medieval Logic"*; J. Bendiek, *On the Logical Structure of Proofs for the Existence of God*; K. Policki, *On the formalization of the Proof ex motu for Existence of God*; J. Herbut, *Jan Salamucha's Efforts Towards the Methodological Modernization of Theistic Metaphysics*; F. Vandamme, *Logic, Pragmatics and Religion*; E. Nieznański, *Logical Analysis of Thomism. The Polish Programme that Originated in 1930s*. **Bibliography.**

VOLUME 78 (2003)

Jacek Juliusz Jadacki
FROM THE VIEWPOINT OF THE LVOV-WARSAW SCHOOL

Preface; *Introduction: Philosophy and Precision*. **Part I. Being and Essence** - *On What Seems Not to Be*; *On the Controversy about Universals*; *On Forms of Objects*; *On Good, Necessity, and Sufficiency*; *On Essence*. **Part II. Truth and Nonsense** - *On the Definition and Criteria of Truth*; *On Linguistic Categories*; *On Questions*; *On Semiotic Function of Conditionals*; *On Nonsense*. **Part III. Understanding and Silence** - *On Misunderstandings about Understanding*; *On Definition, Explication, and Paraphrase*; *On Reasoning*; *On Simplicity*; *On Silence*; *Conclusion: Science and Creation*; *References*; *Index of Names*; *Index of Subjects*.

A Thomistic Tapestry
Essays in Memory of Étienne Gilson

Edited by Peter A. Redpath

Amsterdam/New York, NY 2003. XX, 243 pp.
(Value Inquiry Book Series 142)

ISBN: 90-420-0875-X   € 52,-/US $ 62.-

This book, written by well-known students of Étienne Gilson and especially dedicated to Armand A. Maurer, helps inaugurate a long-overdue special series in philosophy honoring Gilson's legendary scholarship. It presents wide-ranging expositions of Thomist realism in the tradition of Gilsonian humanism covering themes related to philosophy in general, historical method, aesthetics, metaphysics, epistemology, and politics.

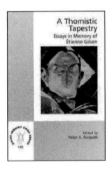

USA/Canada: One Rockefeller Plaza, Ste. 1420, New York, NY 10020,
Tel. (212) 265-6360, Call toll-free (U.S. only) 1-800-225-3998,
Fax (212) 265-6402
All other countries: Tijnmuiden 7, 1046 AK Amsterdam, The Netherlands.
Tel.   31 (0)20 611 48 21, Fax   31 (0)20 447 29 79
Orders-queries@ rodopi.nl   **www.rodopi.nl**
Please note that the exchange rate is subject to fluctuations

Explorations in Contemporary Continental Philosophy of Religion.

Edited by Deane-Peter Baker and Patrick Maxwell

Amsterdam/New York, NY 2003. XIII, 219 pp.
(Value Inquiry Book Series 143)

ISBN: 90-420-0995-0   € 46,-/US $ 55.-

This book is an exploration of the content and dimensions of contemporary Continental philosophy of religion. It is also a showcase of the work of some of the philosophers who are, by their scholarship, filling out the meaning of the term "Continental philosophy of religion."

PHILOSOPHY AND RELIGION (PAR) is dedicated to a critical study of religious attitudes, values, and beliefs. PAR welcomes a wide variety of philosophical approaches to general and specific topics arising from the whole spectrum of religious traditions.

USA/Canada: One Rockefeller Plaza, Ste. 1420, New York, NY 10020,
Tel. (212) 265-6360, Call toll-free (U.S. only) 1-800-225-3998,
Fax (212) 265-6402
All other countries: Tijnmuiden 7, 1046 AK Amsterdam, The Netherlands.
Tel. ++ 31 (0)20 611 48 21, Fax ++ 31 (0)20 447 29 79
Orders-queries@rodopi.nl   www.rodopi.nl
Please note that the exchange rate is subject to fluctuations.

Process Pragmatism
Essays on a Quiet Philosophical Revolution

Edited by Guy Debrock

Amsterdam/New York, NY 2003. XIV, 185 pp.
(Value Inquiry Book Series 137)
ISBN: 90-420-0985-3   € 40,-/US$ 48.-

This book discusses Process Pragmatism, the view that whatever is, derives from interactions. The contributors examine and defend its merits by focusing on major topics, including truth, the existence of unobservables, the origin of knowledge, scientific activity, mathematical functions, laws of nature, and moral agency.

STUDIES IN PRAGMATISM AND VALUES (SPV) promotes the study of pragmatism's traditions and figures, and the explorations of pragmatic inquiries in all areas of philosophical thought.

USA/Canada: One Rockefeller Plaza, Ste. 1420, New York, NY 10020,
Tel. (212) 265-6360, Call toll-free (U.S. only) 1-800-225-3998,
Fax (212) 265-6402
All other countries: Tijnmuiden 7, 1046 AK Amsterdam, The Netherlands.
Tel. ++ 31 (0)20 611 48 21, Fax ++ 31 (0)20 447 29 79
Orders-queries@rodopi.nl   **www.rodopi.nl**
Please note that the exchange rate is subject to fluctuations.

Zur Wissenschaftslehre

Beiträge zum vierten Kongress der Internationalen Johann-Gottlieb-Fichte-Gesellschaftin Berlin vom 03. – 08. Oktober 2000

Hrsg. von Helmut Girndt

Amsterdam/New York, NY 2003. XIII, 284 pp.
(Fichte-Studien 20)
ISBN: 90-420-1184-X € 60,-/US$ 71.-

Inhalt: Vorwort. Nachruf auf Jan Garewicz. Wolfgang JANKE: Vielheit des Seins – Einheit des Ich-existiere. Verwahrung und Vertiefung des transzendentalen Gedankens.
Teil I Zur Wissenschaftslehre 1794
Christian HANEWALD: Absolutes Sein und Existenzgewißheit des Ich. Marina A. PUSCHKAREWA: Der Begriff der nicht offenbaren Tätigkeit und Fichtes *Grundlage der gesammten Wissenschaftslehre*. Frank WITZLEBEN: *Wer* weiß? Eine Re-Interpretation der Theorie der Handlung und des Wissens in Fichtes Wissenschaftslehre von 1794. Ernst-Otto ONNASCH: Ich und Vernunft. Ist J.G. Fichte die Begründung seiner *Grundlage der gesammten Wissenschaftslehre* von 1794/95 gelungen?
Zur Diskussion
Wilhelm METZ: Die produktive Reflexion als Prinzip des wirklichen Bewußtseins
Teil II Zur Wissenschaftslehre von 1801 bis 1805
Virginia LÓPEZ-DOMÍNGUEZ: Die Entwicklung der intellektuellen Anschauung bei Fichte bis zur Darstellung der Wissenschaftslehre (1801-1802). Reinhard LOOCK: Das Schweben des absoluten Wissens. Zur Logik der Einbildungskraft in Fichtes Wissenschaftslehre von 1801/02. Diogo FERRER: Die pragmatische Argumentation in Fichtes Wissenschaftslehre 1801/1802. Ulrich SCHLÖSSER: Entzogenes Sein und unbedingte Evidenz in Fichtes Wissenschaftslehre 1804 (2). Urs RICHLI: Genetische Evidenz – was ist das eigentlich? Berlino D'ALFONSO: Strategien zur Widerlegung des Skeptizismus in Fichtes *Wissenschaftslehre 1804, Zweiter Vortrag*. Peter L. OESTERREICH: Fünf Entdeckungen auf dem Wege zu einer neuen Darstellung der Philosophie Fichtes. Manuel JIMÉNEZ-REDONDO: Der Aporetische Begriff der Erscheinung des Absoluten bei Fichtes WL 1805.
Teil III Zur Wissenschaftslehre von 1811 bis 1814
Alessandro BERTINETTO: Die Grundbeziehung von „Leben" und „Sehen" in der ersten Transzendentalen Logik Fichtes. Hiroshi KIMURA: Sehen und Sagen. Das Sehen sieht das Aussagen seines Grundes. Lu DE VOS: Das Absolute und das Spiel der Modalitäten. Johannes BRACHTENDORF: Der erscheinende Gott – Zur Logik des Seins in Fichtes Wissenschaftslehre 1812. Günter ZÖLLER: „On revient toujours...": Die transzendentale Theorie des Wissens beim letzten Fichte. Hartmut TRAUB: Vollendung der Transzendentalphilosophie.

USA/Canada: One Rockefeller Plaza, Ste. 1420, New York, NY 10020,
Tel. (212) 265-6360, Call toll-free (U.S. only) 1-800-225-3998,
Fax (212) 265-6402
All other countries: Tijnmuiden 7, 1046 AK Amsterdam, The Netherlands.
Tel. ++ 31 (0)20 611 48 21, Fax ++ 31 (0)20 447 29 79
Orders-queries@rodopi.nl www.rodopi.nl
Please note that the exchange rate is subject to fluctuations

Praktische und angewandte Philosophie II

Beiträge zum vierten Kongress der Internationalen Johann-Gottlieb-Fichte-Gesellschaft in Berlin vom 03. – 08. Oktober 2000

Helmut Girndt/Hartmut Traub (Hrsg.)

Amsterdam/New York 2003. VI, 177 pp. (Fichte-Studien 24)

ISBN: 90-420-0855-5 (Bd. 2)   € 36,-/US $ 43.-
ISBN: 90-420-0865-2 (Bde. 1+2)

Inhalt: Thomas Sören HOFFMANN: »... eine besondere Weise, sich selbst zu erblicken«: Zum systematischen Status der Natur nach Fichte. Christian STADLER: Der Transzendentalphilosophische Rechtsbegriff und seine systematische Begründungsleistung. Katja V. TAVER: Fichte und Arnold Gehlen. Fichtes Philosophie des Rechts von 1796 und 1812 im Fokus von Arnold Gehlens philosophischer Anthropologie. Jean-Christophe MERLE: Fichtes Begründung des Strafrechts. Manfred GAWLINA: Verhalten als Synthesis von Recht und Gesinnung. Zur (virtuellen) Auseinandersetzung zwischen Kant, Fichte und Hegel. Carla DE PASCALE: Fichte und die Gesellschaft. Carla AMADIO: Die Logik der politischen Beziehung. Christiana SENIGAGLIA: Die Bestimmung des Bürgers beim späten Fichte. Ferenc L. LENDVAI: Stellung und Spuren einer Sozialethik in Fichtes Philosophie. Teil I: Die Stellung einer Sozialethik in Fichtes Philosophie. Judit HELL: Stellung und Spuren einer Sozialethik in Fichtes Philosophie. Teil II: Die Spuren einer Sozialethik in Fichtes Philosophie. Wladimir Alexejevic ABASCHNIK: Das Konzept des geschlossenen Handelsstaates Fichtes in der Rezeption von Vassilij Nasarovic Karasin. Karl HAHN: Die Relevanz der Eigentumstheorie Fichtes im Zeitalter der Globalisierung unter Berücksichtigung Proudhons und Hegels. Hans HIRSCH: Fichtes Planwirtschaftsmodell als Dokument der Geistesgeschichte und als bleibender Denkanstoß.

USA/Canada: One Rockefeller Plaza, Ste. 1420, New York, NY 10020,
Tel. (212) 265-6360, Call toll-free (U.S. only) 1-800-225-3998,
Fax (212) 265-6402
All other countries: Tijnmuiden 7, 1046 AK Amsterdam, The Netherlands.
Tel. ++ 31 (0)20 611 48 21, Fax ++ 31 (0)20 447 29 79
Orders-queries@rodopi.nl   www.rodopi.nl
Please note that the exchange rate is subject to fluctuations

Praktische und angewandte Philosophie I.

Beiträge zum vierten Kongress der Internationalen Johann-Gottlieb-Fichte-Gesellschaft in Berlin vom 03. – 08. Oktober 2000.

Herausgegeben von Helmut Girndt und Hartmut Traub.

Amsterdam/New York 2003. VII, 232 pp.
(Fichte-Studien 23)

ISBN: 90-420-1025-8 (Bd. 1)   € 50,-/US$ 60.-
ISBN: 90-420-0865-2 (Bde. 1+2)

Inhalt: Rainer ADOLPHI: Weltbild und Ich-Verständnis. Die Transformation des >Primats der praktischen Vernunft< beim späteren Fichte. Jacinto RIVERA DE ROSALES: Das Absolute und die Sittenlehre von 1812. Sein und Freiheit. Marek J. SIEMEK: Fichtes und Hegels Konzept der Intersubjektivität. Ewa NOWAK-JUCHACZ: Das Anerkennungsprinzip bei Kant, Fichte und Hegel. Ronald MATHER: On the Concepts of Recognition. Makoto TAKADA: Verwandlung der Individuumslehre bei Fichte. Hans Georg von MANZ: Deduktion und Aufgabe des individuellen Ich in Fichtes Darstellungen der Wissenschaftslehre von 1810/11. Jürgen STAHL: Zur Kultur in der Vermittlungsrolle zwischen empirischem und absolutem Ich. Christoph ASMUTH: Metaphysik und Historie bei J.G. Fichte. Stephan GNÄDINGER: Vorsehung. Ein religionsphilosophisches Grundproblem bei J.G. Fichte. Johannes HEINRICHS: Die Mitte der Zeit als Tiefpunkt einer Parabel. Fichtes Geschichtskonstruktion und Grundzüge der gegenwärtigen Zeitenwende. Marco M. OLIVETTI: Zum Religions- und Offenbarungsverständnis beim jungen Fichte und bei Kant.

Rodopi

USA/Canada: One Rockefeller Plaza, Ste. 1420, New York, NY 10020,
Tel. (212) 265-6360, Call toll-free (U.S. only) 1-800-225-3998,
Fax (212) 265-6402
All other countries: Tijnmuiden 7, 1046 AK Amsterdam, The Netherlands.
Tel. ++ 31 (0)20 611 48 21, Fax ++ 31 (0)20 447 29 79
Orders-queries@rodopi.nl   **www.rodopi.nl**
Please note that the exchange rate is subject to fluctuations

Scratching the Surface of Bioethics

Edited by Matti Häyry and Tuija Takala.

Amsterdam/New York, NY 2003. XII, 148 pp.
(Value Inquiry Book Series 144)

ISBN: 90-420-1006-1   € 35,-/US $ 42.-

Is bioethics only about medicine and health care? Is it only about law? Is it
only about philosophy? Is it only about social issues? No on all accounts. It
embraces all these and more. In this book, fifteen notable scholars from the
North West of England critically explore the main approaches to bioethics –
and make a scratch on its polished surface.

USA/Canada: One Rockefeller Plaza, Ste. 1420, New York, NY 10020,
Tel. (212) 265-6360, Call toll-free (U.S. only) 1-800-225-3998,
Fax (212) 265-6402
All other countries: Tijnmuiden 7, 1046 AK Amsterdam, The Netherlands.
Tel. ++ 31 (0)20 611 48 21, Fax ++ 31 (0)20 447 29 79
Orders-queries@rodopi.nl   www.rodopi.nl
Please note that the exchange rate is subject to fluctuations

Die Philosophie Karl Leonhard Reinholds.

Herausgegeben von Martin Bondeli und
Wolfgang H. Schrader†.

Amsterdam/New York, NY 2003. XIX, 324 pp.
(Fichte-Studien-Supplementa 16)

ISBN: 90-420-1115-7 € 70,-/US $ 83.-

Die Philosophie Karl Leonhard Reinholds (1757-1823) findet heute vermehrt Beachtung. Während dieser Denker lange Zeit als Popularisator Kants, als Vorläufer Fichtes oder als tatsachenphilosophischer Antipode Schellings und Hegels wahrgenommen wurde und gemeinhin im Ruf eines unsteten und unselbständigen Geistes stand, ist seit einigen Jahrzehnten eine Gegentendenz feststellbar: Reinholds Denkentfaltung wird zunehmend in ihrem gesamten Umfang sowie als eigenwilliger und innovativer Ansatz innerhalb der postkantischen Systemphilosophie zur Kenntnis genommen. Mehr und mehr wird anerkannt, dass Reinhold entscheidende Anstöße zur Entstehung des deutschen Idealismus gegeben und diese Strömung zugleich auf der Grundlage von Einsichten, die in die Richtung der Phänomenologie und Sprachphilosophie des 20. Jahrhunderts weisen, kritisiert hat.

Die in diesem Band vereinigten Beiträge gehen auf die I. Internationale Reinhold-Tagung zurück, die vom 14. bis 18. März 1998 in Bad Homburg stattgefunden hat. Sie geben systematische Fragestellungen und entwicklungsgeschichtliche Kontexte aus Reinholds kantischer Phase der Elementarphilosophie, aus der Periode seiner Zusammenarbeit mit Fichte und Jacobi sowie aus seiner sprachphilosophisch orientierten Spätzeit wieder. Alles in allem spiegeln sie das gegenwärtige Bedürfnis, Reinhold zu entdecken, sein Denken und Schaffen in seinen Wandlungen und in seiner thematischen Breite aufzuarbeiten und nachzuvollziehen. Zudem sind sie repräsentativ für die aktuell bestehenden vielfältigen Zugänge zu diesem produktiven Philosophen.

USA/Canada: One Rockefeller Plaza, Ste. 1420, New York, NY 10020,
Tel. (212) 265-6360, Call toll-free (U.S. only) 1-800-225-3998,
Fax (212) 265-6402
All other countries: Tijnmuiden 7, 1046 AK Amsterdam, The Netherlands.
Tel. 31 (0)20 611 48 21, Fax 31 (0)20 447 29 79
Orders-queries@ rodopi.nl ww.rodopi.nl
Please note that the exchange rate is subject to fluctuations

Graduate Faculty Philosophy Journal

The **Graduate Faculty Philosophy Journal**, published in association with the Department of Philosophy, New School for Social Research, is a forum for historical and contemporary issues in philosophy.

VOL. 24 NO. 2

UPCOMING SPECIAL ISSUES

ARISTOTLE'S ARABIC RECEPTION (25:1, 2)
guest edited by Badr El Fakkek

ESSAYS ON THE HISTORY AND PHILOSOPHY OF MATHEMATICS (26:1, 2)
edited by Alexei Angelides

contributing authors include Lanier Anderson, Juliette Kennedy, Peter Pesic and Danielle MacBeth

All communications should be addressed to the Editor, **Graduate Faculty Philosophy Journal**, Department of Philosophy, New School for Social Research, 65 Fifth Avenue, New York, NY 10003. The *Journal* is biannual. Domestic rates: Individuals: $20.00/year; Students: $12.00/year; Institutions: $45.00/year.

Confidential Relationships
Psychoanalytic, Ethical, and Legal Contexts.

Edited by Christine M. Koggel, Allannah Furlong, and Charles Levin

Amsterdam/New York, NY 2003. XV,263 pp.
(Value Inquiry Book Series 141)

ISBN: 90-420-0835-0   € 50,-/US $ 60.-

This book focuses the collective attention of psychotherapists, the legal community, social scientists, and ethicists on the moral, legal, and clinical problems of confidentiality in psychotherapeutic practice. By providing timely and important interdisciplinary contributions, the book opens the way to understanding, if not resolving, the conflicting interests and values at stake in the debate on confidentiality.

USA/Canada: One Rockefeller Plaza, Ste. 1420, New York, NY 10020,
Tel. (212) 265-6360, Call toll-free (U.S. only) 1-800-225-3998,
Fax (212) 265-6402
All other countries: Tijnmuiden 7, 1046 AK Amsterdam, The Netherlands.
Tel. ++ 31 (0)20 611 48 21, Fax ++ 31 (0)20 447 29 79
Orders-queries@rodopi.nl   **www.rodopi.nl**
Please note that the exchange rate is subject to fluctuations.